Exploitation of Environmental Heterogeneity by Plants

Physiological Ecology
A Series of Monographs, Texts, and Treatises

A complete list of titles in this series appears at the end of this volume.

Exploitation of Environmental Heterogeneity by Plants

Ecophysiological Processes Above- and Belowground

Edited by

Martyn M. Caldwell

Department of Range Science
and the Ecology Center
Utah State University
Logan, Utah

Robert W. Pearcy

Department of Botany
University of California, Davis
Davis, California

Academic Press
A Division of Harcourt Brace & Company
San Diego New York Boston London Sydney Tokyo Toronto

This book is printed on acid-free paper. ∞

Academic Press, Inc.
525 B Street, Suite 1900, San Diego, California 92101-4495

United Kingdom Edition published by
Academic Press Limited
24–28 Oval Road, London NW1 7DX

Library of Congress Cataloging-in-Publication Data

Exploitation of environmental heterogeneity by plants :
  ecophysiological process above- and below ground / edited by Martyn
  M. Caldwell, Robert W. Pearcy.
      p.   cm. -- (Physiological ecology)
    Includes bibliographical references and index.
    ISBN 0-12-155070-2
    1. Plant ecophysiology. I. Caldwell Martyn M.   Date
II. Pearcy, R. W. (Rober W.),   Date   III. Series.
QK905.E96   1994
581.5--dc20
                                                        93-14291
                                                        CIP

PRINTED IN THE UNITED STATES OF AMERICA
94   95   96   97   98   99   BC   9   8   7   6   5   4   3   2   1

Contents

3. Light Gaps: Sensing the Light Opportunities in Highly Dynamic Canopy Environments
Carlos L. Ballaré

4. Canopy Gaps: Competitive Light Interception and Economic Space Filling—A Matter of Whole-Plant Allocation
Manfred Küppers

10. Root–Soil Responses to Water Pulses in Dry Environments
Park S. Nobel

11. Architecture and Biomass Allocation as Components of the Plastic Response of Root Systems to Soil Heterogeneity
A. H. Fitter

12. Exploiting Nutrients in Fertile Soil Microsites
Martyn M. Caldwell

13. Coping with Environmental Heterogeneity: The Physiological Ecology of Tree Seedling Regeneration across the Gap–Understory Continuum

F. A. Bazzaz and P. M. Wayne

14. Spatial Heterogeneity at Small Scales and How Plants Respond to It

Graham Bell and Martin J. Lechowicz

Contributors

Numbers in parentheses indicate the pages on which the authors' contributions begin.

Dennis Baldocchi (21), Atmospheric Turbulence and Diffusion Division, Air Resources Laboratory, National Oceanic and Atmospheric Administration, Oak Ridge, Tennessee 37831

Carlos L. Ballaré (73), Department de Ecología, Facultad de Agronomía, Universidad de Buenos Aires, 1417 Buenos Aires, Argentina

F. A. Bazzaz (349), Department of Organismic and Evolutionary Biology, Harvard University, Cambridge, Massachusetts 02138

Graham Bell (391), Department of Biology, McGill University, Montreal, Quebec, Canada H3A 1B1

Martyn M. Caldwell (325), Department of Range Science and the Ecology Center, Utah State University, Logan, Utah 84322

Robin L. Chazdon (175), Department of Ecology and Evolutionary Biology, University of Connecticut, Storrs, Connecticut 06269

Serge Collineau (21), Atmospheric Turbulence and Diffusion Division, Air Resources Laboratory, National Oceanic and Atmospheric Administration, Oak Ridge, Tennessee 37831

A. H. Fitter (305), Department of Biology, University of York, York YO1 5DD, England

J. P. Grime (1), NERC Unit of Comparative Plant Ecology, Department of Animal and Plant Sciences, The University of Sheffield, Sheffield S10 2TN, England

Katherine L. Gross (237), W. K. Kellogg Biological Station, and Departments of Botany and Zoology, Michigan State University, Hickory Corners, Michigan 49060

Louis J. Gross (175), Department of Mathematics, University of Tennessee, Knoxville, Tennessee 37996

Manfred Küppers (111), Institut für Botanik, Technische Hochschule Darmstadt, D-64287 Darmstadt, Germany

Martin J. Lechowicz (391), Department of Biology, McGill University, Montreal, Quebec, Canada H3A 1B1

Keith A. Mott (175), Department of Biology, Utah State University, Logan, Utah 84322

Park S. Nobel (285), Department of Biology, and Laboratory of Biomedical and Environmental Sciences, University of California, Los Angeles, California 90024

Alma Orozco-Segovia (209), Centro de Ecología, UNAM, Ciudad Universitaria 04510, Mexico

Robert W. Pearcy (145, 175), Department of Botany, University of California, Davis, Davis, California 95616

G. Philip Robertson (237), W. K. Kellogg Biological Station, and Department of Crop and Soil Sciences, Michigan State University, Hickory Corners, Michigan 49060

Daniel A. Sims (145), Department of Botany, University of California, Davis, Davis, California 95616

John M. Stark (255), Department of Biology, Utah State University, Logan, Utah 84322

Carlos Vázquez-Yanes (209), Centro de Ecología, UNAM, Ciudad Universitaria 04510, Mexico

P. M. Wayne (349), Department of Organismic and Evolutionary Biology, Harvard University, Cambridge, Massachusetts 02138

Preface

Considerable interest in the role played by temporal and spatial heterogeneity in ecological organization and biodiversity has emerged in the past decade. Pertinent aspects of heterogeneity are involved in the scaling of patterns and operational organization among different levels of time and space. Yet, the functional responses of organisms to heterogeneity in different environments has received much less attention. This book examines a synthesis of plant response to temporal and spatial heterogeneity, the exploitation of resources from pulses and patches by plants, and their competition with neighbors in the face of this variability. Approximately half of this volume is directed to the aboveground environment, addressing the nature of canopy patchiness and light transitions—the mechanisms by which plants perceive, acclimate and exploit this patchiness. The remainder explores analogous questions of the belowground environment, heterogeneity in the soil environment and how root systems adjust and acquire nutrients and water in the context of soil temporal and spatial variability.

While the importance of scale in addressing temporal and spatial heterogeneity has long been recognized in ecology, the quantification of pattern and scale of heterogeneity is still an evolving field. Even the definition of gaps, sunflecks and fertile soil microsites, though seemingly straightforward, is more complicated upon further inspection. Geostatistical approaches to assessing the scale and structure of variability are gaining acceptance in ecological study. Scale-dependent autocorrelation, rather than continuous autocorrelation, may emerge as a common phenomenon and be linked with plausible causal agents. Applications of wavelet theory in quantifying light variability may eventually experience the same adoption as geostatistics belowground. Although physical measures of heterogeneity receive the most attention and probably will continue to be emphasized, using the physiological responses of plants to determine meaningful scales of variability is clearly pertinent (e.g., At what level are light transitions simply integrated by plants and at what level are they perceived as significant?). At longer and larger scales, the growth and reproductive success of individual plants or local populations is arguably the most meaningful barometer of heterogeneity in a particular case study. However, comparative studies across different environments will still likely depend on physical characterizations.

Plant plasticity is clearly a central theme in plant exploitation of resources in the face of environmental heterogeneity. Altered biomass allocation of shoots toward canopy gaps or by root proliferation in fertile soil patches are obvious and well known. However, architectural plasticity may be more important than biomass allocation in root exploitation of fertile patches. Plastic adjustment of physiological processes of both individual leaves and roots occurs. However, physiological plasticity is usually closely coupled with growth and development, both above- and belowground. While the time scale of biochemical regulatory events, such as induction of the carboxylating enzyme, may be on the scale of minutes, acclimatizing physiological changes of both leaves and roots appear to operate on the scale of days and involve developmental alterations.

Costs of acclimation or foraging are, of course, very pertinent. These include investments in new structure and in new metabolic capacities that are well understood at the leaf and whole-plant level, but poorly for belowground organs. There are clearly linkages between investments in different parts of the plant in order to maintain functional balance— more assimilatory capacity of the shoots demands more support in water and nutrient supply.

Many counterintuitive findings are emerging: Precision in allocating new root mass preferentially into nutrient-rich patches may be a property not of the dominant, fast growing plants, but of the subordinate community members; slow growing plants may be more adept in capitalizing on nutrient pulses of short duration than fast growing species; sun-acclimated leaves do not reap a great return on investment in high light; high respiration rates in leaves does not necessarily indicate high maintenance costs; root shrinkage during pulses of water deprivation may be an important regulator of water transport between plant and soil. When water does move from roots to soil, there can be benefits to the entire root system in acquiring both water and nutrients.

An effective, timely response to opportunities presented by heterogeneous environments requires perception of the opportunities and of present or impending competition for these resources by neighbors. Seeds have evolved many responses to environmental cues, including germination after exposure to millisecond light flashes, and the ability to differentiate between passing light flecks during the day and opportunities presented by gap formation. Plants can detect the presence of neighbors long before appreciable shading develops.

Finally, although environmental heterogeneity involves a sizeable element of stochasticity, evolutionary molding of plasticity is most likely driven by this heterogeneity. Yet, little is known about this evolution, especially at the genetic level. While the importance of environmental heterogeneity is widely appreciated in modern ecology, the processes and mechanisms by which plants cope and exploit resources, and how these mechanisms evolve, remain a pertinent challenge in physiological ecology.

MARTYN M. CALDWELL
ROBERT W. PEARCY

1

The Role of Plasticity in Exploiting Environmental Heterogeneity

J. P. Grime

In recent years, theories of the functioning and evolution of vascular plants, such as the resource-ratio hypothesis (Huston and Smith, 1987; Tilman, 1988), have placed heavy emphasis on the relative abundance of resources above and below ground and trade-offs in the allocation of captured resources between roots and shoots.

This chapter reviews the results of experimental studies of plant responses to resource heterogeneity conducted over the period 1958 to the present. It is concluded that the resource-ratio hypothesis underestimates the interdependence of roots and shoots and, in particular, does not sufficiently allow for the expenditure of assimilate necessary to allow the extension of roots from the localized zones of depletion that are an inevitable consequence of the rapid rates of mineral nutrient in-flow achieved by plants of productive habitats. It is suggested that in fast-growing species exploiting fertile soil, the swift incorporation of carbon and mineral nutrients into plant tissue and the relative constancy of plant chemical composition will dictate strong covariance between root and shoot function.

In the slow-growing plants of unproductive habitats, the relationship between root and shoot activity may relax considerably; here resource heterogeneity in time rather than space is often critical and capture of resources both above and below ground often depends on long-lived tissues that remain viable under extreme conditions. These tissues exploit pulses of resource enrichment that may be of insufficient duration to reward those foraging mechanisms that rely on growth responses.

Evidence is presented of a trade-off between the scale and precision

Exploitation of Environmental Heterogeneity by Plants

of resource foraging by leaf canopies and root systems. It is suggested that this trade-off is relevant to theories of species coexistence in plant communities. Future studies of this trade-off must take account of phylogenetic constraints and will need to recognize the modifying effect of moisture stress on the development and evolution of root morphology.

I. Introduction

In the majority of animals and in most fungi, bacteria, algae, and bryophytes, all essential resources are absorbed through the same surfaces. This contrasts strongly with the trophic design of vascular land plants, in most of which photons and carbon dioxide are captured by leaves whereas mineral nutrients and water are intercepted by roots. Such specialization in function above and below ground has prompted the hypothesis that the most severe challenges of environmental heterogeneity to the fitness of vascular plants growing in natural environments will arise from the competing claims of roots and shoots on the synthetic capacity of the plant. Hence it has been argued (Newman, 1973, 1983; Iwasa and Roughgarden, 1984; Huston and Smith, 1987; Tilman, 1988, 1989) that, on both an evolutionary time scale and within the life span of an individual phenotype, trade-offs in allocation of captured resources between shoot and root will be of paramount importance. Theories that recognize a pivotal role for root–shoot partitioning in the evolution of flowering plants have become known collectively as the "resource-ratio hypothesis" and they have been formally expressed in models of functional types (Huston and Smith, 1987; Tilman, 1988) in which plants with relatively large shoots (assumed to be superior competitors for light) are distinguished from those with relatively large roots (assumed to be strong competitors for below-ground resources).

  In this chapter various sources of evidence are assembled to support an alternative interpretation of the role of plasticity in exploiting environmental heterogeneity. Here, in marked contrast to Huston and Smith (1987) and Tilman (1988), recognition is given to (1) differences between habitats with respect to spatial and temporal patchiness in resource supply, (2) the capacity of fast-growing plants in productive habitats to generate local resource gradients both above *and below* ground, (3) limits to root–shoot trade-offs imposed by the physiological interdependence of photosynthesis and mineral nutrient capture and the constancy of plant chemical composition across the plant kingdom, (4) covariance in the resource foraging characteristics of leaf canopies and root systems of annual and perennial plants of productive habitats, and (5) genetic differences between dominant and subordinate components of plant communities in the scale and precision of resource interception.

II. Distribution of Resources within Habitats

A. Grassland

It is frequently suggested (e.g., Newman, 1973; Tilman, 1988) that grasslands of productive and unproductive habitats are distinguishable by a difference in the nature of the resources limiting plant production. Few would argue with the conclusion that mineral nutrients are often the primary cause of the sparse vegetation cover frequently observed on many shallow and infertile soils. To a surprising extent, however, scientists disagree about the identity of the resources that limit production and determine species composition on fertile soils.

A popular interpretation of the role of resource limitations in dense stands of productive herbaceous vegetation on fertile soil is that intense shade close to the soil surface is acting as a powerful selective mechanism, whereas mineral nutrients are nonlimiting and have minimal influence (Huston and Smith, 1987; Tilman, 1988). On first inspection, this interpretation is attractive, but it is not supported by an extensive literature accumulated over the past 30 years. A more accurate account of the circumstances prevailing in productive vegetation is that both light and mineral nutrients are sufficient to sustain high productivity but that resource interception by the plants causes depletion above and below ground with the result that competitive suppression and even rapid failure may occur in those individuals with leaves and roots confined to the depletion zones. The existence of rapidly expanding depletion zones within the main areas of fine-root concentration in fertile soils has been well documented (Bhat and Nye, 1973) and is predictable (Passioura, 1963; Nye and Tinker, 1977) from the very high rates of nutrient capture achieved by fast-growing crops and perennials and the rapid proliferation and redistribution of the fine roots (see Fig. 11.3 in Chapter 11, this volume) observed in nutritionally heterogeneous media (Drew *et al.*, 1973; Drew, 1975; Crick and Grime, 1987; Campbell and Grime, 1989b; Granato and Raper, 1989; Caldwell, Chapter 12, this volume). Further evidence of the importance of competition for mineral nutrients on fertile soils is provided by the results of the classic experiment of Donald (1958), which used root and shoot partitions to demonstrate the major contribution of belowground competition in the interactions between two perennial grasses growing on fertile soil. Aerts *et al.* (1991) have conducted a similar experiment; the results again demonstrate that competition for mineral nutrients is a major component of species interactions on fertile soils.

We may conclude that the distribution of resources within productive grassland habitats is exceedingly dynamic and strongly determined by the capacity of the leaf canopies and root systems of the dominant plants to generate localized zones of resource depletion. These spatial patterns will change with the seasonal activities of the plants and will be modified by litter-fall, decomposition events, and management interventions such as fertilizing, burning, and ploughing.

In contrast to the role of spatial patchiness of resources in productive grasslands, there is a scattered literature (Davison, 1964; Davy and Taylor, 1975; Gupta and Rorison, 1975; Taylor *et al.*, 1982) indicating that temporal variation in mineral nutrient supply assumes a key significance when we turn our attention to infertile grassland habitats. Here, chronic shortages of mineral nutrients such as nitrogen and phosphorus frequently pervade the entire rooting volume for most of the year; this situation arises because mineral nutrients tend to be sequestered in the living biomass of flowering plants and microorganisms or reside in relatively intractable organic residues, many of which appear to be relicts of the anti-herbivore defenses that protect the living tissues of many of the plants of nutrient-limited vegetation (Grime, 1988; Heal and Grime, 1991). In such conditions, the opportunities for mineral nutrient capture appear to be brief and unpredictable and depend on disruptive events such as drying–wetting cycles and episodes of freeze and thaw. The duration of many of the pulses of nutrient release on infertile soils are likely to be a critical determinant of the success of the various possible root system responses (Grime *et al.*, 1986; Jackson and Caldwell, 1989).

It is quite conceivable that as a result of rapid uptake of pulses by soil microbial populations (Shields *et al.*, 1973; Ritz and Griffiths, 1987) some pulses may be too infrequent and too short to be exploited effectively by the potentially dynamic but relatively short-lived roots of plants normally associated with fertile soils. In two recent experiments (Crick and Grime, 1987; Campbell and Grime, 1989a) plants characteristic of fertile and infertile soils have been compared with respect to their ability to capture nitrogen from pulses of various durations. The results (Figure 1) indicate that where pulses are short, an advantage is enjoyed by slow-growing species of infertile soils such as *Festuca ovina;* these plants appear to have low rates of tissue turnover and have roots that remain functional despite exposure to long periods of mineral nutrient stress.

B. Woodland

The most conspicuous feature of the distribution of resources within woodland is the presence of a shaded stratum extending for a considerable height above the layers occupied by herbs, bryophytes, small shrubs, and tree seedlings. Woodland plants under deciduous tree canopies exhibit a diversity of strategies, many of which are associated with seasonal opportunities to maximize light interception (Salisbury, 1916; Grime, 1966; Givnish, 1982). However, it is vitally important to recognize that the surge (and subsequent rapid decline) of leaf growth observed in many

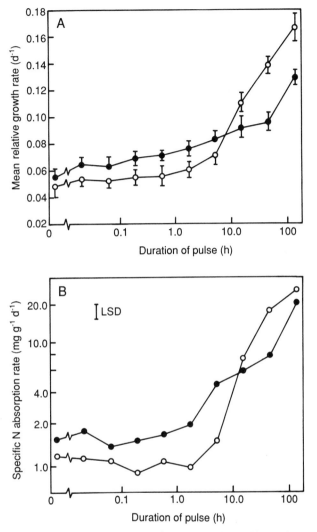

Figure 1   (A) Mean relative growth rate of the potentially rapid-growing *Arrhenatherum elatius* (○) and slow-growing *Festuca ovina* (●) plants exposed once every 6 days to pulses of nutrient enrichment of differing durations. Vertical bar in (B) is L.S.D. ($P < 0.05$) for comparing means on logarithmic scale. Means in (A) are shown ±95% confidence limits. (B) Mean specific nitrogen absorption rate of *Arrhenatherum elatius* (○) and *Festuca ovina* (●) plants exposed once every 6 days to pulses of nutrient enrichment of differing duration. Vertical bar is L.S.D. ($P < 0.05$) for comparing means on logarithmic scale.

spring plants of temperate woodlands (Al-Mufti *et al.*, 1977) coincides not only with the "vernal window" of high irradiance but also with the spring peak of mineral nutrient mobilization from litter decomposition.

A quite different circumstance of resource supply prevails beneath mature canopies of evergreen trees and in deciduous woodland ground floras composed of summergreen or evergreen species. For these species low irradiance is a critical limiting factor and sunflecks are likely to provide a crucial supplement to the energy budget (Pearcy *et al.*, 1988; Kirschbaum *et al.*, 1988; Rincon and Grime, 1989; Pearcy *et al.*, Chapter 6, this volume). It would be a mistake, however, to interpret the physiology of shade plants as a sole consequence of natural selection under low irradiance. The large forest trees that cause deep shade also provide major sinks for mineral nutrients and through intensive mycorrhizal networks may tightly close the mineral nutrient cycle and subject ground flora plants to severe nutrient stress. On this basis, it may be unwise to regard either shade tolerance or a dependence on sunflecks as a simple evolutionary response to shade. Following the argument of Grime (1979, pp. 26,41) it can be postulated that, on infertile soils in deciduous woodland, mineral nutrient constraints preclude the "expensive" development of a temporary canopy during the vernal window and instead favor the slow dynamics (*sensu* Chapin, 1980), low leaf turnover, and conservative use of carbon and mineral nutrients characteristic of many evergreen shade plants (Monk, 1966).

III. Interdependence of Photosynthesis and Mineral Nutrient Capture

Support for the hypothesis that attributes a primary role to evolutionary trade-offs between root and shoot allocation has been drawn from experiments (Brouwer 1962a,b, 1963; Corré, 1983a,b; Hunt and Nicholls, 1986) in which predictable alterations in partitioning of dry matter between shoots and roots have been induced by exposing plants to either low mineral nutrient concentrations or shade treatments. In a wide range of plants, these manipulations have been shown to be capable of modifying root–shoot ratios to a profound extent and this has led to the proposition that there are effective homeostatic mechanisms that, through modifying allocation between root and shoot, preserve a balance between photosynthesis and mineral nutrient capture. Such results encouraged the prediction that plants of fertile and infertile soils or shaded and unshaded habitats will differ in root–shoot ratio. Experimental evidence in support of this hypothesis is lacking; in fact consistent falsification (Olff *et al.*, 1990; Berendse and Elberse, 1989; Tilman and Cowan, 1989; Gleeson and Tilman, 1990; Aerts *et al.*, 1991; Campbell *et al.*, 1991; Shipley and Peters, 1990) has been a feature of the recent literature.

Why has the resource-ratio hypothesis failed to match reality? Several explanations can be suggested:

1. As explained in the preceding section, many habitats cannot be simply classified with respect to single limiting factors. Although shading, for example, is often a conspicuous feature of grasslands and woodlands, mineral nutrients also frequently limit plant production at particular sites.

2. Models such as those of Huston and Smith (1987) and Tilman (1988) and experiments such as those of Brouwer (1963) and Hunt and Nicholls (1986) involve circumstances in which resource depletion is imposed uniformly within the aerial or the rooting environment. This fails to simulate the "patch" and "pulse" phenomena that characterize resource supply in natural environments. In particular, omission of the depletion zones that surround the root surfaces of fast-growing plants growing on fertile soils has led to serious underestimation of the expenditure of assimilate required to sustain the continuous process of root growth necessary to escape from the local but expanding zones of nutrient exhaustion that are an inescapable consequence of the physics of nutrient uptake (Nye and Tinker, 1977) from the rhizosphere.

3. There are biochemical limits to the trade-off between root and shoot. Autotrophy involves the assembly of components derived from both parts of the plant. Chemical analyses of plants reveal that the ratio of root- and shoot-derived elements remains relatively constant across a wide range of ecologies. Interdependence is further enforced by the carbon and energy demand of roots and the mineral nutrient demand of leaves. Scope for variation in root–shoot ratio across plant functional types is restricted by the fact that species with the capacity for high rates of photosynthesis and dry matter production have higher concentrations of leaf nitrogen (Sharkey, 1985; Field and Mooney, 1986) and leaf phosphorus (Figure 2) and in consequence are dependent on high rates of nutrient capture by the root system.

IV. Covariance in Root and Shoot Size and Function

So far in this chapter, strong emphasis has been placed on the interdependence of roots and shoots. It is pertinent to ask, therefore, to what extent this interdependence will dictate a similar partitioning of resources between leaves and roots across a wide range of vascular plants. Greatest uniformity in this respect might be expected to occur during the early exponential phase of seedling growth in both ephemeral species and

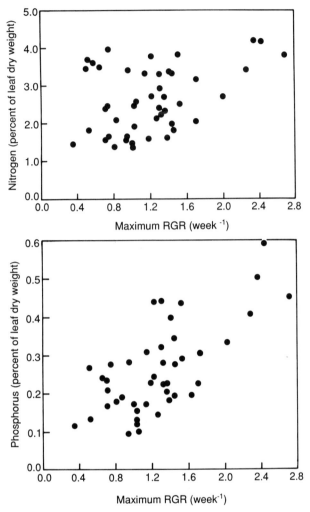

Figure 2 The relationship between maximum potential relative growth rate (RGR; data from Grime and Hunt, 1976) in the seedling phase and the average concentration of nitrogen and phosphorus in the leaf. For nitrogen, $r^2 = 0.339$; for phosphorus, $r^2 = 0.564$; in both cases d.f. = 39 and $P<0.05$. Each determination of nutrient concentration refers to mature, nonsenescent leaves and represents the mean of populations, sampled from a wide range of natural habitats distributed within a 2400-km^2 area in northern England. (Data from Band and Grime, 1981.)

perennials of productive habitats. This is because these plants attain rapid seedling growth by immediate commitment of captured resources to new leaves and roots and will therefore, at each point in time, tend to absorb and utilize carbon and mineral elements in proportions similar to those in which these elements occur in the plant tissues. The elemental composition of plant tissue would not be expected to exercise such a tight constraint on root–shoot allometry in the slow-growing, long-lived perennials of unproductive habitats. In these plants where there may be substantial storage pools and strong reliance on pulse interception (e.g., woodland evergreens, desert succulents), resource capture is usually uncoupled from growth and, in the short term, there may be considerable independence between root and shoot functioning.

V. Resource Heterogeneity and Community Structure

The concept of the plant community arises not merely from the fact that populations of the same or functionally similar species frequently recur together in specific circumstances. It is also of vital concern that consistent patterns are usually observed with respect to the relative abundance of the component populations (Grubb *et al.*, 1982). In a systematic review of the characteristics of common British herbaceous plants (Grime, *et al.* 1988) it is apparent that in terms of their status in vegetation most species conform to a "pecking order." The constancy with which species play either a dominant or subordinate role in vegetation has major implications for both evolutionary theory and physiological ecology. It is not immediately clear why natural selection should drive so many species into subordinate roles; elsewhere (Grime, 1987) an attempt has been made to explain this apparent paradox. More relevant to the subject matter of this volume is the question "Are there identifiable and genetically determined differences between dominants and subordinates in mechanism of response to resource heterogeneity?"

In an attempt to answer this question, experiments have been conducted (Campbell *et al.*, 1991) using two new techniques to assay the responses of the leaf canopies and root systems of a range of plants grown in isolation under standardized patchy conditions simulating aspects of the conditions experienced in a perennial community. Both techniques (Figure 3) present patches of resource without the use of partitions or barriers that could impede growth between patches. In both the root and the shoot assays, measurement was made of the partitioning of dry matter allocation between depleted and undepleted sectors imposed after an initial growth period in a uniform, productive environment. The experiment involved eight common British herbaceous species, widely contrasted in morphology and ecology. To test the predictive value of the assays, comparison was made with the status achieved in a conventional competition experiment in which the eight species were grown together

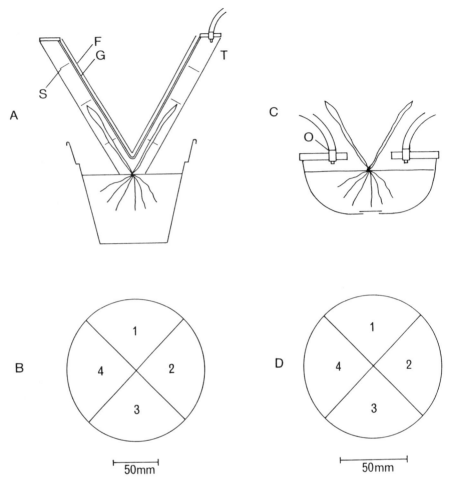

Figure 3 Imposing standardized patches of resource depletion on growing shoot and root systems. Two new techniques have been developed to assay the "resource-foraging" attributes of the leaf canopies and root systems of individual plants grown in isolation under standardized conditions simulating those experienced during competition. An important feature of both techniques is that plants are presented with standardized patches of resource depletion created without the use of partitions or barriers that could impede growth between patches.

Light Patches (A and B). (A) Section through cone-shaped chamber for imposing partial shading on developing shoot system. A transparent glass upper surface (G) is covered with filters (F) to produce standardized patches of shade. Fine struts (S) support leaves and the chamber is supplied with compressed air (T). (B) View from above of shading pattern. Quadrants 1 and 3 are fully illuminated and quadrants 2 and 4 are shaded by filters.

Nutrient Patches (C and D). (C) Section through bowl used to impose patches of nutrient depletion on developing root system. Nutrient solution fed by peristaltic pumps is dripped continuously onto the surface at symmetrically arranged outlets (O). (D) View from above of nutrient distribution pattern. Quadrants 1 and 3 are supplied with nutrient-rich solution and quadrants 2 and 4 are supplied with nutrient-poor solution.

in an equiproportional mixture under productive glasshouse conditions for 16 weeks.

The results (Figure 4) reveal a consistent relationship between the increment of dry matter to the undepleted sectors in both assays and the capacity for dominance in the competition experiment. It is interesting to note, however, that the difference in status achieved in the experimental community by the dominant and subordinate species was not correlated with plasticity (i.e., the precision with which the increment of root and shoot was concentrated in the undepleted sectors). A statistically significant inverse relationship was detected in which the greatest precision of allocation within shoot and within root was evident in the subordinate species. From this we may conclude that the capacity for dominance in a species mixture growing on fertile soil is not related to plasticity per se. Under such conditions dominance appears to depend rather more on a high relative growth rate and a massive architecture above and below ground. These attributes allow coarse-grained foraging of a large volume of canopy space and soil but they are not compatible with precise location of shoots and roots in local parts of the resource mosaic; this is more easily achieved by the smaller leaves and roots of subordinate species.

These results have the important implication that plant community structure under productive conditions is the product of a balance between the resource foraging characteristics of dominant and subordinate components. It seems likely that there is an oscillating equilibrium between (1) the tendency, in the absence of perturbation, for potential dominants to monopolize resource capture and drive the community toward monoculture and (2) the tendency, following perturbation and temporary debilitation of the dominants, for subordinates to generate diversity by exploiting local parts of the resource mosaic. Crucial to this hypothesis is the idea that during phases in which dominance is strongly expressed, the survival of subordinate species will depend on their ability for precise location within any local patches of sunlight and mineral nutrients unexploited by the coarse-grained resource interception of the dominants.

VI. Complicating Factors

A. Phylogeny

Although useful generalizations concerning plant design and community structure can be drawn from the kind of broad principles applied in this paper, it is necessary to recognize the existence of additional factors that may be important in particular circumstances. This can be illustrated by

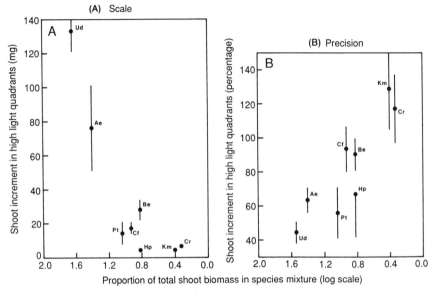

Proportion of total shoot biomass in species mixture (log scale)

Figure 4   (Top panel) Allocation of shoot biomass to unshaded patches within a heterogeneous light environment plotted against species ranking in an experimental community; (A) increment to shoot biomass (mg) in unshaded quadrants; (B) increments to shoot biomass in unshaded quadrants as percentage of total shoot increment. Note that values can exceed 100% because of the capacity of some of the shoots formed prior to the introduction of shading to move phototropically. Ae, *Arrhenatherum elatius;* Be, *Bromus erectus;* Cf, *Cerastium fontanum;* Cr, *Campanula rotundifolia;* Hp, *Hypericum perforatum;* Km, *Koeleria macrantha;* Pt, *Poa trivialis;* Ud, *Urtica dioica.* Each value is the mean of five replicates; vertical bars are 95% confidence limits. For (B), $r = 0.42$, $P<0.01$, $n = 8$.

Species ranking was determined by growing a mixture of the eight species in a heated glasshouse (temperature range 15–25°C) from November 1986 to March 1987 with natural light supplemented for 16 h per day with mercury vapor lamps supplying a PPFD of 70 μmol m^{-2} s^{-1}. Equal numbers of seedlings of each species were planted 25 mm apart in 160-mm-diameter pots of sand irrigated every 10 days with a nutrient solution three times the concentration of Rorison solution. Live shoot biomass was harvested after 16 weeks. Species ranking was calculated as $\log_{10}$ (species biomass × 100/total biomass).

(Opposite panel) Allocation of root biomass to undepleted patches of a nutritionally heterogeneous environment plotted against species ranking in an experimental community; (A) increment to root biomass (mg) in nutrient-rich quadrants; (B) nitrogen gain (mg) to whole plant; (C) increment of dry matter to root biomass in nutrient-rich quadrants as percentage of total root increment. Total plant nitrogen was determined using the micro-Kjeldahl technique and ammonia gas diffusion method. Each value is the mean of five replicates; vertical bars are 95% confidence limits. For (C), $r = 0.48$, $P<0.01$, $n = 8$.

Scale

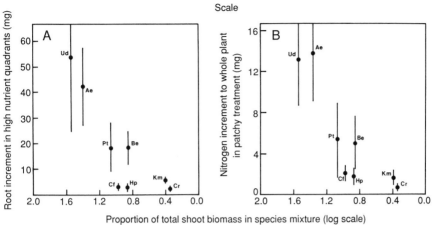

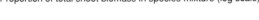

Proportion of total shoot biomass in species mixture (log scale)

Precision

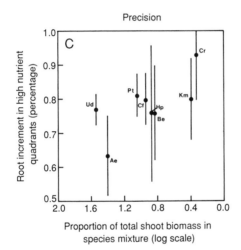

Figure 4   *(Continued)*

reference to Figure 4C (bottom panel), which reveals a tendency for the grasses (Km, Be, Pt, and Ae) to exhibit less precise foraging in response to shading than the four broad-leaved herbs (Cr, Hp, Cf, and Ud). Here we may suspect the operation of a phylogenetic constraint arising from the lack of internode extension in many grasses.

A similar complication is suggested in Figure 5, which compares the precision of root foraging (assessed by the assay method described in Figure 3) in a large number of British plants classified by morphological criteria (canopy height, lateral spread, and litter accumulation) known to be strongly correlated with their capacity for community dominance (Grime, 1973; Grime, *et al.* 1988). These data confirm that there is a general tendency for root foraging to be more precise in subordinate species. It is also apparent, however, that regardless of their potential status in communities, most of the grasses are less precise than the dicotyledons; further research is required to identify the basis of this phylogenetic constraint on root development.

B. Water Supply

So far in this chapter, root and shoot responses to resource heterogeneity have been considered with an exclusive emphasis on light and mineral nutrients. As other chapters in this volume testify, in some habitats seasonal and spatial heterogeneity in moisture supply has exercised strong selection pressures, not least through restriction of opportunities for photosynthesis and mineral nutrient capture.

An example of such constraint is apparent in Figure 5, where it is evident that, in comparison with the majority of subordinates, two dicotyledons, *Hieracium pilosella* and *Leontodon hispidus*, exhibit low precision in the mineral nutrient foraging assay. Examination of the morphology of the roots harvested from the tests revealed that, in these species, the weak discrimination between nutrient-rich and nutrient-poor patches was correlated with the appearance of a taproot at an early stage of seedling development. In their natural habitats, both species exploit crevices in shallow, droughted soils (Anderson, 1927; Grime, *et al.* 1988); experiments on a wider range of species are needed to examine the theoretical possibility of a trade-off between early allocation to taproots in crevice exploiters and early allocation to fine roots in plants of mesic conditions.

VII. Conclusions

Although the concept of limiting factors (Liebig, 1840) retains its value in diagnosing the causes of low yield in agriculture and horticulture,

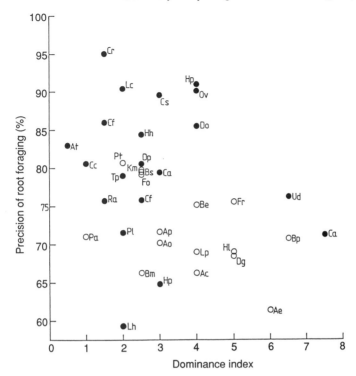

Figure 5 Comparison of the precision of root allocation to undepleted patches of a standardized nutritionally heterogeneous environment (see Figs. 1.3C and 1.3D) by a range of common British plants, classified by reference to a dominance index related to shoot morphology (see Grime, 1979, p. 129). Open circles are grasses, solid circles are dicotyledons. Ac = *Agrostis capillaris,* Ae = *Arrhenatherum elatius,* Ao = *Anthoxanthum odoratum,* Ap = *Avenula pratensis,* AT = *Arabidopsis thaliana,* Be = *Bromus erectus,* Bm = *Briza media,* Bp = *Brachypodium pinnatum,* Bs = *Bromus sterilis,* Ca = *Chenopodium album,* Can = *Chamerion angustifolium,* Cc = *Conyza canadensis,* Cf = *Cerastium fontanum,* Cfl = *Carex flacca,* Cr = *Campanula rotundifolia,* Cs = *Centaurea scabiosa,* Dg = *Dactylis glomerata,* Do = *Dryas octapetala,* Dp = *Digitalis purpurea,* Fo = *Festuca ovina,* Fr = *Festuca rubra,* Hl = *Holcus lanatus,* Hn = *Helianthemum nummularium,* Hp = *Hypericum perforatum,* Hpi = *Hieracium pilosella,* Km = *koeleria macrantha,* Lc = *Lotus corniculatus,* Lh = *Leontodon hispidus,* Lp = *Lolium perene,* Ov = *Origanum vulgare,* Pa = *Poa annua,* Pl = *Plantago lanceolata,* Pt = *Poa trivialis,* Ra = *Rumex acetosella,* Tp = *Thymus praecox,* Ud = *Urtica dioica.*

ecological analyses are placing an increasing emphasis on the interactive nature of resource capture by leaves and roots. The interdependence of photosynthesis and mineral nutrition limits the scope for evolutionary attunement to fertile and infertile soils by trade-offs in allocation between root and shoot. Comparisons of the root–shoot ratios of plants of contrasted ecology support this view. Assays measuring the plasticity of root and shoot systems in standardized patchy conditions strongly suggest that, under productive conditions, resource capture and fitness depend primarily on plasticity in allocation within both the root and shoot systems rather than modifications in root–shoot partitioning.

Interspecific comparisons of root and shoot responses to resource patchiness such as those described here (Figures 4 and 5) suggest that a trade-off exists between the scale (high in dominants) and the precision (high in subordinates) in resource foraging. This trade-off is highly relevant to the development of theories of the structure and dynamics of plant communities. Refinements of these theories will require recognition of phylogenetic constraints on root and shoot plasticity. There is also an urgent need to examine the interplay of mineral nutrition and water supply as determinants of root structure and function.

Acknowledgments

Most of the original data presented in this paper were obtained in collaboration with Mrs. J. M. L. Mackey and Dr. B. D. Campbell as part of the Integrated Screening Programme, a UCPE initiative funded by the Natural Environment Research Council.

References

Aerts, R., Boot, R. G. A., and van der Aart, P. J. M. (1991). The relation between above- and below-ground biomass allocation patterns and competitive ability. *Oecologia* **87,** 551–559.

Al-Mufti, M. M., Sydes, C. L., Furness, S. B., Grime, J. P., and Band, S. R. (1977). A quantitative analysis of shoot phenology and dominance in herbaceous vegetation. *J. Ecol.* **65,** 759–791.

Anderson, V. L. (1927). The water economy of the chalk flora. (Studies of the vegetation of the English chalk 5). *J. Ecol.* **15,** 72–129.

Band, S. R., and Grime, J. P. (1981). Chemical composition of leaves. *Annual Rep. Univ. Sheffield: Unit Comp. Plant Ecol. (NERC)* pp. 6–8.

Berendse, F., and Elberse, W. T. (1989). Competition and nutrient losses from the plant. *In* "Causes and Consequences of Variation in Growth Rate and Productivity of Higher Plants" (H. Lambers, H. Konings, M. L. Cambridge, and T. L. Pons, eds.), pp. 269–284. SPB Academic Publishing, The Hague, The Netherlands.

Bhat, K. K. S., and Nye, P. H. (1973). Diffusion of phosphate to plant roots in soil. 1. Quantitative autoradiography of the depletion zone. *Plant Soil* **38,** 161–175.

Brouwer, R. (1962a). Distribution of dry matter in the plant. *Neth. J. Agric. Sci.* **10,** 361–376.

Brouwer, R. (1962b). Nutritive influences on the distribution of dry matter in the plant. *Neth. J. Agric. Sci.* **10,** 399–408.

Brouwer, R. (1963). Some aspects of the equilibrium between overground and underground plant parts. *Inst. Biol. Scheikd. Onderz. Landbouwgewassen, Wageningen, Jaarb.*, pp. 31–39.

Campbell, B. D., and Grime, J. P. (1989a). A new method of exposing developing root systems to controlled patchiness in mineral nutrient supply. *Ann. Bot. (London)* [N.S.] **63,** 395–400.

Campbell, B. D., and Grime, J. P. (1989b). A comparative study of plant responsiveness to the duration of episodes of mineral nutrient enrichment. *New Phytol.* **112**, 261–267.

Campbell, B. D., Grime, J. P., and Mackey, J. M. L. (1991). A trade-off between scale and precision in resource foraging. *Oecologia* **87**, 532–538.

Chapin, F. S., II (1980). The mineral nutrition of wild plants. *Annu. Rev. Ecol. Syst.* **11**, 233–260.

Corré, W. J. (1983a). Growth and morphogenesis of sun and shade plants. I. The influence of light intensity. *Acta Bot. Neerl.* **32**, 49–62.

Corré, W. J. (1983b). Growth and morphogenesis of sun and shade plants. III. The combined effects of light intensity and nutrient supply. *Acta Bot. Neerl.* **32**, 277–294.

Crick, J. C., and Grime, J. P. (1987). Morphological plasticity and mineral nutrient capture in two herbaceous species of contrasted ecology. *New Phytol.* **107**, 403–414.

Davison, A. W. (1964). Some factors affecting seedling establishment in calcareous soils. PhD Thesis, University of Sheffield.

Davy, A. J., and Taylor, K. (1975). Seasonal changes in the inorganic nutrient concentrations in *Deschampsia caespitosa* (L.) Beauv. in relation to its tolerance of contrasting soils in the Chiltern Hills. *J. Ecol.* **63**, 27–39.

Donald, C. M. (1958). The interaction of competition for light and for nutrients. *Aust. J. Agric. Res.* **9**, 421–432.

Drew, M. C. (1975). Comparison of the effect of a localized supply of phosphate, nitrate, ammonium and potassium on the growth of the seminal root system and the shoot in barley. *New Phytol.* **75**, 479–490.

Drew, M. C., Saker, L. R., and Aschley, T. W. (1973). Nutrient supply and the growth of the seminal root system in barley. III. The effect of nitrate concentration on the growth of axes and laterals. *J. Exp. Bot.* **24**, 1189–1202.

Field, C., and Mooney, H. A. (1986). The photosynthesis–nitrogen relationship in world plants. *In* "On the Economy of Plant Form and Function" (T. V. Givnish, ed.), pp. 25–55. Cambridge Univ. Press, Cambridge, UK.

Givnish, T. J. (1982). On the adaptive significance of leaf height in forest herbs. *Am. Nat.* **120**, 353–381.

Gleeson, S. K., and Tilman, D. (1990). Allocation and the transient dynamics of succession on poor soils. *Ecology* **71**, 1144–1155.

Granato, T. C., and Raper, C. D. (1989). Proliferation of maize (*Zea mays* L.) roots in response to localised supply of nitrate. *J. Exp. Bot.* **40**, 263–275.

Grime, J. P. (1966). Shade avoidance and shade tolerance in flowering plants. *In* "Light as an Ecological Factor" (R. Bainbridge, G. C. Evans, and O. Rackham, eds.), pp. 187–207. Blackwell, Oxford.

Grime, J. P. (1973). Competitive exclusion in herbaceous vegetation. *Nature (London)* **242**, 344–347.

Grime, J. P. (1979). "Plant Strategies and Vegetation Processes." Wiley, Chichester.

Grime, J. P. (1987). Dominant and subordinate components of plant communities: Implications for succession, stability and diversity. *In* "Colonization, Succession and Stability" A. J. Gray, M. J. Crawley, and P. J. Edwards, eds.), pp. 413–428. Blackwell, Oxford.

Grime, J. P. (1988). Fungal strategies in ecological perspective. *Proc.—R. Soc. Edinburgh, Sect. B: Biol. Sci.* **94**, 167–169.

Grime, J. P., and Hunt, R. (1976). Relative growth rate: Its range and adaptive significance in a local flora. *J. Ecol.* **63**, 393–422.

Grime, J. P., Crick, J. C., and Rincon, J. E. (1986). The ecological significance of plasticity. *In* "Plasticity in Plants" (D. H. Jennings and A. J. Trewavas, eds.), pp. 5–19. Company of Biologists, Cambridge, U. K.

Grime, J. P., Hodgson, J. G., and Hunt, R. (1988). "Comparative Plant Ecology: A Functional Approach to Common British Species" Unwin Hyman, London.

Grubb, P. J., Kelly, D., and Mitchley, J. (1982). The control of relative abundance in communities of herbaceous plants. *In* "The Plant Community as a Working Mechanism" (E. I. Newman, ed.), Spec. Publ. Ser. Br. Ecol. Soc., No. 1, pp. 77–97. Blackwell, Oxford.

Gupta, P. L., and Rorison, I. H. (1975). Seasonal differences in the availability of nutrients down a podzolic profile. *J. Ecol.* **63**, 521–534.

Heal, O. W., and Grime, J. P. (1991). Comparative analysis of ecosystems: Past lessons and future directions. *In* "Comparative Analyses of Ecosystems: Patterns, Mechanisms and Theories" (Cole, J., Lovett, G. S., and Findlay S. R., eds.) pp. 7–23. Springer-Verlag, New York.

Hunt, R., and Nicholls, A. O. (1986). Stress and the coarse control of root–shoot partitioning in herbaceous plants. *Oikos* **47**, 149–158.

Huston, M. A., and Smith, T. M. (1987). Plant succession: Life history and competition. *Am. Nat.* **130**, 168–198.

Iwasa, Y., and Roughgarden, J. (1984). Shoot/root balance of plants: optimal growth of a system with many vegetative organs. *Theor. Popul. Biol.* 25, 78–104.

Jackson, R. B., and Caldwell, M. M. (1989). The timing and degree of root proliferation in fertile-soil microsites for three cold desert perennials. *Oecologia* **81**, 149–153.

Kirschbaum, M. U. F., Gross, L. J., and Pearcy, R. W. (1988). Observed and modelled stomatal responses to dynamic light environments in the shade plant *Alocasia macrorrhiza*. *Plant, Cell Environ.* **11**, 111–121.

Liebig, J. (1840). "Chemistry and Its Applications to Agriculture and Physiology." Taylor & Walton, London.

Monk, C. D. (1966). An ecological significance of evergreeness. *Ecology* **47**, 504–505.

Newman, E. I. (1973). Competition and diversity in herbaceous vegetation. *Nature (London)* **244**, 310.

Newman, E. I. (1983). Interactions between plants. Physiological Plant Ecology III: Responses to the Chemical and Biological Environment. Encyclopedia of Plant Physiology: New Series. Vol. 12C (O. L. Lange, P. S. Nobel, C. B. Osmond, and H. Zeigler, eds.). pp 679–710. Springer-Verlag, Berlin.

Nye, P. H., and Tinker, P. B. (1977). *In* "Solute movement in the soil system" Blackwell Scientific Publications, Oxford.

Olff, H., Van Andel, J. and Bakker, J. P. (1990). Biomass and shoot/root allocation of five species from a grassland succession series at different combinations of light and nutrient supply. *Funct. Ecol.* **4**, 193–200.

Passioura, J. B. (1963). A mathematical model for the uptake of ions from the soil solution. *Plant Soil* **18**, 225–228.

Pearcy, R. W., Osteryoung, K., and Calkin, H. W. (1988). Photosynthetic responses to dynamic light environments by Hawaiian trees. *Plant Physiol.* **79**, 896–902.

Rincon, E., and Grime, J. P. (1989). Plasticity and light interception by six bryophytes of contrasted ecology. *J. Ecol.* **77**, 439–446.

Ritz, K., and Griffiths, B. S. (1987). Effects of carbon and nitrate additions to soil upon leaching of nitrate, microbial predators and nitrogen uptake by plants. *Plants and Soil* **102**, 229–237.

Salisbury, E. J. (1916). The emergence of the aerial organs of woodland plants. *J. Ecol.* **4**, 121–218.

Sharkey, T. D. (1985). Photosynthesis in intact leaves of C_3 plants: Physics, physiology and rate limitations. *Bot. Rev.* **51**, 53.

Shields, J. A., Paul, E. A. and Lowe, W. E. (1973). Turnover of microbial tissue in soil under field conditions. Soil Biology and Biochemistry **5**, 753–764.

Shipley, B., and Peters, R. H. (1990). A test of the Tilman model of plant strategies: Relative growth rate and biomass partitioning. *Am. Nat.* **136,** 139–153.

Taylor, A. A., DeFelice, J., and Havill, D. C. (1982). Seasonal variation in nitrogen availability and utilization in an acidic and calcareous soil. *New Phytol.* **92,** 141–152.

Tilman, D. (1988). "Plant Strategies and the Structure and Dynamics of Plant Communities." Princeton Univ. Press. Princeton, NJ.

Tilman, D. (1989). Competition, nutrient reduction and the competitive neighbourhood of a bunchgrass. *Funct. Ecol.* **3,** 215–219.

Tilman, D., and Cowan, M. L. (1989). Growth of old field herbs on a nitrogen gradient. *Funct. Ecol.* **3,** 425–438.

2

The Physical Nature of Solar Radiation in Heterogeneous Canopies: Spatial and Temporal Attributes

Dennis Baldocchi and Serge Collineau*

I. Introduction

When tramping through the woods or a field an observant hiker will notice a plethora of bright and dark light patterns throughout the vegetation and on the ground. The shape, lifetime, and amount of radiation in these sun and shade light patches are of great importance to plants. For example, light energy directly drives many fundamental plant and biophysical processes (photosynthesis, stomatal conductance, transpiration, and leaf temperature). Light energy also indirectly influences many secondary plant processes. A list of these processes includes plant growth, seedling regeneration (Waring and Schlesinger, 1985; Chazdon, 1988), the vertical structure and crown shape of forest stands (Terborgh, 1985; Oker-Blom, 1986), and the uptake and emission of trace gases that participate in biogeochemical cycling and atmospheric chemistry (Baldocchi, 1991).

The qualitative nature of the light environment within a plant canopy can be described with ease. The flux density of light energy received at a particular location inside a plant canopy consists of beam and diffuse solar radiation that penetrates through gaps in the canopy. It also contains complementary radiation that is generated by the interception and the consequent (wavelength-dependent) transmission through leaves and

* Permanent address: Institut National de la Recherche Agronomique (INRA), Station de Bioclimatologie, 78850, Thiverval-Grignon, France.

reflection by leaves and soil (Lemeur and Blad, 1974; Ross, 1976; Myneni *et al.*, 1989). On the other hand, it is very difficult to quantify the light environment in a plant canopy because the light environment exhibits much spatial and temporal variability. This variability is associated with structural and environmental heterogeneity on a variety of space and time scales. Key factors causing heterogeneity in the canopy light environment include: (1) the clumping and gapping of foliage, (2) gaps in canopy crowns due to treefall or cultivation practices, (3) spatial variations in leaf orientation angles, (4) penumbra, (5) leaf flutter, (6) clouds, (7) topography, (8) seasonal trends in plant phenology, and (9) seasonal and diurnal movement of the sun. Determining the existence and extent of the cited factors is a critical component of any theoretical or experimental study on radiative transfer in heterogeneous plant canopies.

The ecological and environmental importance of light transfer through plant canopies has generated an extensive body of research reports; authoritative reviews of this research typically cite many hundreds of works (Anderson, 1964, 1971; Lemeur and Blad, 1974; Norman, 1975; Ross, 1975, 1981; Myneni *et al.*, 1989; Oker-Blom *et al.*, 1991). Yet, the main body of work on this topic has concentrated on measuring and modeling radiative transfer through *uniform* crop canopies, instead of through *heterogeneous* vegetation stands (which are of prime interest in this volume). In this chapter we describe the spatial and temporal characteristics of solar radiation in heterogeneous plant canopies. To accomplish this goal, we describe measurement and sampling requirements and discuss theoretical models and analysis methods that are used to measure, interpret, and calculate radiative transfer through heterogeneous canopies.

Before we proceed, several terms used in this chapter must be defined. Employing the vernacular of the radiative transfer field, we define a heterogeneous canopy as one that *cannot* be abstracted as a turbid medium with randomly positioned leaves. There is also imprecision in the literature concerning usage of the terms solar radiation and light. Solar radiation brackets wave bands between 0.15 and 4 μm. Light is visible electromagnetic radiation, which has wave bands between 0.4 and 0.7 μm and drives photosynthetic reduction reactions. The term "light" typically refers to a flux density of photons, whereas the term "solar radiation" refers to a flux density of energy. These two terms can be used interchangeably when the factors governing the transfer of photons and energy are wavelength independent, such as the probability of beam transmission. The terms are not interchangeable when differential absorption and scattering, which are wavelength dependent, are involved.

II. Measurement and Sampling Principles

To interpret the literature on light transfer in heterogeneous plant canopies properly, one must understand the measurement and sampling
requirements that are needed to assess the canopy light environment.
First, there are intrinsic differences in the principles and relative merits
of thermal detectors (they measure the flux density of radiant energy)
and photon detectors (they measure the flux density of quanta within a
selected spectral wave band); several excellent reviews cover this topic
(Anderson, 1971; Pearcy, 1989; Kubin, 1971; Fritschen and Gay, 1979;
Sheehy, 1986), so this information is not repeated here. Second, the
reader must recognize that many early and contemporary works in the
phytoactinometric literature are unsuitable for detailed analysis (e.g.,
Turton, 1985). Light measurements are of questionable quality if:
(1) the spectral response of the instrument is not adequate; (2) variations
in the spectral composition of light within and above the canopy are not
considered; (3) the time constant of the sensor is too long to properly
account for changes in irradiance as the sun moves across the sky or as
a sensor moves through sunflecks; (4) the sensor is not sampled with
sufficient frequency to account for high-frequency fluctuations or for
variations as the sun moves across the sky; and (5) inadequate spatial
sampling procedures are used (Anderson, 1964, 1971; Kyle *et al.*, 1977;
Salminen *et al.*, 1983; Herrington *et al.*, 1972). Instead of presenting an
exhaustive review, we have gleaned the relevant literature and only present information that does not suffer from these limiting attributes.

When conducting radiative transfer experiments in heterogeneous
canopies, problems 1 through 4 can be avoided with proper experimental
and sensor design. On the other hand, inadequate spatial sampling is a
tough problem to circumvent and merits further discussion. Sampling
problems arise because the light environment is highly variable in space
and time (Anderson, 1966; Norman and Jarvis, 1974; Gay *et al.*, 1971).
Consequently, enough samples are needed to ensure representative estimates of the population.

The complex spatial variability of light in a plant canopy can cause
sampling requirements to be severe. For example, Reifsnyder *et al.* (1971/
1972) suggest that 18 and 412 sensors are needed, respectively, to estimate the instantaneous field of direct radiation in a deciduous and conifer
forest within a standard error of 7 W m^{-2}. On the other hand, only
two instruments are needed to measure the more uniform diffuse light
environment in these disparate forest types.

On the basis of elementary statistical theory, and a liberal assumption
that the probability distribution of light is Gaussian, we calculate how

many sensors are required to estimate the sample mean of radiation within a plant canopy. Table I shows that the number of samples needed to measure the light regime within 10% of the population mean increases from 3 to 690 as the spatial coefficient of variation (CV; the ratio between the sample standard deviation and the mean) of the population increases from 10 to 150%. To give the reader a better perspective on these sampling requirements, the spatial CV of light above the floor of deciduous, tropical, and conifer forests often exceeds 100% (Reifsnyder *et al.*, 1971/1972; Gay *et al.*, 1971; Baldocchi *et al.*, 1986; Oberbauer *et al.*, 1988). Therefore, it is more often the rule than the exception to implement a large number of instruments to sample the short-term, radiation field near a forest floor correctly. On the other hand, the coefficient of variation is less than 30% in the crown of a fully leafed deciduous forest or throughout a leafless forest (Baldocchi *et al.*, 1986) and is between 20 and 40% in chrysanthemum and Scots pine canopies (Acock *et al.*, 1970; Smolander, 1984). Hence, fewer sensors are warranted in these circumstances. For experimental design purposes, the spatial variability of radiation in vegetation stands generally increases with depth in the canopy and increases as the proportion of beam to diffuse radiation increases (Acock *et al.*, 1970).

Fewer instruments are needed to evaluate the daily average radiation environment. For example, Reifsynder *et al.* (1971/1972) recommend only 1 and 10 instruments to measure daily-averaged direct radiation in deciduous and coniferous forests, respectively. Gay *et al.* (1971) recommend only five sensors to study the daily mean light environment in a uniform loblolly pine plantation. In this case, fewer sensors are needed to measure long-term averages because the coefficient of variation of

Table I A Statistical Estimate of the Number of Sensors Needed to Define the Light Environment with a Given Percent of the Population Mean ($n = t^2s^2/d^2$) [a]

CV, coefficient of variation (percent)	n, number of samples (within 10% of the population mean)	n, number of samples (within 5% of the population mean)
150	609	865
100	270	382
50	68	96
25	17	24
10	3	4

[a] These computations are based on the simple assumption of a Gaussian distribution. t is Student's t statistic, s is the standard deviation, and d is the difference from the mean.

the flux density of light at the floor of a pine forest decreases from about 100% to about 20% as the averaging interval increases from 5 to 120 minutes.

Our analysis of sampling requirements does not provide information on the area that must be sampled. The above computations implicitly assume that the CV of the population has been ascertained over a representative area. The minimum sampling area to be considered in studying radiation in orchards or crops is the scale of the repeating unit of the planting pattern (Jackson, 1980). For canopies that are discontinuous in two directions, Jackson (1980) recommends evenly spacing sensors in a grid. If this is not possible, distribute the sensors in space to account for the diverse components of the light regime. Our analysis also says nothing about individual sensor size in comparison to domain size. Many workers, however, recommend small sensors to avoid the smearing of small sunflecks (Anderson, 1971; Herrington *et al.*, 1972; Kyle *et al.*, 1977).

In row crops, line sensors are often used to measure light across a row (Szeicz, 1974; Szeicz *et al.*, 1964; Pearcy, 1989; Anderson, 1971). The orientation of these sensors must account for row orientation, solar azimuth angle, and azimuthal dependence of the instruments. Anderson (1971) recommends that two orthogonally aligned instruments should be used. The reader should appreciate that using a line sensor to measure mean radiation flux densities may be useful only if one is correlating radiation with canopy scale flux density measurements of heat, water vapor, and CO_2. On the other hand, Anderson (1971) cautions that reporting mean values may be misleading. Separate measurements of light in sunflecks and shade are needed to calculate photosynthesis, stomatal conductance, or other nonlinear, light-dependent process. An alternative to using line-averaged sensors to measure light in plant canopies is the "mouse." This instrument traverses through a canopy and measures sunfleck area and fractional interception of light and (Matthews and Saffell, 1987; Ross, 1981).

As can be deduced from Table I, the cost of implementing the proper amount of stationary instruments to measure short-term radiation below a canopy can be expensive. Alternatively, many workers use a sensor that traverses through vegetation to measure the solar radiation regime (Mukammal, 1971; Brown, 1973; Sinclair and Lemon, 1974; Clegg *et al.*, 1974; Norman and Jarvis, 1974; Sheehy and Chapas, 1976; Kyle *et al.*, 1977; Sinclair and Knoerr, 1982; Baldocchi *et al.*, 1986; Cohen and Fuchs, 1987; Pech, 1986; Black and Kelliher, 1989). Not only do such systems allow a spatial estimate of the radiation field to be made, but they reduce the absolute error that would otherwise arise by implementing multiple instruments. The limitation of this approach involves the mixing of spatial and temporal information by the moving sensor, making it difficult to separate one factor from the other.

To ensure high-quality radiation measurements, one must design a traversing path that is not too short (Norman and Jarvis, 1974; Cohen and Fuchs, 1987). Figure 1 shows that a sampling length of 10 to 12 m is needed within a Sitka spruce forest and orange grove to converge the

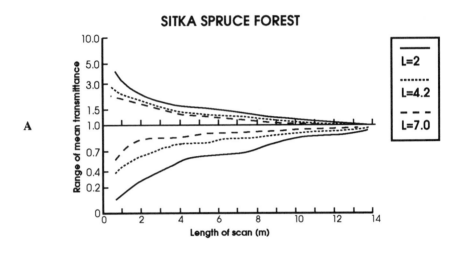

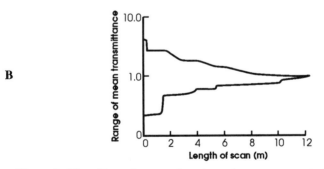

Figure 1 The effect of transect length on the measurement of light transmittance. These transmittances have been normalized by the mean transmittance that was measured over the total transect. (A) Sitka spruce for various cumulative leaf area indices (L) (after Norman and Jarvis, 1974). (B) An orange hedgerow orchard (after Cohen and Fuchs, 1986).

range of mean transmittance measurements (Norman and Jarvis, 1974; Cohen and Fuchs, 1987). As a broad guideline for designing the length of a traversing system, one should account for the length scale of the largest repeating light pattern (Jackson, 1980). Such scales are associated with plant spacing, size of gaps, or the diameter of tree crowns.

The rate that a sensor moves through the canopy and the rate at which it is sampled are other design issues to be considered. To avoid smearing of sensor signals as a sensor moves through a sequence of sunflecks, it is necessary to use a sensor with a short time constant or one that moves slowly through the canopy. Herrington *et al.* (1972) report that the velocity of the sensor should be less than the product of the minimum wavelength being resolved times the data system cutoff frequency. Consequently, the time response of the sensor (T) should be less than the ratio between the minimum wavelength and the velocity of the moving sensor. For example, if the minimum wavelength to be detected is 0.02 m and the system is being sampled at 10 Hz, then the translation velocity of the sensor should be less than 0.03 m s^{-1} and the sensor time constant should be less than 0.6 s.

III. Modeling and Analysis Methods

The complex nature of light in both uniform and heterogeneous canopies forces us to rely on geometrical and statistical models to describe and interpret spatial and temporal patterns of light and to characterize the orientation and position of foliage elements through which light passes (Lemeur and Blad, 1974; Ross, 1981; Myneni *et al.*, 1989). In the following we outline some modeling and statistical approaches that are used to evaluate light data in heterogeneous plant canopies.

A. Models for Predicting Light in Heterogeneous Plant Canopies

Light transmission through canopies that are heterogeneous in the horizontal dimension can be modeled by considering discrete arrays of vegetation envelopes (Norman and Welles, 1983; Lang and Xiang, 1986). Foliage envelopes have been abstracted as hedges with either rectangular or triangular cross sections (Jackson and Palmer, 1972, 1979; Cohen and Fuchs, 1987; Allen, 1974; Gijzen and Goudriaan, 1989), as arrays of ellipsoids (Mann *et al.*, 1980; Norman and Welles, 1983; Whitfield, 1986; Wang and Jarvis, 1990a,b; Grace *et al.*, 1987), as cones (Li and Strahler, 1987; Pukkula *et al.*, 1991; Kuuluvaninen and Pukkala, 1987; Oker-Blom and Kellomaki, 1982; Oker-Blom *et al.*, 1991), cubes (Fukai and Loomis, 1976; Roberts and Miller, 1977; Arkin *et al.*, 1978; Myneni

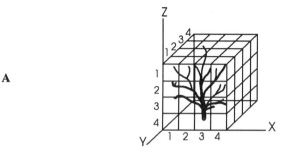

A

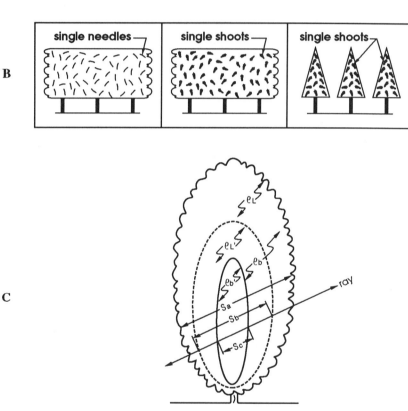

B

C

Figure 2 Various abstractions of plant canopies used to model radiative transfer in heterogeneous scenes. (A, Adapted from Roberts and Miller, 1975; B, from Oker-Blom, 1986; C, from Norman and Welles, 1983.)

et al., 1991), and cylinders (Brown and Pandolfo, 1969). Figure 2 shows examples of canopy abstractions that are used to model radiative transfer.

The simplest modeling approaches consider the shadows cast by opaque shapes (Jackson and Palmer, 1972; Arkin *et al.*, 1978; Li and

Strahler, 1987; Brown and Pandolfo, 1969; Grant, 1985). More complicated models calculate the probability of beam penetration through the
foliage envelope (e.g., Norman and Welles, 1983; Grace *et al.*, 1987;
Wang and Jarvis, 1990a,b). An attraction of foliage envelope models is
their inherent flexibility. They can be used to estimate light transmission
in closed canopies, row crops, or canopies with widely spaced individuals
(Norman and Welles, 1983).

Ellipsoid and cube array models have been tested by Roberts and
Miller (1977), Grace *et al.* (1987), Norman and Welles (1983), and Wang
and Jarvis (1990a) for shrubs, conifers, and crops. Tests of foliage envelope models demonstrate that they are capable of calculating the canopy
light environment well. However, many envelope models have distinct
weaknesses that merit discussion. For example, Grace *et al.* (1987) and
Wang and Jarvis (1990a) report that agreement between calculated and
measured transmittance values in pine and spruce canopies is good over
short time intervals, but they caution that small errors in measuring
crown shape, position, and leaf area can cause errors in predicted transmission values. When transmittance values are averaged over a daily
basis the consequence of these structural errors is usually small. Across
corn rows, an ellipsoid model does not perfectly mimic the radiation
field, but nearly all predictions fall within one standard deviation of
measurements (Norman and Welles, 1983).

Most envelope radiation transfer models neglect penumbra, clumping
of foliage, the content of woody biomass, and scattering of light (e.g.,
Pukkala *et al.*, 1991; Li and Strahler, 1987; Allen, 1974; Wang and Jarvis,
1990a,b). The omission of these processes may restrict the use of these
models under certain conditions. Candidates for restricted use include
conditions when whorl and branch geometry allows distinct foliage gaps
between internodes, foliage is clumped, and penumbra and light scattering are significant.

Procedural and Monte Carlo models are methods capable of evaluating
radiation transfer in three-dimensional canopies (Oker-Blom, 1985,
1986; Kimes *et al.*, 1980; Kimes and Kirchner, 1982; Myneni and Impens,
1985). The procedural approach uses information on the probability
density functions of foliage position, shape, size, and inclination to construct a numerical plant canopy within a defined region. Then shadow
projection of leaves are computed to determine if an "observer" is in
full sun, shade, or penumbra (Myneni and Impens, 1985; Myneni *et al.*,
1986); the computation of these projections depends on the leaf angle
and angle of incoming sunlight.

Monte Carlo models use stochastic theory to follow the transfer of an
ensemble of photons; typically, the transfer of 10^6 photons is followed
to obtain reliable results (Spanier and Gelbard, 1969; Myneni *et al.*,
1989). Monte Carlo approaches are powerful for they do not depend

on simplifying assumptions, which are required to solve analytical models for light transfer in heterogeneous canopies. Consequently, they can be made as complicated as necessary, as long as information on the distribution and dispersion of foliage and its optical properties is available; for example, these models can calculate the light environment within heterogeneous canopies with arbitrary leaf orientations and spatial dispersions and can consider scattering and penumbra (e.g., Oker-Blom, 1985). The method was pioneered by Tanaka (see Monsi *et al.*, 1973) and Szwarcbaum and Shaviv (1976). A review of Monte Carlo modeling of radiative transfer in plant canopies is presented by Ross and Marshak (1991).

One of the most comprehensive models for calculating radiative transfer in a three-dimensional vegetation scene was published recently by Myneni (1991) and Myneni *et al.* (1991). They solve the differential equation for photon transport in turbid media using the simplified discrete ordinates scheme; the angular direction of photon transport is discretized into a small number of directions. Tests show that the model performs well under field conditions.

B. Probability Statistics for Describing Light Transmission through Plant Canopies

The probability statistics are a useful tool for evaluating the probability of beam transmission through foliage envelopes (see Nilson, 1971; Lemeur and Blad, 1974; Myneni *et al.*, 1989). Probability density functions are classically derived by considering a plant canopy as a horizontally homogeneous turbid medium and by dividing the canopy into a number of statistically independent layers (N). The optical thickness of a layer is defined by the ratio between the cumulative leaf area index (L) and N. The probability that a ray of light passes through a foliage layer, without interception, is a function of the layer's leaf area that is projected in the direction perpendicular to the incoming ray (see Monsi and Saeki, 1953; Nilson, 1971; Mann *et al.*, 1977). Algebraically, the probability of a light ray passing through a foliage layer without interception is

$$P_0 = 1 - \frac{G \, \Delta L}{\cos \theta}. \tag{1}$$

ΔL is the leaf area index of the discrete layer and θ is the zenith angle of the light ray. G is the leaf orientation function. The G function defines the mean projection of unit leaf area onto a plane perpendicular to an incoming light ray (Ross, 1975, 1981). When the solar elevation and the leaf inclination angles are orthogonal, the probability of beam penetration is minimized. On the other hand, when solar and leaf elevation angles are similar, the probability of beam penetration is maximized (Lemeur and Blad, 1974; Ross, 1975).

From elementary calculus, one can derive the Poisson distribution for calculating zero contacts (the probability of beam penetration) from Eq. (1) by taking the limit of N approaching infinity:

$$P_0 = \exp\left(\frac{-GL}{\cos\theta}\right). \tag{2}$$

When applying this classic equation several assumptions must hold. Leaves must be small and opaque, they must be distributed randomly in space, they must not overlap, and only one interception event is allowed in any layer.

Although the Poisson distribution was derived for homogeneous canopies, a variation of Eq. (2) is used widely to describe the transmission of light through foliage envelopes in heterogeneous scenes (Norman and Welles, 1983; Wang and Jarvis, 1990b; Grace *et al.*, 1987; Lang and Xiang, 1986):

$$P_0 = \exp\left(-G\, a\,(z)\, s\right), \tag{3}$$

where $a(z)$ is leaf area density of the foliage envelope and s is the path length that a pencil of radiation must traverse through the envelope. When applying Eq. (3), the cited assumptions must hold only inside the foliage envelope.

The negative and positive binomial distributions can be used to calculate the probability of beam penetration through foliage envelopes with clumped or regularly dispersed foliage (Nilson, 1971; Acock *et al.*, 1970; Baldocchi *et al.*, 1985). The negative binomial probability distribution is defined as

$$P_0 = \left(1 + \frac{G\,g}{\cos\theta}\right)^{-L/g} = \exp\left[-\frac{L}{g}\ln\left(1 + \frac{G\,g}{\cos\theta}\right)\right]. \tag{4}$$

It is derived on the assumption that more than one contact can occur in a layer. Unfortunately, this distribution introduces a clumping factor, g, that is unknown and must be determined empirically. So far, attempts to define this parameter mechanistically have failed (Acock *et al.*, 1970; Lemeur and Blad, 1974).

The positive binomial probability distribution is defined as

$$P_0 = \left(1 - \frac{G\,g}{\cos\theta}\right)^{L/g} = \exp\left[\frac{L}{g}\ln\left(1 - \frac{G\,g}{\cos\theta}\right)\right]. \tag{5}$$

Its derivation allows none or one contact per layer. Both the negative and positive binomial distributions approach the Poisson distribution as the constituent clumping factor approaches zero.

The fourth approach for estimating the probability of beam penetration through a heterogeneous canopy is the Markov model (Nilson, 1971). Contrary to assumptions used to derive the Poisson and binomial models, the Markov model assumes that adjacent layers are dependent on each other. Hence, the future probabilistic behavior depends on the present status of the system. This model allows either one or no contacts and invokes a conditional probability, depending on whether or not a contact occurred:

$$P_0 = \exp\left[\frac{-\lambda_0 G L}{\cos \theta}\right]. \qquad (6)$$

λ_0 is the Markov factor. The Markov factor equals one if leaves are randomly distributed, is less than one if they are clumped, and is greater than one if leaves are distributed in a regular pattern. Hence, Eq. (6) becomes identical to the Poisson distribution when leaves are randomly dispersed.

The probability of diffuse radiation transmission is computed by integrating the probability of beam transmission (P) over the sky's hemisphere (Lemeur and Blad, 1974; Ross, 1981). If the sky brightness is not uniform, then the integrand must also be multiplied by a normalized sky brightness function (Lemeur and Blad, 1974).

How the probability of beam penetration varies with increasing clumping or regularity is shown in Figure 3. In all cases the probability of beam transmission decreases logarithmically with a linear increase in cumulative leaf area. Clumping, however, allows more light to penetrate through a given thickness of canopy, in comparison to canopies with randomly distributed leaves (the Poisson case). On the other hand, regularly dispersed foliage allows less light transmission than that transmitted through a canopy with randomly dispersed foliage.

Probabilistic Theory on Sunfleck Size Distributions   While many theories describe the probability that a beam of light will pass through vegetation without interception, only Norman and colleagues (Miller and Norman, 1971a,b; Norman *et al.*, 1971; Norman and Jarvis, 1975) have developed a theory to describe the size distribution of sunflecks. For simple convex leaves that are randomly distributed at a given height and have similar size, shape, and azimuthal orientation, the sunfleck segment length distribution $f(X)$ along a linear transect is

$$f(X) = \exp(-L)\exp(-\rho w X)(1 + \rho w X), \qquad (7)$$

where ρ is the number density of leaves per unit area and w is the width of leaves perpendicular to the linear transect. The function $f(X)$ represents the fraction of a transect that is occupied by gaps greater

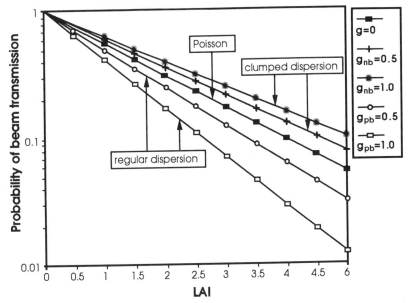

Figure 3   Theoretical influence of increasing leaf area and variations in leaf dispersion on the probability of beam penetration through vegetation. g is the clumping factor (see Eqs. (4) and (5)). g equal to zero is indicative of a Poisson probability distribution. g_{nb} and g_{pb} represent clumping factors for the negative and positive binomial distributions, respectively.

than X. In subsequent papers, Miller and Norman (1971b) and Norman *et al.* (1971) consider the role of penumbra on intensity of light on the sunfleck segments. This theory has been extended to heterogeneous forest canopies, with clumped foliage, by Norman and Jarvis (1975).

C. Defining the Structural and Optical Properties of a Canopy

The structural properties of a heterogeneous canopy must be quantified to model light transmission through its foliage. As can be deduced from Eqs. (1) through (6), the amount of leaf or plant area per unit volume and the leaf orientation distribution function, G, are the most essential factors needed to calculate the light environment within homogeneous and heterogeneous plant canopies. More information is needed if other structural heterogeneities arise.

If the plants are heliotropic or if their leaf distribution is asymmetrical, data on the azimuthal orientation of leaves are needed (Lemeur and Blad, 1974; Verstraete, 1987); sunflower, Jerusalem artichoke, corn, soybeans, and *Quercus coccifera* are examples of crops and trees that exhibit asymmetry in their leaf azimuthal distribution (Lemeur, 1973; Caldwell *et al.*, 1986). Many plant stands also exhibit vertical variations in leaf inclination angles (broadleaf forests: Stoner *et al.*, 1978; Miller, 1969a; Hutchison *et al.*, 1986; Caldwell *et al.*, 1986; Hollinger, 1989; woodland herbs: Anderson, 1971; alpine grasses: Tappeiner and Cernusca, 1989; salt marsh grasses: Morris, 1989; wheat: Denmead, 1976). Leaf angles are easily measured with a compass and protractor and the effect of their variation on radiative transfer can be accommodated by varying G with height in Eqs. (2) through (6).

The clumping of foliage is an attribute of many native plants stands (broadleaf forests: Baldocchi, 1989; Kruijt, 1989; Neumann *et al.*, 1989; conifer forests: Oker-Blom *et al.*, 1991; Norman and Jarvis, 1974; alpine shrubs: Tappeiner and Cernusca, 1989). Clumping is difficult to quantify. An estimate of clumping can be obtained by inverting the negative binomial or Markov models [Eqs. (4) and (6)] or by measuring the ratio between the total leaf area of the clump and its orthogonal projected area (Carter and Smith, 1985; Leverenz and Hinckley, 1990; Kruijt, 1989; Chen *et al.*, 1991).

Penumbra arises in tall plant stands and in canopies with narrow leaves (Miller and Norman, 1971b; Denholm, 1981a,b; Oker-Blom, 1985; Baldocchi, 1989); penumbra occurs when the angular diameter of a leaf, above a reference vantage point, does not fully obscure the sun (Miller and Norman, 1971b; Denholm, 1981a,b). Its evaluation requires information on leaf size and the vertical distribution of leaves.

In discontinuous canopies structural information is needed on the vertical and horizontal distribution of leaf area and the spacing and the dimensions of foliage envelopes (Norman and Welles, 1983; Grace *et al.*, 1987; Wang and Jarvis, 1990a; Myneni *et al.*, 1991) to evaluate Eq. (3). Unfortunately, few detailed data sets are available that describe the canopy architecture this completely (Whitehead *et al.*, 1991; Wang and Jarvis, 1990b).

It is often desirable to simplify the structural description of the canopy under study as much as possible. The degree that one can simplify the architectural description of a canopy depends on its inherent complexity. Short closed crop canopies are often the simplest to describe. In general, they tend to possess randomly dispersed leaves that have spherical inclination angles (Uchijima, 1976; Norman, 1979; Ross, 1981); this greatly simplifies the solution of Eq. (2). Of course, other leaf angle and spatial distributions occur (see de Wit, 1965). Rice, for example, has an erectophile leaf angle distribution, whereas corn has a plagiophile orientation (Uchijima, 1976).

Orchards and immature or widely spaced crop canopies do not form a continuous canopy, but vary in either two or three dimensions. A description of the foliage envelope and the leaves it contains is often satisfactory for describing radiative transfer through such managed canopies (Jackson and Palmer, 1972; Jackson, 1980; Allen, 1974; Cohen and Fuchs, 1987).

Temperate broadleaf deciduous and evergreen forests tend to form closed canopies, except for narrow gaps between crowns (Horn, 1971; Terborgh, 1985). Consequently, a three-dimensional description of the canopy is not necessary to successfully model radiative transfer through these heterogeneous stands (Miller, 1969a; Stoner *et al.*, 1978; Baldocchi *et al.*, 1985; Baldocchi, 1989). On the other hand, several structural heterogeneities must be considered when modeling radiative transfer through broadleaf forests. These include the clumping of foliage within crowns and vertical variations in leaf inclination angles (Miller, 1969a; Stoner *et al.*, 1978; Baldocchi, 1989; Chason *et al.*, 1991; Neumann *et al.*, 1989). Penumbra is also significant in broadleaf forests (Baldocchi, 1989) because relatively small leaves (~0.1m) are distributed over tens of meters (Hutchison *et al.*, 1986). Therefore, information on leaf size and the vertical distribution of leaves is required to compute penumbra (Denholm, 1981a,b).

Conifer stands may be either open or closed due to contributions from fires, climate, soil nutrition, moisture and temperature, topography, elevation, and latitude (Bonan and Shugart, 1989). Conceptually, the architectural description of a conifer canopy requires information on leaf area density and foliage inclination angles, plus a description of the shape and position of crown envelope and the geometry of shoots (Oker-Blom, 1986; Wang and Jarvis, 1990a; Norman and Jarvis, 1975; Carter and Smith, 1985; Leverenz and Hinckley, 1990; Whitehead *et al.*, 1991). If one must economize on the amount of structural data needed to model radiative transfer in a conifer forest, Wang and Jarvis (1990a) report that information on total leaf area and its distribution within the crown is more important than information on leaf inclination and crown shape.

Tropical forests are a mosaic of patches at different stages of development (Chazdon and Fetcher, 1984; Torquebiau, 1988). Seven kinds of eco-units, including treefall gaps, small and tall trees with respectively short or long boles, and tall dead trees, have been used to define components of the mosaic (Torquebiau, 1988). The eco-units vary in size from a few square meters to several hundred square meters. To account for the spectrum of gaps that may occur in a tropical forest, information on crown spacing and dimensions may be necessary to model radiative transfer successfully. Field tests are needed to demonstrate the applicability of simpler one-dimensional models in tropical forest stands.

Optical properties (reflection and transmission coefficients) of leaves are needed to calculate scattering of intercepted radiation (Ross, 1975;

Myneni *et al.*, 1989). Preferred absorption and scattering of selected wave bands act to alter the composition of solar radiation with depth into a canopy (Smith, 1980). Absorption is high (e.g., 90%) in the photosynthetically active wave band, and low (15 to 20%) in the near-infrared bands (wavelengths $> 7 \mu$m) (Ross, 1975; Norman, 1979).

D. Statistical Tools for Describing Spatial and Temporal Variability

1. Global Statistics: Evaluating Properties of the Radiation Regime   Most studies on light in plant canopies evaluate mean quantities. Yet, mean values of light flux density and transmittance are often useless statistics (Anderson, 1971; Miller and Norman, 1971a). Probability density functions (pdf) are a useful tool for studying light transfer through heterogeneous plant canopies; pdf's describe the proportion of transmitted radiation that is associated with different absolute or normalized flux density classes. This conclusion is drawn because the frequency distribution of light inside vegetation is often bimodal or skewed (Section IV,C).

Second-, third-, and fourth-order statistics, such as the variance (σ^2), skewness ($Sk = \overline{X'^3}/\sigma_x^3$), and kurtosis ($Kr = \overline{X'^4}/\sigma_x^4$), describe how dispersed, skewed, and peaked a population of data are. Statistical moments are also needed to calculate nonnormal, probability distributions ($P(x)$), such as the Gram-Charlier probability density function (von Mises, 1964):

$$P(x) = \frac{1}{\sqrt{2\pi}}\exp\left(-\frac{x^2}{2}\right)\left[1 + Sk\frac{(x^3-3x)}{6} + (Kr-3)\frac{(x^4-6x^2+3)}{24}\right]. \tag{8}$$

2. Spectral and Autocorrelation Analysis   From time series measurements at a single point or spatial measurements over a given domain, it is desirable to extract features such as the periodicity, duration, and size of dominant light patches. Two categories exist for evaluating these statistics from time- or space-dependent light data. Global methods calculate the periodicity of scales contained in the whole record and can be used to evaluate *integral* length and time scales. Local methods, on the other hand, detect the onset and end of nonperiodic events from a large set of data, as is needed to determine the duration and size of characteristic patches. Spectral and autocorrelation analyses belong to the first family. Wavelet analysis and conditional sampling schemes belong to the second family (Liandrat and Moret-Bailly, 1990; Antonia, 1981). Since neither family of analysis methods is used often by ecologists, a brief overview will be given.

Spectral analysis provides information on the contribution of different time or space scales to the variance of a series of data. Numerical transforms, such as those named after Fourier, Hartley, or Laplace, can be applied to convert time or space information into either frequency or wave number domains (Bracewell, 1990). The Fourier transform is the most common method used for this purpose. The Fourier transform $(\hat{F})$ of a time-dependent function, $f(t)$, into angular frequency domain (ω) is

$$\hat{F}(\omega) = \frac{1}{\sqrt{2\pi}} \int_{-\infty}^{+\infty} f(t)\, e^{-i\omega t} dt, \tag{9}$$

where ω equals 2π times natural frequency (n, cycles per second). This theory is derived from the concept that a time- or space-dependent data set can be approximated as the sum of a series of sine and cosine functions, whose periods vary from minus to plus infinity. Fast algorithms (Fast Fourier Transform or FFT) have made this computation affordable in time and allow its wide use (Press *et al.*, 1986). Detailed discussions on Fourier transforms are provided in Bracewell (1990), Press *et al.* (1986), and Hamming (1983).

Spectral densities ($S(\omega)$) are computed by squaring the modulus of the Fourier transform ($\hat{F}(\omega)$). Evaluating spectral densities, instead of Fourier transform coefficients, has distinct advantages because the spectral density at a particular angular frequency can be interpreted as the contribution of that particular frequency to the total variance. Thereby, examining the peaks in spectral plots can be used to identify the periodicity of the dominant time or length scale associated with the process under investigation.

Autocorrelation analysis can describe the time or length scales associated with the persistence or the repetition of a given event. In other words, it's a measure of how well a given function maintains its original value as one deviates in time, or space, from the initial point (see Becker and Smith, 1990). The basic definition of the autocorrelation of a function, R_f, is

$$R_f(\tau) = \frac{1}{T} \int_{O}^{T} f(t)\, f(t + \tau)\, dt, \tag{10}$$

where τ is the time lag and T is the integral time. Equation (10) can also be reposed to examine spatial correlations, by substituting τ with a spatial lag. Direct computation of the autocorrelation function is computer-time-consuming. An inverse FFT can be used to streamline this operation:

$$R_f(\tau) = \int_{-\infty}^{+\infty} \hat{F}(\omega)\, e^{i\omega\tau}\, d\omega. \tag{11}$$

Integrating Eq. (10) with respect to τ (or a spatial lag) and dividing by the variance of $f(t)$ gives information on time (or length) scales that represent the persistence of a periodic event.

3. Local Approaches: Event Detection   Many empirical methods exist in the turbulence literature for detecting events from time series or spatial measurements of light in plant canopies. These methods include conditional sampling and variable interval time averaging (VITA) schemes (Antonia, 1981; Subramanian *et al.*, 1982; Pearcy *et al.*, 1990). Each scheme has its own strengths and weaknesses. For example, most conditional sampling methods depend on the choice of a detection threshold value and a hold time. Consequently, Subramanian *et al.* (1982) states, "it is the onus of the experimentalist to demonstrate that the structure recognized is not an artifact of the detection method used."

Subramanian's statement is particularly relevant because few studies exist that examine sunflecks with event detection methods. To provide some guidance we will focus the following discussion on the simple, conditional sampling method proposed by Pearcy *et al.* (1990). A sunfleck is said to occur in a fluctuating record when the photon flux density exceeds a given threshold value (e.g., 50 μmol m^{-2} s^{-1}). Then the duration and corresponding proportion of the photon flux density are deduced for each sunfleck.

Unfortunately, the capability of any conditional sampling scheme depends on the definition of sunflecks, which is often arbitrary and author dependent (Chazdon, 1988). Moreover, it is difficult to propose a universal definition among both crops and forest canopies because of the relative effect of penumbra in the understory. For example, quick changes from shade to near-full sun tend to dominate in crops, whereas penumbra contributes significantly to the total photosynthetic photon flux density (PPFD) within forests and prevent "sunflecks" from attaining flux densities equal to the value incident at the top of the canopy.

Using photosynthetically active radiation (PAR) data measured below a deciduous forest, near Oak Ridge, Tennessee, we observe that the number of sunflecks detected is very sensitive to the choice of the threshold photosynthetic photon flux density (PPFD) (Figure 4); the number of events detected decreases monotonically as the threshold value decreases. Based on these data it can be concluded that a more sophisticated method is necessary to analyze sunflecks in heterogeneous canopies where penumbral effects are significant. The wavelet transform method is a new and objective way to identify sunflecks.

4. Wavelet Analysis   Wavelet analysis allows a user to identify and classify patterns, or singularities, in a data record according to position, size, and shape (Argoul *et al.*, 1989; Liandrat and Moret-Bailly, 1990; Collineau, 1992). Applying wavelet transforms to analyze a series of data can be compared with using a microscope to examine an object; both tools probe features at different magnifications and are able to translate

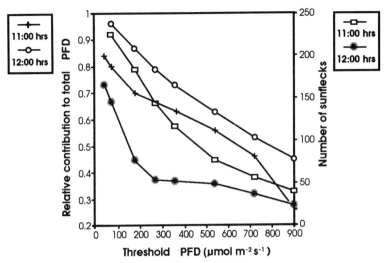

Figure 4   The effect of varying detection thresholds on recognizing sunflecks. These data are from PFD measurements made below a fully leafed deciduous forest growing near Oak Ridge, Tennessee, on Day 277 at 1100 and 1200 EST. The cross (+) and circle (O) refer to the relative contribution of sunflecks at a given threshold to the total PFD at the given times. The square and arterisk refer to the number of sunflecks at a given threshold.

the position of the focal point (Argoul *et al.*, 1989). A major advantage of this method, over Fourier analysis, is that the original record need not be periodic.

Information on position and scale of a given event is obtained by translating a wavelet function

$$g\left(\frac{t-b}{a}\right),$$

possessing various window widths (*a*), across a series of data (*h(t)*). A "wavelet" is a particular function (*g(t)*) that has a zero mean. Functionally, it is zero everywhere except in the domain of a window, where it exhibits one to several canceling oscillations. The width of the wavelet window is denoted *a* and its translational position is *b*. Both *a* and *b* are real numbers, and *a* must be greater than zero.

Wavelet transform (*T(a,b)*) are defined as a convolution between the original function (*h(t)*) and a wavelet function (*g(t − b)/a*)):

$$T(a,b) = \frac{1}{a}\int\limits_{-\infty}^{+\infty} h(t)\, g\left(\frac{t-b}{a}\right)\, dt. \tag{12}$$

The computation of $T(a,b)$ can be streamlined using Fourier transforms (denoted by capital letters with hats, $\hat{F}$) by taking advantage of a key property of convolutions—the Fourier transform of a convolution of two functions is the product of their Fourier transforms (Hamming, 1983). Thus, $T(a,b)$ can also be expressed by taking the inverse Fourier transform of the product of these two transformed functions:

$$T(a,b) = \int_{-\infty}^{+\infty} \hat{H}(\omega)\hat{G}(a\omega) \exp(i\,b\,\omega)\, d\omega. \qquad (13)$$

Since the transform of a wavelet function consists of a distinct banded window within some given frequency domain, this transform can be interpreted as a filtering function that is applied to the original data record (Liandrat and Moret-Bailly, 1990). $T(a,b)$ approaches a maximum value when the translation position and the scale of the wavelet match the characteristic scale of the data under examination.

A wavelet variance, $W(a)$, can be defined by integrating the square of $T(a,b)$ with respect to the translation position (b):

$$W(a) = \frac{1}{L} \int_{0}^{L} |T(a,b)|^2 db. \qquad (14)$$

By performing this operation for various dilation coefficients, b, we can obtain a spectral characterization of the events that match the wavelet dilation scale. The integration of the wavelet variance over a does not directly retrieve energy conservation. However, it reveals the scales that are best correlated with the wavelet shape. Energy conservation can be attained by applying wavelet-dependent coefficients to Eq. (14) (Collineau, 1992).

Figure 5 shows the capability of the wavelet transform method to identify sunflecks measured across a 30-m transect at the floor of a deciduous forest floor. For this analysis the "Mexican hat" wavelet (the second derivative of a Gaussian probability density function) is used because it is especially adept at detecting a sharp rise or decrease in the data record, through zero-crossings in the wavelet transform. These features can be used to determine starting and ending instances of sunflecks. In a deciduous forest, wavelet analysis detects sunflecks as events that exhibit both small and large excursions above the background diffuse radiation regime. These data support our assertion that sunflecks in a deciduous forest do not possess a distinct threshold flux density because penumbral shading diminishes the energy contained within them (Baldocchi et al., 1986). Consequently, the "Mexican hat" wavelet is pre-

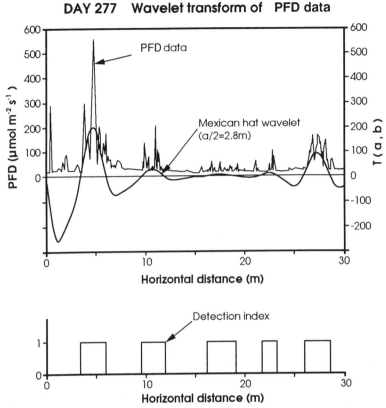

Figure 5   The application of wavelet analysis to detect sunflecks of a given size across a horizontal transect of PFD measured below a fully leafed deciduous forest (Day 277, 1989, 1100 to 1200 EST). A Mexican hat wavelet, whose dilation size is 2.8 m, is translated through the data record. The detection index denotes the detection and horizonal extent of sunflecks, detected with the noted wavelet.

ferred over others because it does not depend on some threshold value to detect events.

IV. Spatial Variability of Solar Radiation in Plant Canopies

Vertical, horizontal, and angular (elevational and azimuthal) distribution of leaves and the horizontal and vertical distribution of individual plants affect the manner by which pencils of radiation are able to penetrate through heterogeneous plant canopies. In this section we survey the literature and describe how the radiation regime varies vertically and horizontally in heterogeneous herb and forest stands.

A. Vertical Transmission of Beam Radiation through Plant Canopies

Data are compiled in Table II to ascertain which probability distribution describes the beam penetration through a variety of crop, grass, forest, and shrub plant stands. Several concepts can be deduced here. First, the Poisson probability density function describes beam penetration through many crop canopies. These data support the contention that vegetative elements within foliage envelopes of crops are often dispersed randomly. However, it is noteworthy that crops such as wheat, sorghum, and corn possess distinct heterogeneities. Multiple breakpoints occur in the Poisson distribution that are caused by spikelets, tassels, and seedheads. It must be emphasized, however, that the vegetation within each distinct foliage zone tends to be dispersed randomly.

The probability of beam penetration through most closed forest and shrub stands is described often by negative binomial or Markov probability density functions. Clumped foliage and vertical variations in leaf inclination angles in native and closed plant stands enhance the probability of beam penetration, as opposed to light transmission through a stand with randomly distributed foliage. In other words, these architectural features act to redistribute light throughout the canopy.

In conifer stands, negative binomial or Markov probability density functions best describe probability of beam penetration when the fundamental descriptor of the canopy is needle area index. However, data suggest that the probability of beam penetration is Poisson if one considers the basic foliage unit to be the projected shoot area index (Oker-Blom, 1986; Gower and Norman, 1991; Oker-Blom *et al.*, 1991; Chen *et al.*, 1991). This is a fortuitous and simplifying finding, for it suggests that shoots on conifers are distributed randomly in space. This observation is supported theoretically by Mann *et al.* (1977) and Norman *et al.* (1971), who show that the projection of shoots, not necessarily the individual foliage elements, must be randomly distributed for the Poisson distribution to be useful.

B. Mean Vertical Transmission of Total Radiation through Plant Stands

Most studies on light in plant canopies have focused on the vertical transmission of total radiation (beam plus diffuse and scattered) in some selected wave band. In practice, a simple Beer's Law relation adequately describes radiation transmission through many crop and forest canopies (Szeicz, 1974; Uchijima, 1976; Rauner, 1976; Jarvis and Leverenz, 1983; Baldocchi *et al.*, 1984a). Hence, the mean amount of radiation received at the canopy floor is most dependent on the canopy's leaf area index. Secondary factors include leaf inclination angle distributions, site lati-

Table II   A Listing of the Probability Density Functions That Describe the Probability of Beam Penetration into Plant Canopies[a]

Crop	Spatial distribution	Clumping parameter	Sources
Maize, sparse	Random: Poisson (1)		Ross (1981)
Maize, dense	Random: Poisson (3)		Ross (1981)
Maize, hexagonal planting	Random: Poisson		Sinclair and Lemon (1974)
Maize, 76-cm-wide rows	Clumped: Markov	$\lambda_0 = 0.4$ to 0.9.	Sinclair and Lemon (1974)
Sunflower	Random: Poisson (1)		Saugier (1976)
Sorghum	Random: Poisson (3)		Ross (1981)
Cotton	Random: Poisson		Stanhill (1976)
Oranges	Clumped: negative binomial	$g = 1.0$	Cohen and Fuchs (1987)
Deciduous forest full leafed (oak, maple, aspen)	Clumped: negative binomial or Markov	$\lambda = 0.53$ to 0.67; $\Delta L = 2.65$	Miller (1969b); Baldocchi *et al.* (1985); Neumann *et al.* (1989)
Rice	Random: Poisson (3)		Uchijima (1976)
Pinus radiata	Clumped		Whitehead *et al.* (1991)
Wheat	Random: Poisson (2)		Ross (1981)
Deciduous forest: leafless	Regular: positive binomial		Chason *et al.* (1991)
Pinus taeda	Clumped: Markov ($G = 0.47$)	$\lambda = 0.6$ to 1.0	Sinclair and Knoerr (1982)
Ryegrass	Regular: positive binomial ($L = 2.03$)		Acock *et al.* (1970)
Ryegrass	Random: Poisson ($L = 1.16$)		Acock *et al.* (1970)
Ryegrass	Clumped: negative binomial ($L = 0.82$)		Acock *et al.* (1970)
Lucerne	Clumped: negative binomial ($L = 2.94$)		Acock *et al.* (1970)
Pinus resinosa	Random: Poisson		Reifsnyder *et al.* (1971/1972)

(continues)

Table II   *(Continued)*

Crop	Spatial distribution	Clumping parameter	Sources
Grass			Sheehy and Chapas (1976)
Conifers: Sitka spruce	Grouped		Norman and Jarvis (1975)
Douglas fir	Clumped: Markov	0.36–0.45	Chen *et al.* (1991)
Sorghum	Clumped: Markov	0.8	Niilisk *et al.* (1979)
Cotton	Clumped: Markov	0.65	Niilisk *et al.* (1970)

[a] Number in parentheses indicates number of discrete zones that occur inside the canopy. Data in table are from studies investigating beam transmission, point quadrats interception, or gap probability along a horizontal transect.

tude, and season. Peculiar features and similarities that are associated with the transmission of light through a variety of cultivated and native plant stands are examined in the following sections.

1. Herbaceous Crops and Grasses   Light in the PPFD, infrared (IR), and shortwave radiation wave bands decreases monotonically with increasing leaf area in corn (Allen and Brown, 1965; Kyle *et al.*, 1977) and sorghum (Clegg *et al.*, 1974) and log-linearly in sugar beet, wheat, and bean (Szeicz, 1974). Attenuation with canopy depth is most rapid for PPFD (due to preferential absorption) and least for IR. Most crops surveyed had leaf area indices exceeding four and allowed less than 10% of PPFD to penetrate to the ground.

The amount of light transmitted through grass canopies is quite variable. Some stands allow up to 40% mean transmission, at leaf area indices exceeding seven, whereas other stands intercept almost all incoming light (Sheehy and Chapas, 1976). The amount of light transmitted through a grass canopy also depends on sky conditions. Light transmission through grass from an overcast sky is attenuated more slowly than light from a clear sky (Sheehy and Chapas, 1976). Opposite results have been reported for sorghum (Clegg *et al.*, 1974) and chrysanthemum (Acock *et al.*, 1970). When evaluating light transmission through grass canopies, it must be recognized that live and dead leaves intercept photons.

Saltmarsh grass communities (*Spartina alterniflora*) possess leaves that tend to be more erect near the top of the canopy than near the bottom. This architecture causes light to be attenuated less rapidly in the upper canopy than in the lower portion (Morris, 1989). At midday, only 5% of incoming PPFD penetrates through the canopy.

2. Orchards   Orchards consist of arrays of trees or hedges separated by major gaps. This arrangement is necessary to allow passage of cultivation equipment and to maintain high light levels on individual fruit, which is needed to attain proper ripening and color (Jackson, 1970). Consequently, the area-averaged leaf area and light interception values are often lower than in other canopy types. For example, Jackson (1970) reports that light transmission through an apple tree, integrated over a 2-week period, is about 18% at the base of trees and increases up to 78% as one radiates 3 m from tree bases. Theoretically, more than 80% interception is possible when half the ground area is covered with hedgerow and the hedgerow height is equal to or greater than the alleyway width (Jackson and Palmer, 1972). However, under certain cultivation schemes, extremely high interception values can be attained. For example, Cohen and Fuchs (1987) report that 1% transmittance occurs in a hedgerow orange canopy with a leaf area index (LAI) ranging between 6.1 and 8.6.

3. Temperate Broadleaf Forests   In most deciduous forest stands, solar radiation decreases with depth in a monotonic or log-linear fashion (Hutchison and Matt, 1977a; Baldocchi *et al.*, 1984a; Hicks and Chabot, 1985). Kinks, or inversions, in radiation profile measurements have been reported in some oak–hickory and aspen stands (Miller, 1969b; Thompson and Hinckley, 1977). These kinks are attributed to the reception of scattered beam radiation from upper leaves near gaps. However, it is noteworthy that only in cases when the light regime is undersampled are kinked profiles observed.

In some cases, 9 to 25% of incoming sunlight penetrates to the floor of fully leafed deciduous forests at midday (Miller, 1969b; Reifsnyder *et al.*, 1971/1972; Thompson and Hinkley, 1977; Rauner, 1976; Horn, 1971). On the other hand, another group of studies reports that less than 5% of incoming light is transmitted through broadleaf forests at high sun angles (Hutchison and Matt, 1977b; Baldocchi *et al.*, 1984a,b; Hicks and Chabot, 1985; Caldwell *et al.*, 1986; Terborgh, 1985; Horn, 1971). No clear consensus arises on why the amount of light transmitted through a temperate broadleaf forest ranges so widely. However, factors such as climate, site fertility, latitude, species, and elevation affect the amount of leaf area established by temperate broadleaf forests and, in turn, the amount of light they intercept.

4. Tropical Broadleaf Forests   Light gradients are severe in the upper portion of tropical forests. About 60% of incoming PPFD is absorbed in the upper 20% of the canopy, where foliage is densest (Torquebiau, 1988). Inversions in the PPFD interception profile have also been observed. PPFD profile inversions are thought to occur because crown gaps allow upper sources of scattered radiation to penetrate deeper into the canopy. Typically less than 1% of incoming radiation is transmitted through Amazonian, Malaysian, Indonesian, and Costa Rican tropical forests (Shuttleworth, 1984; Yoda, 1974; Torquebiau, 1988; Terborgh, 1985; Chazdon and Fetcher, 1984).

The occurrence of clouds also affects mean light transmission through tropical forests. Chazdon and Fetcher (1984) report that a greater percentage of light is transmitted through a canopy on a cloudy day than on a clear one.

5. Conifer Forests   Despite the economic importance and accessibility to researchers, relatively few studies measure light transmission through conifer stands with proper sampling procedures. Furthermore, despite the paucity of data, several studies focus on red pine (*Pinus resinosa*) (Mukkamal, 1971; Reifsnyder *et al.*, 1972; Pech, 1986). In one pioneering study, Mukkamal (1971) reports that mean transmission through a red pine forest in Canada ranges between 10 and 25% over the course of a clear day and ranges between 18 and 20% on a cloudy day. Mukkamal also reports that the irregularity of the canopy causes occasions when light transmission recorded at 1.8 m height exceeded values measured at 10 m. High values for light transmission (30%) through a red pine stand have also been reported by Reifsnyder *et al.* (1971/1972), whereas Pech (1986) reports a lower range of transmission values (5 to 7%) through another Canadian red pine stand.

The transmission of PPFD through loblolly pine (*Pinus taeda*) decreases logarithmically with linear increases in leaf area (Sinclair and Knoerr, 1982). Near noon about 20% of incoming radiation is transmitted through a canopy with a leaf area index of four. At lower sun angles (e.g., 10 degrees) only 4% is transmitted.

Sitka spruce stands, growing in a moist Scottish milieu, maintain high leaf area indices (8.2). Consequently, they allow less light transmission (10%) (Norman and Jarvis, 1974) than do loblolly pine, which grows in drier and warmer climates, and red pine, which grows in colder and harsher environments.

6. Comments on Light Transmission   Among all canopy types, it can be concluded that tropical forests allow the least amount of light to penetrate to the canopy floor. Terborgh (1985) has hypothesized why tropical forests allow three to four times less light to penetrate to the forest floor than temperate forests. Tropical forests can support twice as much leaf area as temperate forests because high sun angles and their planar or dome-shaped crowns allow enough sunflecks to penetrate gaps and support multiple-storied canopies. In contrast, conical-shaped tree crowns living at high latitudes, where sun angles are low, quickly extinguish light. Consequently, there is insufficient light in high-latitude forests to support the development of an understory (Terborgh, 1985). Mathematical analysis by Oker-Blom and Kellomaki (1982) supports the idea that disclike and conical tree crowns are effective absorbers of light at low and high latitudes.

C. Vertical Variation in the Probability Distribution of Light

The shape of the light transmission probability density function typically varies with depth in plant canopies. When the sun is high in the sky, the frequency histogram (an approximation of the probability density function) is negatively skewed near the top of the canopy, is bimodal within the foliage crown, and is positively skewed below the foliage crown (Figure 6). High flux density contributions are from sunflecks and low flux density contributions embody shade patches, consisting of diffuse

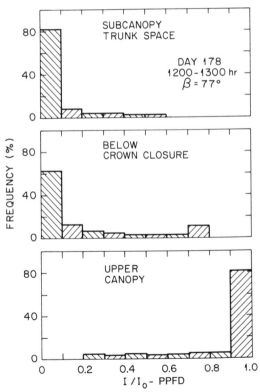

Figure 6 Frequency distributions of PFD measured in three distinct layers within a fully leafed deciduous forest (after Baldocchi *et al.*, 1986).

and scattered radiation (crops and herbs: Lemeur and Blad, 1974; Ross, 1981; Sinclair and Lemon, 1974; Acock *et al.*, 1970; Sheehy and Chapas, 1976; temperate forests: Hutchison and Matt, 1977a; Baldocchi *et al.*, 1986; conifer forests: Norman and Jarvis, 1974; Sinclair and Knoerr, 1982; tropical forests: Yoda, 1974; Chazdon and Fetcher, 1984). In tall forest canopies and canopies with narrow leaves, penumbra smears sunflecks and redistributes their energy into intermediate bins (Sinclair and Knoerr, 1982; Norman and Jarvis, 1974; Sheehy and Chapas, 1976; Baldocchi *et al.*, 1986). Sky conditions are another factor affecting the shape of light transmission pdf's. Sheehy and Chapas (1976) and Hutchison and Matt (1977a) report that the frequency distribution of light under a grass and a broadleaf forest stand was more uniformly spread across light classes when received from an overcast sky than from a clear sky.

Specific or unique features associated with pdf's for light transmission under various canopy classes are explored in the following sections. As a word of warning, most of the data to be discussed involve some convolution between space and time—pdf's are typically derived using a moving sensor or an array of sensors over a given sampling interval, such as an hour.

1. Crops and Grasses   The vertical variation in light probability density functions has been examined for several crops and grasses (sunflower: Lemeur and Blad, 1974; chrysanthemum: Acock *et al.*, 1970; corn: Niilisk *et al.*, 1970; Kyle *et al.*, 1977; Sinclair and Lemon, 1974; Allen and Lemon, 1972; grass: Sheehy and Chapas, 1976; beans: Impens *et al.*, 1970; soybeans: Pearcy *et al.*, 1990). Several distinct features are noticed. First, intermediate flux densities contribute little to the light transmission pdf of many crops. This feature is attributed to the lack of penumbral shade. Penumbral shade is minimal in crops and grasses because of their short stature and relatively broad leaves (Pearcy *et al.*, 1990). Consequently, sunflecks generally have flux densities equal to that of the sunlight incident on the canopy. However, we do not want to mislead the reader and suggest that penumbral shading is nonexistent in all short-stature vegetation. Sheehy and Chapas (1976) report that 11 to 16% of the area beneath grass canopies with unit leaf area is penumbra. But it must be recognized that grasses have much narrower leaves than beans, corn, and sunflowers. They therefore have a greater potential for casting penumbral shade.

Second, the patterns of the pdf are not identical for all crops. When the sun was near its zenith no sunflecks are observed near the bottom of a grass canopy ($L = 4$), whereas sunflecks are observed below corn (LAI $= 3+$). Near the top of a corn and sunflower canopy, about 20%

of measurements consist of full-intensity sunflecks, whereas less than 5% of observations at lower levels ($L = 2.5$, sunflower; $L = 3.2$, corn) are in full sunflecks (Lemeur and Blad, 1974; Sinclair and Lemon, 1974).

While the focus of this paper has been on the transmission of solar radiation and visible light, one study on net radiation (the balance between incoming and outgoing solar and terrestrial radiation) merits discussion. In a corn canopy, the net radiation flux density in sunflecks can exceed the net radiation balance above the canopy (Allen and Lemon, 1972). Furthermore, negative radiation balances can be observed in the shade. These two effects are due to the respective absorption and emission of infrared radiation from neighboring and constituent leaves.

2. *Temperate Broadleaf and Conifer Forests*   A distinct feature associated with light pdf's measured in temperate broadleaf and conifer forests is that penumbra smears sunflects significantly. This effect causes intermediate contributions to the canopy light environment and reduces the energy flux density within sunflecks located below midcrown (Norman and Jarvis, 1974; Hutchison and Matt, 1977a; Sinclair and Knoerr, 1982; Baldocchi *et al.*, 1986). For example, Figure 6 shows that between 20 and 30% of light measurements below crown closure of a deciduous forest are comprised of smeared sunflecks, by penumbra. Only above canopy closure do sunflecks contain full incident energy, whereas at midcanopy the maximum energy in sunflecks is about 75% of the incident level.

While it has been qualitatively established that light transmission pdf's are skewed, only two studies in the literature quantify the magnitude of this skewness. Within a Finnish Scots pine stand, the skewness coefficient rarely exceeds 2.5 over a wide range of incident light conditions (Smolander, 1984). In a fully leafed, broadleaf forest, skewness values at midday progressively increase with depth (Baldocchi *et al.*, 1986). Skewness values in the crown, below crown closure, and in the stem space are about -3, 2, and 3, respectively.

3. *Tropical Forests*   The shape of PPFD pdf's in tropical forests depends on whether one is in the understory or in a small or large gap. In a Costa Rican forest small (200 m^2) and large (400 m^2) gaps have a fairly uniform distribution of light flux densities in the range between 0 and 1000 μmol m^{-2} s^{-1} (Chazdon and Fetcher, 1984). In the dense understory of Costa Rican, Mexican, and Malaysian forests, pdf's are positively skewed (Chazdon and Fetcher, 1984; Chazdon *et al.*, 1988; Yoda, 1974). Between 74 and 100% of light measurements below a Costa Rican forest are less than 10 μmol m^{-2} s^{-1} during the wet and dry season, respectively, and fewer than 5% of measurements exceed 50 μmol m^{-2} s^{-1} (Chazdon and Fetcher, 1984). In another study at the same site,

Oberbauer *et al.* (1988) report that less that 2% of the understory light environment contained full sunlight and 90% of PPFD were less than 25 μmol m^{-2} s^{-1}. A slightly different picture is painted for the light environment of understory plants below a *Piper* species forest in Mexico. In this case, 97 to 99% of light (PPFD) is below 50 μmol m^{-2} s^{-1} (Chazdon *et al.*, 1988). Furthermore, the shaded environment is quite uniform, having a spatial coefficient of variation of 1.5%. The light environment within sunflecks, on the other hand, was quite variable—coefficients of variation ranged between 39 and 102%.

D. Horizontal Variations in the Canopy Light Environment

Horizontal variations in light within heterogeneous canopies can be as great as vertical variations. Such variation occurs because there are distinct locations where light does and does not penetrate through the foliage. Next we examine characteristics of the horizontal light regime in a number of canopies.

1. Crops and Orchards   Huge horizontal gradients in light can occur in row crops (Figure 7; Denmead *et al.*, 1962; Luxmoore *et al.*, 1970; Charles-Edwards and Thorpe, 1976). As one traverses from the base of individual stems to the row's midpoint, a distance often less than 0.40 m, light transmission values between 5 and 100% are observed. In general, the magnitude of light gradients depends on the ratio between canopy height and alley width, row orientation, latitude, leaf area density of the foliage envelope, direct to diffuse radiation ratio, solar elevation, and azimuth angles (Charles-Edwards and Thorpe, 1976; Jackson, 1980; Allen, 1974; Mutsaers, 1980).

Examples of how light varies across a row of immature corn and a hedgerow apple orchard are given in Figure 7. When the sun azimuth is nearly aligned with a row, light transmission varies symmetrically across the row; minimum values occur at both legs of the row where foliage density is greatest and maximum transmission occurs at midrow where beam interception is negligible. When the sun's azimuth is less than the row orientation azimuth angle, transmission is greatest on the face looking at the sun and is least on the posterior position where shading is greatest. The converse occurs when the solar azimuth is greater than the row's azimuth angle. When the sky is diffuse, light transmission varies symmetrically and unimodally across the row, reaching a maximum at midrow regardless of row orientation.

From a crop management perspective, east–west-oriented rows, at high latitudes, intercept less light than north–south rows (Allen, 1974; Charles-Edwards and Thorpe, 1976); on a daily basis, Allen (1974) reports that E–W-oriented wide-row sorghum intercepts 37% of incoming

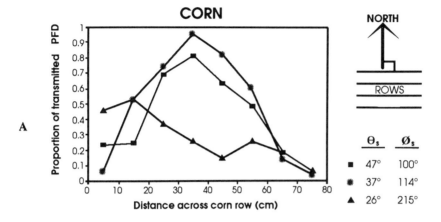

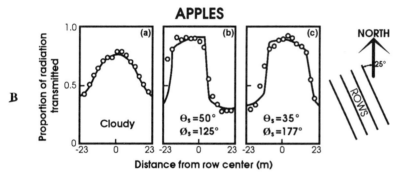

Figure 7   Horizonal variations of PFD across rows in (A) an immature corn canopy growing in eastern Oregon (D. Baldocchi, unpublished) and (B) an English apple orchard (after Charles-Edwards and Thorpe, 1976). θ_s is the solar zenith angle and ϕ_s is the solar azimuth angle. Insets show the orientation of the rows with respect to north.

radiation whereas N–S-oriented rows intercept 44% of incoming radiation; Charles-Edwards and Thorpe (1976) report that E–W rows absorb 13% less beam radiation in an apple orchard (LAI = 1.8, 4.5-m-wide rows, 2-m-high trees, crown width 1.8 m) than do N–S rows. East–west rows allow greater light penetration during the morning and afternoon hours than north–south rows because the solar azimuth angle is nearly parallel with the row orientation. Near noon the solar elevation angle is high; thus, the effects of row orientation on light transmission are minimal. At low latitudes north–south-oriented hedges intercept more light than east–west rows (Mutsaers, 1980; Charles-Edwards and Thorpe, 1976).

Small-scale spatial heterogeneity is a predominant feature of the light environment in many crops (oats, bean, sunflower, corn; Impens *et al.*, 1970). Spatial correlation coefficients between two adjacent miniature net radiometers are typically low, between 0.4 and 0.5. Only under cloudy skies, when the light environment is more uniform, are higher spatial correlation coefficients (0.8) observed.

2. Broadleaf Temperate Forests   Few studies describe the detailed features of horizontal variations in temperate broadleaf forests. Consequently, we obtained additional PPFD data below a forest near Oak Ridge, Tennessee, to examine horizontal features of the canopy light environment in greater detail; information on the canopy and experiment setup are presented by Baldocchi *et al.* (1984a,b, 1986) and Chason *et al.* (1991).

Figures 8A and B show a power spectrum and a spatial autocorrelation distribution of PPFD measured below the forest. The power spectrum (Figure 8A) demonstrates that light patches possessing wave numbers between 0.04 and 1 m^{-1} contribute most to the PPFD variance. The accompanying spatial autocorrelation analysis (Figure 8B) suggests that the integral length scale of these light patches is quite small; correlation coefficients drop rapidly with increasing lag and are decorrelated when the distance between two sensor readings exceeds 3m.

The threshold event detection and the wavelet analysis methods are used to study the size characteristics and energy contribution of sunflecks for 1 h in full-leaf (Day 277) and post-leaf-fall (Day 317) stages (Figure 9). The threshold even detection technique quantitatively shows that most sunflecks are small (over 70% are less than 0.5 m in length). Yet, these methods reveal that most energy observed under a foliated canopy is contained in two larger size classes, corresponding to lengths of sunflecks of 0.85 and 2.5 m. After leaf fall, the shape of the wavelet variance spectrum becomes broader, but peak energy is still contained in sunflecks 2 m in length. The plethora of small sunflecks explains why the spatial autocorrelation function is decorrelated over small lag scales.

3. Conifers   Horizontal transects of light in Sitka spruce and Scots pine stands indicate a diverse light environment punctuated by shade and sun patches (Norman and Jarvis, 1974; Smolander, 1984). Overall, one may conclude that sunflecks are relatively small. For example, a theoretical analysis of sunfleck segment size, by Norman and Jarvis (1975), shows that most sunflecks in a sparse canopy (LAI = 2) are less than 1.8 m long and that most sunflecks in the dense canopy (LAI = 6) are less than

DECIDUOUS FOREST

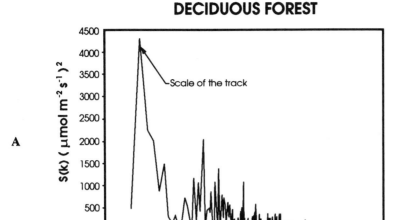

A

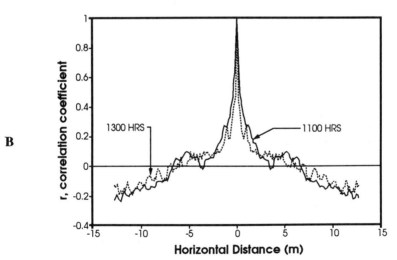

B

Figure 8 (A) Wave number power spectrum and (B) spatial autocorrelation analysis of PFD. Data were measured across a 30-m transect beneath a fully leafed deciduous forest, growing near Oak Ridge, Tennessee. Data are presented for two periods near midday on Day 277, 1989.

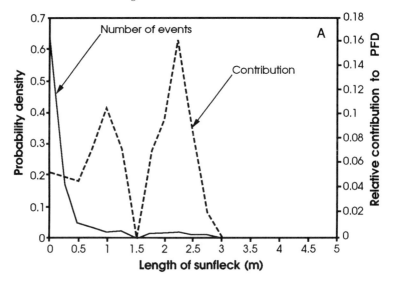

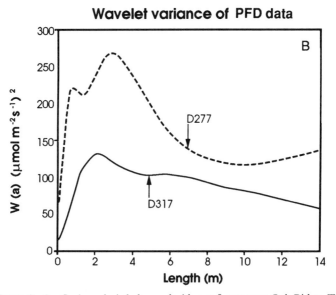

Figure 9 Sunfleck analysis below a deciduous forest near Oak Ridge, Tennessee. (A) Threshold detection spectrum. This method evaluates the number of sunflecks and relative energy contained within them in different size classes. The forest was fully leafed (Day 277,1100 to 1300 EST). A detection threshold of 43 μmol m^{-2} s^{-1} was used. (B) Wavelet variance spectrum. This figure yields information on the relative contribution of different-sized sunflecks to the variance of PFD measured below a fully leafed (Day 277) and leafless deciduous forests (Day 317).

0.3 m. The spatial autocorrelation function of light in Scots pine exhibits zero correlations at separation exceeding 0.18 m, for short averaging periods (5 min) (Smolander, 1984). On the other hand, spatial correlation coefficients are near one for a wider range of separation distances under overcast skies.

4. Tropical Forests Horizontal heterogeneity of radiation in tropical forests is dominated by gap dynamics. Consequently, an extensive effort has been made to quantify spatial patterns in tropical plant canopies. Typically, the amount of radiant energy in the canopy diminishes as one progresses from a gap to canopies under pioneer trees and to a dense mature forest (Chazdon and Fetcher, 1984; Pearcy, 1988; Torquebiau, 1988). For example, Torquebiau (1988) shows that mean daily transmission values through an Indonesian forest are 20, 3, and 0.8% for the three cited classes. In one other study, Chazdon and Fetcher (1984) report that mean daily light transmission observed in the understory and in a 200-m^2 and a 400-m^2 gap are 1–2%, 9%, and 20–35%, respectively.

The light environment within crowns is also highly variable. Oberbauer *et al.* (1988), for example, reports that the CV of light within crowns of saplings below a Costa Rican forest is 135%.

Only two studies report on spatial autocorrelation coefficients in a tropical forest. Becker and Smith (1990) report that radiation under a Panamanian forest, during a typical year, is only weakly correlated at distances out to 2.5 m and no correlation is exhibited between two sensors at spatial separations exceeding 5 m. During drought periods, when leaf fall occurs, the light environment becomes more homogeneous. Positive spatial correlation coefficients are observed for separation distances between 2 and 20 m. Chazdon *et al.* (1988) report even smaller-scale sunflecks in the heterogeneous light environment below a *Piper* forest. Correlation coefficients between two sensors only 0.2 m apart were below 0.4 and at a 0.5-m separation distance the correlation coefficient was less than 0.03 (Figure 10). Only in clearings are spatially separated light measurements highly correlated over a range of distances.

Chazdon *et al.* (1988) also report that the sunfleck size distribution is bimodal, with dominant contributions from 0.15 to 0.30 m and 1.05 to 1.2 m size classes. The scale of the sunfleck patches coincides with crown diameters of *Piper* (0.5 to 2 m). On the basis of their data, Chazdon *et al.* conclude that leaf orientation and within-crown shading contributed most to spatial heterogeneity in light, not species differences.

5. Isolated Plants Isolated groups of foliage such as bunchgrass intercept light on the sides and tops of tussocks (Caldwell *et al.*, 1983). Hence, the interception fraction is a function of the zenith angle of incoming light, the foliage angle distribution, and the spatial pattern of both live

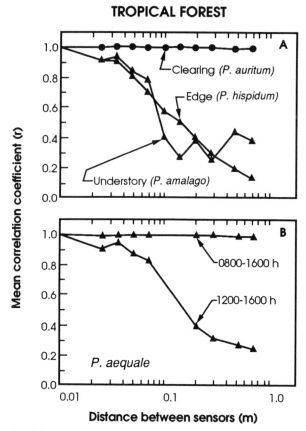

Figure 10   Spatial autocorrelation between spatially separated sensors that were located below a tropical forest, *Piper* species. The sensors were recorded every 2 min and were averaged for (A) 8 h and (B) 4 h (cloudy morning and sunny afternoon) (after Chazdon *et al.*, 1988).

and dead leaves. Considerable self-shading is observed in the tussock. Yet, the widely spaced architecture of the bunches allows as much beam radiation to be intercepted by the tussock as would otherwise be intercepted by a rhizomatous grass occupying a ground area six times greater.

V. Temporal Variability of Light in Plant Canopies

Temporal variability of light at a particular point in a plant stand occurs on a variety of time scales. Variability arises from leaf flutter, the occurrence, type and distribution of clouds, seasonal and diurnal movement in the sun's position, topographic shading, and seasonal changes in canopy phenology and leaf pigment. An expanded discussion of these factors is presented in the following sections.

A. Short-Term Fluctuations

How well plants capture and use light for photosynthesis and stomatal mechanics depends on the frequency and magnitude of available light (Pearcy, 1990). For example, fluctuating periods of dark and light on the order of minutes to hours can cause induction phenomena to occur, which results in a slow increase of photosynthesis when light increases. Despite a need for fast-response light data, few temporal spectra of light fluctuations are reported in the literature. In one brief study, Norman and Tanner (1969) report significant frequency contributions to the temporal variance of light in several crop canopies (soybeans, corn, and alfalfa) from 0.1 to 30 Hz fluctuation. They conclude that high-frequency fluctuations were attributed to leaf flutter.

In a more complete study, Desjardins *et al.* (1973) evaluated the spectra of light fluctuations in a corn canopy using traversing sensors. They report that four factors affect the frequency distribution of light. These factors are spatial position, sensor height in the canopy, crop structure, and wind speed. Peak spectral densities shift from higher to lower frequencies (0.30 to 0.06) as one progresses from the top to the bottom of a canopy (Figure 11). This shift occurs because the size of shade patches increases with depth. For example, near the top of the canopy, fluctuations up to 5 Hz are observed, while as one progresses to the bottom of the canopy, few fluctuations occur with frequencies exceeding 0.3 Hz. Increasing wind increases leaf flutter and causes higher-frequency modes, as lightflecks penetrate through transient gaps.

In the upper third of an alfalfa stand, peak spectral densities of PFD fluctuations shift from 0.30 to 1 Hz as wind speed increases from 1.3 to 6.3 m s^{-1} (Hongliang and Hipps, 1991). Furthermore, 10-Hz contributions change from negligible to significant as wind increases. The integral lifetime of sunflecks in alfalfa is short-lived because the time-lag autocorrelation coefficient decreases rapidly. Values below 0.2 are observed as the time lag approaches 1 s. At greater lag times, periodic fluctuations in the correlation coefficient are observed. These oscillations may be due to waving of the foliage in the wind, but more work is needed to substantiate this speculation.

In a fourth study, Pearcy *et al.* (1990) report that light fluctuations having frequencies greater than 1 Hz contribute little to the PPFD variance in soybeans. Dominant frequencies are across a broad spectral plateau, between 0.6 and 0.005 Hz. Pearcy *et al.* (1990) also used a conditional sampling scheme to examine the duration and the energy contained

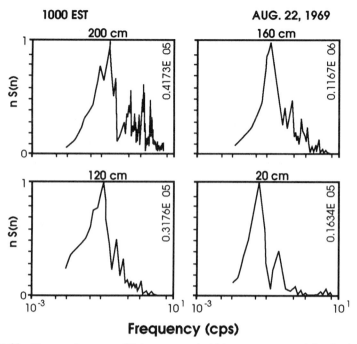

Figure 11   Temporal spectra of light measured within a corn canopy (after Desjardins *et al.,* 1973).

within sunflecks at a particular location in a soybean crop. They found that the most frequently occurring class of sunflecks lasts between 0.4 and 0.8 s. These short-lived sunflecks are inconsequential in determining the canopy energy input. Sunfleck durations less than 1.6 s contribute only 6.7% of the total daily photosynthetic energy, whereas longer-living sunflecks (those exceeding 12 s) account for 77% of the energy but last less than 10% of the time.

The dominant time scales of light fluctuations in tropical and conifer forests are longer than those observed in crops. Pearcy (1988) reports that the mean duration of sunflecks below an Australian tropical forest is about 50 s. Smolander (1984) reports that the temporal lag autocorrelation coefficient in Scots pine decreases linearly with time lag, reaching about 0.6 after 200 s.

B. Daily Scale

Diurnal variability in light transmission mainly results from variations in the altitudinal and azimuthal position of the sun; routines for calculating these angles are presented in Walraven (1978), among others. In humid regions, daytime buildup and dissipation of convective clouds can also force a diurnal modulation on the solar radiation regime in plant canopies.

In general, mean light transmission in crops and forests increases with increasing solar elevation (Ross, 1975). The theoretical role of varying sun angles on light transmission has been treated in detail by many workers (Lemeur, 1973; Ross, 1975; Anderson, 1966) and will not be repeated here. Empirically, light transmission through many crop stands (wheat, beans, kale: Szeicz, 1974; sorghum: Clegg *et al.*, 1974) is maximal, but invariant, during the 4- to 6-h period surrounding midday and decreases afterward. On the other hand, light transmission through a pine stand only exhibits midday temporal invariance on cloudy days (Mukkamal, 1971). On clear days, light transmission peaks at midday and is associated with lower transmission readings preceding and succeeding noon (Mukkamal, 1971).

The shape of the light transmission pdf's varies with time of day. Qualitatively, light transmission pdf's in the crown of crop and forest canopies transcend from a bimodal to a unimodal, but negatively skewed, distribution as the sun descends from the zenith and the probability of beam penetration becomes nil (Sinclair and Lemon, 1974; Sinclair and Knoerr, 1982; Baldocchi *et al.*, 1986).

Higher statistical moments associated with the pdf for PAR transmission through a temperate broadleaf forest have been observed to vary over the course of a day (Baldocchi *et al.*, 1986). The data show that skewness values are positive (they range between 2 and 4) below crown closure and do not depend on solar elevation. In the upper crown skewness values decrease linearly from 0 to −4 as solar elevation increases from the horizon toward the zenith. Kurtosis values suggest that the PAR pdf distribution below crown closure is peaked, but the peakedness depends on sun angle; kurtosis decreases from about 20 to 0 with linear increases in solar elevation. Kurtosis values are independent of solar elevation about crown closure. There, they range between 0 and 8, suggesting a flatter probability distribution.

Diurnal patterns of light at single points have been observed below crops, grasses, and tropical and temperae forests (Yoda, 1974; Pearcy *et al.*, 1990; Pearcy 1987, 1988; Tang *et al.*, 1988; Chazdon and Fetcher, 1984; Chazdon *et al.*, 1988; Weber *et al.*, 1985). In the upper canopy, the diurnal course is punctuated with many short-lived sunflecks, whereas in the understory periods of frequent sunflecks are followed by long periods with few or no sunflecks. Diurnal variations in the sun's position also cause the maximum flux density of energy in the sunflecks to vary sinusoidally with time, peaking at midday.

The overarching conclusion drawn from studies under grass and tropical and deciduous forest canopies is that a disproportionate amount of radiation reaching the canopy floor over the course of a day is due to sunflecks. Below grass stands, Tang *et al.* (1988) report that sunflecks exceeding 100 μmol m^{-2} s^{-1} occur only 6 to 54% of the time during a day, but contribute 25 to 82% of the daily energy in the PPFD wave band. Between 10 and 78% of total photon flux at the floor of a tropical forest is attributed to sunflecks (Pearcy, 1987, 1988; Chazdon, 1988). In one specific case, sunflecks under an Australian forest only lasted 16 min over the course of a day, but contributed 38% of the light received (Pearcy, 1987). In another study, Pearcy (1988) reports that the total daily flux of 1.57 mol m^{-2} d^{-1} was mainly due to 120 lightflecks, each exceeding 50 μmol m^{-2} s^{-1}. The total duration of sunflecks was about 100 min, with a 50-s mean duration. Penumbral effects, however, drastically reduce the flux density of light in the sunflecks from being equal to the full-sun values (Pearcy, 1988); only 1% of sunflecks exceeded 1200 μmol m^{-2} s^{-1}. In contrast, the diffuse light regime of an Australian tropical forest ranges between 10 and 20 μmol m^{-2} s^{-1}. Under a deciduous forest, Weber *et al.* (1985) report that sunflecks with a PPFD greater than 100 μE m^{-2} s^{-1} occur less than 25% of the day but contribute 45–55% of daily PPFD.

Topographic Effects   Topographic shading affects the canopy light environment in predictable ways as a result of (1) blockage of direct beam radiation on steep, sloping sites and (2) the generation of anisotropic diffuse radiation through the topographic shading of this input (Flint and Childs, 1987). On sloping terrain the direct beam incident on the surface depends on the solar zenith angle and the angle defined between the normal to the surface and the incoming beam (Nuñez, 1980). Requirements for dealing with topography include slope, aspect, latitude, longitude, and time. Sunlight available on slopes with topographical shading can be estimated using algorithms published by Nuñez (1980) or Garnier and Ohmura (1968) in conjunction with a topographical data base.

C. Seasonal Scale

Seasonal variations in light transmission are strongly dependent on changes in solar position. Seasonal changes in leaf area are important in crops and deciduous forests. As the growing season starts and proceeds, spaces between rows diminish (crops) and bare limbs become vegetated (deciduous trees and shrubs). At the end of the growing season pigmentation changes and leaves senesce and drop. Seasonal variations in cloud cover will also affect the light environment in plant canopies.

1. Crops   In wheat, a gradual decrease in light transmission occurs with age (between June and August), even though leaf area declines after mid-July (Szeicz, 1974). Szeicz proposes that heading of wheat compensates for the reduction in leaf area and causes a continual seasonal increase in interception.

The amount of light penetrating through a cotton canopy decreases from 100% at the start of the growing season (Day 140) to about 10% near the end (Day 260) at a rate of about 1.3% per day, when penetration again increases as a result of leaf senescence (Marani and Ephrath, 1985). This seasonal decrease in light transmission was not monotonic, but exhibited several kinks due to changes in canopy structure, clouds, and leaf angles.

Seasonally integrated light interception will vary with hedgerow configurations and latitude (Jackson and Palmer, 1972; Mutsaers, 1980). With north–south orientation little seasonal change occurs. On the other hand, a marked seasonal pattern occurs with east–west-oriented rows. For example, at 25 degrees latitude, beam interception for a given configuration oriented north–south varies from 70 to 65% between January and June, whereas light interception by a canopy oriented east–west varies 65 to 30% over this time interval (Mutsaers, 1980).

2. Temperate Broadleaf Forests   Hutchison and co-workers (Hutchison and Matt, 1977a,b; Baldocchi *et al.*, 1984a,b, 1986) have conducted several seasonal studies on variations of light in deciduous forest. Maximum radiation penetrates the forest in early spring, prior to leaf expansion, as solar paths rise higher in the sky. After leaf expansion average radiation received inside the forest diminishes despite continual increases in solar radiation. Full leaf is obtained in early June and canopy structure remains static until leaf abscission near the autumnal equinox. A reduction in solar angles after the summer solstice reduces light transmission. Transmission increases somewhat with leaf fall, but then diminishes as winter approaches and sun paths through the leafless canopy increase with lower sun angles. Light transmission through a leafless forest near the winter solstice is relatively low (0.20 to 0.60) because the arc of the solar path is obscured by woody biomass of optically thick trunks throughout most of the day and there are large contributions by penumbra (Hutchison and Matt, 1977a; Baldocchi *et al.*, 1986; Federer, 1971).

Baldocchi *et al.* (1986) and Hutchison and Matt (1977a) also show that frequency distributions of light in a deciduous forests (Oak-hickory and Tulip poplar) vary seasonally. In the winter leafless period, pdf's are unimodal and negatively skewed at all levels. During the leafing period, light transmission pdf's are negatively skewed. (Full-leaf period pdf's have already been described.) Fall pdf's had low kurtosis due to low sun angles, penumbra, and senescing leaves.

Seasonal variations in phenology also affect the penetration of diffuse radiation through a deciduous forest. Hutchison and Matt (1977a) report that in the winter the proportion of radiation penetrating through a leafless forest is greater under cloudy rather than clear skies. In the summer, when the canopy is fully leafed, the reverse relationship is true.

3. Conifer Stands   Schomaker (1968) compared the difference in solar radiation transmission in a conifer spruce stand and a deciduous birch stand between May and June (leafless to full leaf). Transmission through a birch stand decreased from 59 to 17%, whereas transmission through a conifer stand stayed constant at about 8%. In a red pine stand, light transmission decreased from 43 to 33% from the beginning of May to mid-June, as boughs grew and the few constituent deciduous trees leafed out (Mukkamal, 1971). Further reduction in mean transmission, to 18%, continued until September.

4. Broadleaf Evergreen Forests   In broadleaf evergreen forests leaf fall occurs continuously throughout the year. However, there are several periods through the year when leaf fall is not met with concurrent replacement of emerging leaves (Lowman, 1986). For example, cool temperate rain forests lose half their leaves in the autumn and they are not replaced until spring. Lowman (1986) examined whether seasonal patterns of leaf fall affected the light environment within three distinct Australian rain forests, occupying subtropical, cool temperate, and warm temperate climates. He found that Australian evergreen rain forests did not exhibit any seasonal change in the amount of unobscured visible sky despite the fact that the cool temperate canopy lost half of its leaves and the subtropical canopy contained deciduous species.

Seasonality may also be expected in some tropical forests because of distinct dry and wet periods as the amount of light in a clearing significantly varies. Yet, no seasonal trend was detected for the amount of light transmitted through a Costa Rican forest (Chazdon and Fetcher, 1984).

VI. Concluding Remarks

The light environment below closed plant canopies consists of a low-level background diffuse radiation regime that is punctuated by seemingly random, but deterministic patches of sunlight. Sunfleck patches possess a variety of length scales. The dimensions of these sunflecks vary because of the position, number, size, and orientation of overhead leaves, size of gaps, and the location of the sun. Below closed plant canopies, sunflecks possessing small length scales are most frequent, but those with moderate length scales contain the most energy. Since the canopy light environment is characterized by great spatial variability, much care is needed to properly measure and representatively sample this environment. A simple means of sampling the spatial light regime in heterogeneous canopies, involving sensors, is available for routine use.

Many temporal scales contribute to fluctuations of the light environment at a single point. It can be concluded that sunflecks are short-lived as a result of movement of the sun, passage of clouds, and leaf flutter. Although sunflecks are relatively small and short-lived, they account for a disproportionate amount of radiation received over the course of a day at a given location, because the photon flux density in a sunfleck can be 100 times greater than that residing in the background diffuse regime (2000 versus 20 μmol m^{-2} s^{-1}). In forests, penumbra act to reduce the flux density of energy in sunflecks, but the concept just proposed still holds.

The smearing of sunflecks by penumbra makes defining sunflecks a difficult exercise. Technically, a sunfleck should be defined as an event in which the flux density of light exceeds the background diffuse radiation regime due to the receipt of photons from the sun, as detected by wavelet methods. However, if we hold to this definition, threshold values for determining sunfleck could range from about 25 to 2000 μmol m^{-2} s^{-1}. Though a conservative definition of a sunfleck is preferred, a practical definition of a threshold to detect sunflecks may need to vary according to the intended use. For example, because photosynthetic capacity varies with depth in a plant canopy (Field, 1990; Pearcy, 1987) physiologically relevant thresholds for sunflecks should also vary with depth in the canopy. In the case of *Argyrodendron peralatum* leaves, light saturation and compensation points for understory leaves occur at about 250 and 4 μmol m^{-2} s^{-1}, respectively, whereas in the upper canopy these values are at about 750 and 24 μmol m^{-2} s^{-1} (Pearcy, 1987). Obviously, using the same definition of a sunfleck in the understory and canopy crown will lead to a misinterpretation of the data; in the crown a sunfleck threshold of 500 μmol m^{-2} s^{-1} would be meaningful, whereas it would be too high in the understory. Obviously, more work is needed on defining and interpreting sunflecks in various plant stands. Wavelet theory may be one way of objectively quantifying sunfleck thresholds, but more work on this method is needed before it can be applied routinely in a wide range of canopies.

Many modeling frameworks exist for calculating radiative transfer in discontinuous canopies. However, more work is needed to account for many complicating factors that tend to be ignored, such as scattering and penumbra. If we hope to routinely operate three-dimensional radiative transfer models in heterogeneous canopies, more data will be needed on the architectural features and radiation field of heterogeneous canopies in order to test these detailed models and to develop simpler parameterization schemes.

Acknowledgments

This work was partially supported by the National Oceanic and Atmospheric Administration and the U.S. Department of Energy. The senior author is grateful to Dr. Boyd Hutchison, for introducing him to the field of canopy radiative transfer and for providing a helpful review of this manuscript. We are also grateful to comments provided by Dr. Larry Hipps. Ms. Lala Chambers' help in acquiring many of the offprints used in this paper is also appreciated.

References

Acock, B., Thornley, J. H. M., and Wilson, J. W. (1970). Spatial variation of light in the canopy. *In* "Proceedings of the IBP/PP Technical Meeting, Trebon, Czechoslovakia," pp. 91–102. Pudoc, Wageningen, The Netherlands.

Allen, L. H., Jr. (1974). Model of light penetration into a wide-row crop. *Agron. J.* **66**, 41–47.

Allen, L. H., Jr., and Brown, K. W. (1965). Shortwave radiation in a corn crop. *Agron. J.* **57**, 575–580.

Allen, L. H., Jr., and Lemon, E. R. (1972). Net radiation frequency distribution in a corn crop. *Boundary-Layer Meteorol.* **3**, 246–254.

Anderson, M. C. (1964). Light relations of terrestrial plant communities and their measurement. *Biol. Rev. Cambridge Philos. Soc.* **39**, 425–486.

Anderson, M. C. (1966). Stand structure and light penetration. II. A theoretical analysis. *J. Appl. Ecol.* **3**(1), 41–54.

Anderson, M. C. (1971). Radiation and crop structure. *In* "Plant Photosynthetic Production: Manual of Methods" (Z Sestak, J. Catsky, and P. G. Jarvis, eds.), pp. 412–466. Dr. Junk. The Hague.

Antonia, R. A. (1981). Conditional sampling in turbulence measurement. *Annu. Rev. Fluid Mech.* **13**, 131–156.

Argoul, F., Arneodo, A., Grasseau, G., Gagne, Y., Hopfinger, E. J., and Frisch, U. (1989). Wavelet analysis of turbulence reveals the multifractal nature of the Richardson cascade. *Nature (London)* **338**, 51–53.

Arkin, G. F., Ritchie, J. T., and Maas, S. J. (1978). A model of calculating light interception by a grain sorghum canopy. *Trans. ASAE* **21**(2), 303–308.

Baldocchi, D. D. (1989). Turbulent transfer in a deciduous forest. *Tree Physiol.* **5**, 357–377.

Baldocchi, D. D. (1991). Canopy control of trace gas emission. *In* "Trace Gas Emission by Plants" (T. Sharkey, E. Holland, and H. Mooney, eds.), pp. 293–333. Academic Press, San Diego.

Baldocchi, D. D., Matt, D., Hutchison, B. A., and McMillen, R. T. (1984a). Solar radiation within an oak–hickory forest: An evaluation of the extinction coefficients for several radiation components during fully-leafed and leafless periods. *Agric. For. Meteorol.* **32**, 307–322.

Baldocchi, D. D., Hutchison, B., Matt, D., and McMillen, R. (1984b). Seasonal variations in the radiation regime within an oak–hickory forest. *Agric. For. Meteorol.* **33**, 177–191.

Baldocchi, D. D., Hutchison, B., Matt, D., and McMillen, R. (1985). Canopy radiative transfer models for spherical and known leaf inclination angle distributions: A test in an oak–hickory forest. *J. Appl. Ecol.* **22**, 539–555.

Baldocchi, D. D., Hutchison, B., Matt, D., and McMillen, R. (1986). Seasonal variation in the statistics of photosynthetically active radiation penetration in an oak–hickory forest. *Agric. For. Meteorol.* **36**, 343–361.

Becker, P., and Smith, A. P. (1990). Spatial autocorrelation of solar radiation in a tropical moist forest understory. *Agric. For. Meteorol.* **52**, 373–379.

Black, T. A., and Kelliher, F. M. (1989). Process controlling understorey evapotranspiration. *Proc. R. Soc. London, Ser. B* **324**, 207–231.

Bonan, G. B., and Shugart, H. (1989). Environmental factors and ecological processes in boreal forests. *Annu. Rev. Ecol. Syst.* **20**, 1–28.

Bracewell, R. N. (1990). Numerical transforms. *Science* **248**, 697–703.

Brown, G. W. (1973). Measuring transmitted global radiation with fixed and moving sensors. *Agric. Meteorol.* **11**, 115–121.

Brown, P. S., and Pandolfo, J. P. (1969). An equivalent obstacle model for the computation of radiative flux in obstructed layers. *Agric. Meteorol.* **6**, 407–421.

Caldwell, M. M., Dean, T. J., Nowak, R. S., Dzurec, R. S., and Richards, J. H. (1983). Bunchgrass architecture, light interception, and water-use efficiency: Assessment by fiber optic point quadrants and gas exchange. *Oecologia* **59**, 178–184.

Caldwell, M. M., Meister, H. P., Tenhunen, J. D., and Lange, O. L. (1986). Canopy structure, light microclimate and leaf gas exchange of *Quercus coccifera* a Portuguese macchia: Measurements in different canopy layers and simulations with a canopy model. *Trees* **1**, 25–41.

Carter, G. A., and Smith, W. K. (1985). Influence of shoot structure on light interception and photosynthesis in conifers. *Plant Physiol.* **79**, 1038–1043.

Charles-Edwards, D. A., and Thorpe, M. R. (1976). Interception of diffuse and direct-beam radiation by a hedgerow apple orchard. *Ann. Bot. (London)* [N.S.] **40**, 603–613.

Chason, J., Baldocchi, D., and Huston, M. (1991). A comparison of direct and indirect methods for estimating forest canopy leaf area. *Agric. For. Meteorol.* **57**, 107–128.

Chazdon, R. L. (1988). Sunflecks and their importance to forest understorey plants. *Adv. Ecol. Res.* **18**, 1–63.

Chazdon, R. L., and Fetcher, N. (1984). Photosynthetic light environments in a lowland and tropical rain forest in Costa Rica. *J. Ecol.* **72**, 553–564.

Chazdon, R. L., Williams, K., and Field, C. B. (1988). Interactors between crown structure and light environment in five rain forest *Piper* species. *Am. J. Bot.* **75**(10), 1459–1471.

Chen, J. M., Black, T. A., and Adams, R. S. (1991). Evaluation of hemispherical photography for determining plant area index and geometry of forest stand. *Agric. For. Meteorol.* **56**, 129–144.

Clegg, M. D., Biggs, W. W., Eastin, J. D., Maranville, J. W., and Sullivan, C. Y. (1974). Light transmission in field communities of sorghum. *Agron. J.* **66**(4), 471–475.

Cohen, S., and Fuchs, M. (1987). The distribution of leaf area, radiation, photosynthesis and transpiration in a Shamouti orange hedgerow orchard. Part I. Leaf area and radiation. *Agric. For. Meteorol.* **40**, 123–144.

Collineau, S. (1994). On the detection of coherent motion from single point measurements in forest canopy layers. *Boundary-Layer Meteorol.* (in press).

Denholm, J. V. (1981a). The influence of penumbra on canopy photosynthesis. I. Theoretical considerations. *Agric. Meteorol.* **25**, 145–161.

Denholm, J. V. (1981b). Influence of penumbra on canopy photosynthesis. II. Canopy of horizontal circular leaves. *Agric. Meteorol.* **25**, 167–194.

Denmead, O. T. (1976). Temperate cereals. *In* "Vegetation and the Atmosphere" (J. L. Monteith, ed.), Vol 2, pp. 1–31. Academic Press, London.

Denmead, O. T., Fritschen, L. J., and Shaw, R. H. (1962). Spatial distribution of net radiation in a corn field. *Agron. J.* **54**, 505–510.

Desjardins, R. L., Sinclair, T. R., and Lemon, E. R. (1973). Light fluctuations in corn. *Agron. J.* **65**, 904–908.

de Wit, C. T. (1965). Photosynthesis of leaf canopies. *Inst. Biol. Scheikd. Onderz. Landbouwgewassen*, Wageningen, *Jaarb.*, pp. 1–57.

Federer, C. A. (1971). Solar radiation absorption by leafless hardwood forests. *Agric. Meteorol.* **9**, 3–20.

Field, C. B. (1990). Ecological scaling of carbon gain to stress and resource availability. *In* "Integrated Responses of Plants to Stress" (H. A. Mooney, W. E. Winner, and E. J. Pell, eds.), pp. 1–32. Academic Press, San Diego.

Flint, A. L., and Childs, S. W. (1987). Calculation of solar radiation in mountainous terrain. *Agric. For. Meteorol.* **40**, 233–249. (review of Nuñez, 1980).

Fritschen, L. J., and Gay, L. W. (1979). "Environmental Instrumentation." Springer-Verlag, Berlin.

Fukai, S., and Loomis, R. S. (1976). Leaf display and light environments in row-planted cotton communities. *Agric. Meteorol.* **17**, 353–379.

Garnier, B. J., and Ohmura, A. (1968). A method of calculating the direct shortwave radiation income of slopes. *J. Appl. Meteorol.* **7**, 796–800.

Gay, L. W., Knoerr, K. R., and Braaten, M. O. (1971). Solar radiation variability on the floor of a pine plantation. *Agric. Meteorol.* **8**, 39–50.

Gijzen, H., and Goudriaan, J. (1989). A flexible and explanatory model of light distribution and photosynthesis in row crops. *Agric. For. Meteorol.* **48**, 1–20.

Gower, S. T., and Norman, J. M. (1991). Rapid estimation of leaf area index in conifer and broadleaf plantations using the LICOR LAI-2000. *Ecology* (in press).

Grace, J. C., Jarvis, P. G., and Norman, J. M. (1987). Modelling the interception of solar radiant energy in intensively managed stands. *N. Z. J. For. Sci.* **17**, 193–209.

Grant, R. H. (1985). The influence of the sky radiance distribution on the flux density in the shadow of a tree crown. *Agric. For. Meteorol.* **35**, 59–70.

Hamming, R. W. (1983). "Digital Filters." Prentice-Hall, Englewood Cliffs, NJ.

Herrington, L. P., Leonard, R. E., Hamilton, J. E., and Heisler, G. M. (1972). The response of moving radiometers. *Boundary-Layer Meteorol.* **2**, 395–405.

Hicks, D. J., and Chabot, B. F. (1985). Deciduous forest. *In* "Physiological Ecology of North American Plant Communities" (B. F. Chabot and H. A. Mooney, eds.), pp. 257–277. Chapman & Hall, New York.

Hollinger, D. (1989). Canopy organization and foliage photosynthetic capacity in a broad-leaved evergreen montane forest. *Funct. Ecol.* **3**, 53–62.

Hongliang, T., and Hipps, L. (1991). The properties of sunflecks in a flexible plant canopy and their relationship to turbulence. *Proc. Conf. Agric. For. Meteorol., 20th*, pp. 218–221.

Horn, H. S. (1971). "The Adaptive Geometry of Trees." Princeton Univ. Press, Princeton, NJ.

Hutchison, B. A., and Matt, D. R. (1977a). The distribution of solar radiation within a deciduous forest. *Ecol. Monogr.* **47**(2), 186–207.

Hutchison, B. A., and Matt, D. R. (1977b). The annual cycle of solar radiation in a deciduous forest. *Agric. Meteorol.* **18**, 255–265.

Hutchison, B. A., Matt, D. R., McMillen, R. T., Gross, L. J., Tajchman, S. J., and Norman, J. M. (1986). The architecture of a deciduous forest canopy in eastern Tennessee. *J. Ecol.* **74**, 635–646.

Impens, I., Lemeur, R. and Moermans, R. (1970). Spatial and temporal variation of net radiation in crop canopies. *Agric. Meteorol.* **7**, 335–337.

Jackson, J. E. (1970). Aspects of light climate within apple orchards. *J. App. Ecol.* **7**, 207–216.

Jackson, J. E. (1980). Light interception and utilization by orchard systems. *Hortic. Rev.* **2**, 209–267.

Jackson, J. E., and Palmer, J. W. (1972). Interception of light by model hedgerow orchards in relation in latitude, time of year and hedgerow configuration and orientation. *J. Appl. Ecol.* **9**, 341–357.

Jackson, J. E., and Palmer, J. W. (1979). A simple model of light transmission and interception by discontinuous canopies. *Ann. Bot. (London)* [N.S.] **44**, 381–383.

Jarvis, P. G., and Leverenz, J. W. (1983). Productivity of temperate deciduous and evergreen forests. *In* "Encyclopedia of Plant Physiology" (O. L. Lange, P. S. Nobel, C. B. Osmond, and H. Ziegler, eds.), Vol. 12D, pp. 233–280. Springer-Verlag, Berlin.

Kimes, D. S., and Kirchner, J. A. (1982). Radiative transfer model for heterogeneous 3-D scenes. *Appl. Op.* **21**(22), 4119–4129.

Kimes, D. S., Ranson, K. J., and Smith, J. A. (1980). A Monte Carlo calculation of the effects of canopy geometry on phar absorption. *Photosynthetica* **14**, 55–64.

Knapp, A. K., and Reiners, W. A. (1989). Penumbral effects on sunlight penetration in plant communities. *Ecology* **70**(6), 1603–1609.

Kruijt, B. (1989). Estimating canopy structure of an oak forest at several scales. *Forestry* **62**(3), 269–284.

Kubin, S. (1971). Measurement of radiant energy. *In* "Plant Photosynthetic Production: Manual of Methods." (Z Sestak, J. Catsky, and P. G. Jarvis, eds.), pp. 702–763. Dr. W. Junk Publisher, The Hague, The Netherlands.

Kuuluvainen, T., and Pukkala, T. (1987). Effect of crown shape and tree distribution on the spatial distribution of shade *Agric. For. Meteorol.* **40**, 215–231.

Kyle, W. J., Davies, J. A., and Nuñez, M. (1977). Global radiation within corn. *Boundary-Layer Meteorol.* **12**, 25–35.

Lang, A. R. G., and Xiang, Y. (1986). Estimation of leaf area index from transmission of direct sunlight in discontinuous canopies. *Agric. For. Meteorol.* **37**, 229–243.

Lemeur, R. (1973). A method for simulating the direct solar radiation regime in sunflower, jerusalem artichoke, corn and soybean canopies using actual stand structure data. *Agric. Meteorol.* **12**, 229–247.

Lemeur, R., and Blad, B. L. (1974). A critical review of light models for estimating the shortwave radiation regime of plant canopies. *Agric. Meteorol.* **14**, 255–286.

Leverenz, J. W., and Hinckley, T. M. (1990). Shoot structure, leaf area index and productivity of evergreen conifer stands. *Tree Physiol.* **6**, 135–149.

Li, X., and Strahler, A. H. (1987). Modelling gap probabilities in discontinuous vegetation canopies. *Proc. IGARSS Symp., 1987,* Ann Arbor, MI, pp. 1483–1486.

Liandrat, J., and Moret-Bailly, F. (1990). The wavelet transform: Some applications to fluid dynamics and turbulence. *Eur. J. Mech.* **9**, 1–19.

Lowman, M. D. (1986). Light interception and its relation to structural differences in three Australian rainforest canopies. *Aust. J. Ecol.* **11**, 163–170.

Luxmoore, R. J., Millington, R. J., and Peters, D. B. (1970). Symap and net radiation in a soybean canopy. *Agron. J.* **62**, 830–831.

Mann, J. E., Curry, G. L., Hartfiel, D. J., and DeMichele, D. W. (1977). A general law for direct sunlight penetration. *Math. Biosci* **34**, 63–78.

Mann, J. E., Curry, G. L., DeMichele, D. W., and Baker, D. N. (1980). Light penetration in a row-crop with random plant spacing. *Agron. J.* **72**, 131–142.

Marani, A., and Ephrath, J. (1985). Penetration of radiation into cotton crop canopies. *Crop Sci.* **25**, 309–313.

Matthews, R. B., and Saffell, R. A. (1987). An instrument to measure light distribution in row crops. *Agric. For. Meteorol.* **39**, 177–184.

Miller, E. E., and Norman, J. M. (1971a). A sunfleck theory for plant canopies. I. lengths of sunlit segments along a transect. *Agron. J.* **63**, 735–738.

Miller, E. E., and Norman, J. M. (1971b). A sunfleck theory for plant canopies. II. Penumbra effect: Intensity distribution along sunfleck segments. *Agron. J.* **63**, 739–743.

Miller, P. C. (1969a). Tests of solar radiation models in three forest canopies. *Ecology* **50**(5), 878–885.

Miller, P. C. (1969b). Solar radiation profiles in openings in canopies of aspen and oak. *Science* **164**, 308–309.

Monsi, M., and Saeki, T. (1953). Über den Lichtfaktor in den Pflanzengesellschaften und seine Bedeutung für die Stoffproduktion. *Jpn. J. Bot.* **14**, 22–52.

Monsi, M., Uchijima, Z., and Oikawa, T. (1973). Structure of foliage canopies and photosynthesis. *Annu. Rev. Ecol. Syst.* **4**, 301–327.

Morris, J.T. (1989). Modelling light distribution within the canopy of the marsh grass *Spartina alterniflora* as a function of canopy biomass and solar angle. *Agric. For. Meteorol.* **46**, 349–361.

Mukammal, E. I. (1971). Some aspects of radiant energy in a pine forest. *Arch. Meteorol., Geophys. bioklimatol.* **19**, 29–52.

Mutsaers, H. J. W. (1980). The effect of row orientation, date and latitude on light absorption by row crops. *J. Agric. Sci.* **95**, 381–386.

Myneni, R. B. (1991). Modeling radiative transfer and photosynthesis in three-dimensional vegetation canopies. *Agric. For. Meteorol.* **55**, 323–344.

Myneni, R. B., and Impens, I. (1985). A procedural approach for studying the radiation regime of infinite and truncated foliage spaces. I. Theoretical considerations. *Agric. For. Meteorol.* **33**, 323–337.

Myneni, R. B., Asrar, G., Kanemasu, E. T., Lawlor, D. J., and Impens, I. (1986). Canopy architecture, irradiance distribution on leaf surfaces and consequent photosynthetic efficiencies in heterogeneous plant canopies. I. Theoretical considerations. *Agric. For. Meteorol.* **37**, 189–204.

Myneni, R. B., Ross, J., and Asrar, G. (1989). A review on the theory of photon transport in leaf canopies. *Agric. For. Meteorol.* **45**, 1–153.

Myneni, R. B., Marshak, A., Knyazikhin, V., and Asrar, G. (1991). Discrete ordinates method for photon transport in leaf canopies. *In* "Photon–Vegetation Interactions" (R. B. Myneni and J. Ross, eds.), pp. 45–109. Springer-Verlag, Berlin.

Neumann, H. H., Hartog, G. D., and Shaw, R. H. (1989). Leaf area measurements based on hemispheric photographs and leaf-litter collection in a deciduous forest during autumn leaf-fall. *Agric. For. Meteorol.* **45**, 325–345.

Niilisk, H., Nilson, T., and Ross, J. (1970). Radiation in plant canopies and its measurement *In* "Prediction and Measurements of Photosynthetic Productivity" (I. Setlik, ed.), pp. 165–177. PUDOC.

Nilson, T. (1971). A theoretical analysis of the frequency of gaps in plant stands. *Agric. Meteorol.* **8**, 25–38.

Norman, J. M. (1975). Radiative transfer in vegetation. *In* "Heat and Mass Transfer in the Biosphere" (D. A. deVries and N. H. Afgan, eds.), pp. 187–205. Scripta Book Co., Washington, DC.

Norman, J. M. (1979). Modeling the complete crop canopy. *In* "Modification of the Aerial Environment of Plants" (B. J. Barfield and J. F. Gerber, eds.), pp. 249–277. Am. Soc. Agric. Eng. St. Joseph, MI.

Norman, J. M., and Jarvis, P. G. (1974). Photosynthesis in sitka spruce (*Picea sitchensis* (bong.) carr.). III. Measurements of canopy structure and interception of radiation. *J. Appl. Ecol.* **11**, 375–398.

Norman, J. M., and Jarvis, P. G. (1975). Photosynthesis in sitka spruce (*Picea sitchensis* (bong.) carr.). V. Radiation penetration theory and a test case. *J. Appl. Ecol.* **12**, 839–878.

Norman, J. M., and Tanner, C. B. (1969). Transient light measurements in plant canopies. *Agron. J.* **61**, 847–848.

Norman, J. M., and Welles, J. M. (1983). Radiative transfer in an array of canopies. *Agron. J.* **75**, 481–488.

Norman, J. M., Miller, E. E., and Tanner, C. B. (1971). Light intensity and sunfleck-size distributions in plant canopies. *Agron. J.* **63**, 743–748.

Nuñez, M. (1980). The calculation of solar and net radiation in mountainous terrain. *J. Biogeogr.* **7**, 173–186.

Oberbauer, S. F., Clark, D. B., Clark, D. A., and Quesada, M. (1988). Crown light environments of saplings of two species of rain forest emergent trees. *Oecologia* **75**, 207–212.

Oker-Blom, P. (1985). The influence of penumbra on the distribution of direct solar radiation in a canopy of Scots pine. *Photosynthetica* **19**, 312–317.

Oker-Blom, P. (1986). Photosynthetic radiation regime and canopy structure in modeled forest stands. *Acta For. Fenn.* **197**, 1–44.

Oker-Blom, P., and Kellomaki, S. (1982). Theoretical computations on the role of crown shape in the absorption of light by forest trees. *Math. Biosci.* **54**, 291–311.

Oker-Blom, P., Lappi, J., and Smolander, H. (1991). Radiation regime and photosynthesis of coniferous stands. *In* "Photon–Vegetation Interactions" (R. B. Myneni, and J. Ross, eds.), pp. 469–499. Springer-Verlag, Berlin.

Pearcy, R. W. (1987). Photosynthetic gas exchange responses of Australian tropical forest trees in canopy, gap and understory micro-environments. *Funct. Ecol.* **1**, 169–178.

Pearcy, R. W. (1988). Photosynthetic utilization of lightflecks by understory plants. *Aust. J. Plant Physiol.* **15**, 223–238.

Pearcy, R. W. (1989). Radiation and light measurements. *In* "Plant Physiological Ecology: Field Methods and Instrumentation" (R. W. Pearcy, J. Ehleringer, H. A. Mooney, and P. W. Rundel, eds.), pp. 97–116. Chapman & Hall, New York.

Pearcy, R. W. (1990). Sunflecks and photosynthesis in plant canopies. *Annu. Rev. Plant Physiol. Plant Mol. Biol.* **41**, 421–453.

Pearcy R. W., Roden, J. S., and Gamon, J. A. (1990). Sunfleck dynamics in relation to canopy structure in a soybean (*Glycine max* (L.) merr.) canopy. *Agric. For. Meteorol.* **52**, 359–372.

Pech, G. (1986). Mobile sampling of solar radiation under conifers. *Agric. For. Meteorol.* **37**, 15–28.

Press, W. H., Flannery, B. P., Teukolsky, S. A., and Vetterling, W. T. (1986). "Numerical Recipes." Cambridge Univ. Press, Cambridge, UK.

Pukkala, T., Becker, P., Kuuluvainen, T., and Oker-Blom, P. (1991). Predicting spatial distribution of direct radiation below forest canopies. *Agric. For. Meteorol.* **55**, 295–307.

Rauner, J. L. (1976). Deciduous forests. *In* "Vegetation and the Atmosphere" (J. L. Monteith, ed.), vol. **2**, pp.241–253. Academic Press, London.

Reifsnyder, W. E., Furnival, G. M., and Horowitz, J. L. (1971). Spatial and temporal distribution of solar radiation beneath forest canopies. *Agric. Meteorol.* **9**, 21–37.

Roberts, S. W., and Miller, P. C. (1977). Interception of solar radiation as affected by canopy organization in two Mediterranean shrubs. *Ecol. Plant.* **12**(3), 273–290.

Ross, J. (1975). Radiative transfer in plant communities. *In* "Vegetation and the Atmosphere" (J. W. Monteith, ed.), pp. 13–55. Academic Press, London.

Ross, J. (1981). "The Radiation Regime and Architecture of Plant Stands." Dr. W. Junk Publisher, The Hague, The Netherlands.

Ross, J., and Marshak, A. (1991). Monte Carlo models. *In* "Photon–Vegetation Interactions" (R. B. Myneni and J. Ross, eds.), pp. 441–467. Springer-Verlag, Berlin.

Salminen, R., Hari, P., Kellomaki, S., Korphilahti, E., Kotiranta, M., and Sievanen, R. (1983). A measuring system for estimating the frequency distribution of irradiance within plant canopies. *J. Appl. Ecol.* **20**, 887–895.

Saugier, B. (1976). Sunflower. *In* "Vegetation and the atmosphere" (J. L. Monteith, ed.), Vol. 2, pp. 87–100. Academic Press, London.

Schomaker, C. E. (1968). Solar radiation measurements under a spruce and a birch canopy during May and June. *For. Sci.* **14**(1), 31–37.

Sheehy, J. (1986). Radiation. *In* "Instrumentation for Environmental Physiology" (B. Marshall and F. I. Woodward, eds.), pp. 5–28. Cambridge Univ. Press, Cambridge, UK.

Sheehy, J. E., and Chapas, L. C. (1976). The measurement and distribution of irradiance in clear and conditions in four temperate forage grass canopies. *J. Appl. Ecol.* **13**, 831–840.

Shuttleworth, W. J. (1984). Observations of radiation exchange above and below Amazonian forest. *Q. J. R. Meteorol. Soc.* **110**, 1163–1169.

Sinclair, T. R., and Knoerr, K. R. (1982). Distribution of photosynthetically active radiation in the canopy of a loblolly pine plantation. *J. Appl. Ecol.* **19**, 183–191.

Sinclair, T. R., and Lemon, E. R. (1974). Penetration of photosynthetically active radiation in corn canopies. *Agron. J.* **66**, 201–205.

Sinclair, T. R., Desjardins, R. L., and Lemon, E. R. (1974). Analysis of sampling errors with traversing radiation sensors in corn canopies. *Agron. J.* **66**, 214–217.

Smith, H. (1980). Light quality as an ecological factor. *In* "Plants and Their Atmospheric Environment" (J. Grace, E. D. Ford, and P. G. Jarvis, eds.), pp. 93–110. Blackwell Scientific Publications, Inc., Cambridge, Massachusetts.

Smolander, H. (1984). Measurement of fluctuating irradiance in field studies of photosynthesis. *Acta For. Fenn.* **187**, 1–56.

Spanier, J., and Gelbard, E. M. (1969). "Monte Carlo Principles and Neutron Transport Problems." Addison-Wesley, Reading, MA.

Stanhill, G. (1976). Cotton. *In* "Vegetation and the Atmosphere" (J. L. Monteith, ed.), Vol 2, pp. 121–150. Academic Press, London.

Stoner, W. A., Miller, P. C., and Miller, P. M. (1978). A test of a model of irradiance within vegetation canopies at northern latitudes. *Arct. Alp. Res.* **10**(4), 761–767.

Subramanian, C. S., Rajagopalan, S., Antonia, R. A., and Chambers, A. J. (1982). Comparison of conditional sampling and averaging techniques in a turbulent boundary layer. *J. Fluid Mech.* **123**, 335–362.

Szeicz, G. (1974). Solar radiation in crop canopies. *J. Appl. Ecol.* **11**, 1117–1131.

Szeicz, G., Monteith, J. L., and dos Santo, J. M. (1964). Tube solarimeter to measure radiation among plants. *J. Appl. Ecol.* **1**, 169–174.

Szwarcbaum, I., and Shaviv, G. (1976). A Monte-Carlo model for the radiation field in plant canopies. *Agric. Meteorol.* **17**, 333–352.

Tang, Y., Washitani, I., Tsuchiya, T., and Iwaki, H. (1988). Fluctuation of photosynthetic photon flux density within a *Miscanthus sinensis* canopy. *Ecol. Res.* **3**, 253–266.

Tappeiner, U., and Cernusca, A. (1989). Canopy structure and light climate of different alpine plant communities: Analysis by means of a model. *Theor. Appl. Climatol.* **40**, 81–92.

Terborgh, J. (1985). The vertical component of plant species diversity in temperate and tropical forests. *Am. Nat.* **126**(6), 761–776.

Thompson, D. R., and Hinckley, T. M. (1977). Effect of vertical and temporal variations in stand microclimate and soil moisture on water status on several species in an oak–hickory forest. *Am. Midl. Nat.* **97**(2), 373–380.

Torquebiau, E. F. (1988). Photosynthetically active environment, patch dynamics and architecture in a tropical rainforest in Sumatra. *Aust. J. Plant Physiol.* **15**, 327–342.

Turton, S. M. (1985). The relative distribution of photosynthetically active radiation within four tree canopies, Craigiegurn Range, New Zealand. *Aust. For. Res.* **15**, 383–394.

Uchijima, Z. (1976). Maize and rice. *In* "Vegetation and the Atmosphere" (J. L. Monteith, ed.), Vol. 2, pp. 33–43. Academic Press, New York.

Verstraete, M. M. (1987). Radiation transfer in plant canopies: Transmission of direct solar radiation and the role of leaf orientation. *J. Geophys. Res.* **92**(D9), 10,985–10,995.

von Mises, R. (1964). "Mathematical Theory of Probability and Statisitcs." Academic Press, New York.

Walraven, R. (1978). Calculating the position of the sun. *Sol. Energy* **20**, 393–397.

Wang, Y. P., and Jarvis, P. G. (1990a). Description and validation of an array model—MAESTRO. *Agric. For. Meteorol.* **51**, 257–280.

Wang, Y. P., and Jarvis, P. G. (1990b). Influence of crown structural properties on PAR absorption, photosynthesis, and transpiration in Sitka spruce: Application of a model (MAESTRO). *Tree Physiol.* **7,** 297–316.

Waring, R. H., and Schlesinger, W. H. (1985). "Forest Ecosystems: Concepts and Management." Academic Press, New York.

Weber, J. A., Jurik, T. W., Tenhunen, J. D., and Gates, D. M. (1985). Analysis of gas exchange in seedlings of *Acer saccharum:* Integration of field and laboratory studies. *Oecologia* **65,** 338–347.

Whitehead, D., Grace, J. C., and Godfrey, M. J. S. (1991). Architectural distribution of foliage in individual *Pinus Radiata.* Don crowns and the effect of clumping on radiation interception. *Tree Physiol.* **7,** 135–155.

Whitfield, D. M. (1986). A simple model of light penetration into row crops. *Agric. For. Meteorol.* **36,** 297–315.

Yoda, K. (1974). Three-dimensional distributions of light intensity in a tropical rain forest of West Malaysia. *Jpn. J. Ecol.* **24,** 247–254.

3

Light Gaps: Sensing the Light Opportunities in Highly Dynamic Canopy Environments

Carlos L. Ballaré*

I. Introduction

In natural environments the amount of photosynthetic light energy received at a given point in space is extremely variable over time. Part of this variability is associated with seasonal changes, part is due to the activity of nearby vegetation, and, at a finer time scale, to weather factors and changes in solar angle during the course of the day. Plant fecundity and carbon gain are likely to be positively correlated in most environments. Therefore, in situations where light is limiting growth, the ability of plants to locate and efficiently exploit light opportunities would be a major determinant of evolutionary success.

Plants exhibit a fascinating variety of mechanisms for acquiring information about the light environment. These mechanisms operate at several stages in controlling, *inter alia,* the timing of seed germination, the transition to the photoautotrophic stage, the spatial orientation of branches and leaves, the elongation of stems and, hence, the height of leaf insertion in the canopy, the production of new branches or tillers, and the performance of the photosynthetic apparatus.

This chapter deals with the processes involved in the perception of light opportunities by plants in plant communities characterized by rapid changes of plant cover over time, such as those dominated by herbaceous

* Permanent address: Departamento de Ecología, Facultad de Agronomía, Universidad de Buenos Aires, Buenos Aires, Argentina.

Exploitation of Environmental Heterogeneity by Plants   73 species in early secondary succession. In the first and major part of the chapter I concentrate on the nature of environmental signals and sensory systems that regulate seed germination and plant morphological responses to changes in light environment. I then present an overview of the roles of these information-acquiring systems in the exploitation of patchy and highly dynamic light environments. Finally, I discuss the importance of considering these systems in mechanistic models of plant competition. Most of the examples I use in the discussion are derived from research on herbaceous species from agricultural plant communities.

II. Sensing and Responding to Light Opportunities

A. Environmental Signals and Plant Receptors

Plants use many environmental cues to obtain information about the prevailing light climate. These cues are factors of the natural environment that (1) can be sensed by plants and (2) are either directly associated or correlated with characteristics of the flux of photosynthetically active radiation (PAR), such as its intensity, direction, and likelihood of future change. Examples are the amplitude of the daily fluctuation of soil temperature sensed by a seed, the differential intensity of blue light received at opposite sides of a stem or the ratio between the fluxes of red and far-red radiation. Light signals, perceived by specific receptors, play a fundamental role in the detection of light opportunities. Therefore, a brief description of plant photoreceptors is necessary.

In addition to the two photosystems in the chloroplasts (PSI and PSII), higher plants have at least three families of photoreceptors (e.g., Mohr, 1986) (Figure 1):

1. Phytochromes, which absorb radiation over a wide range from the ultraviolet (UV) to the far-red (FR) with maxima around 660 nm for the Pr form and 730 nm for the Pfr form.
2. One or more blue (B)/UV-A absorbing pigments.
3. A little characterized receptor for UV-B radiation.

These photoreceptors are molecular devices that translate electronic excitations caused by light into specific cellular signals (see Smith, 1982, and Quail, 1991, for more detailed discussion). Through a variety of signal-transduction pathways, the original signal, which conveyed information about the light environment, leads to altered cellular metabolism and ultimately influences plant growth and development.

It should be kept in mind that PSI and PSII can also act as sensors of variations in light climate via the participation of ATP and NADPH in

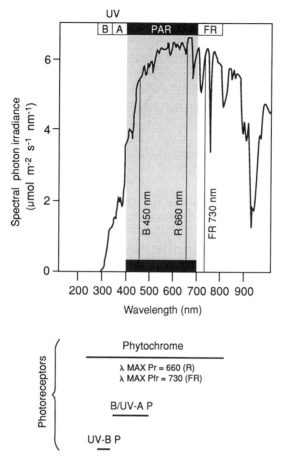

Figure 1 Spectral distribution of daylight and wavelengths absorbed by known plant photoreceptors. The shaded region is the visible portion of the spectrum. Abbreviations: B, blue; B/UV-A P, blue/ultraviolet-A absorbing photoreceptor(s); FR, far-red; PAR, photosynthetically active radiation; Pfr, far-red absorbing form of phytochrome; Pr, red absorbing form of phytochrome; R, red; UV, ultraviolet (B and A); UV-B P, ultraviolet-B absorbing photoreceptor. The spectrum was scanned near noon on a clear summer day at Corvallis, Oregon, USA.

photosynthetic feedback mechanisms, which may lead to altered chloroplast stoichiometry and photosynthetic capacity (e.g., Chow *et al.*, 1990). More indirectly, plants may accommodate to variations in light conditions by sensing and reacting to changes in levels of photosynthetic products, which may act as internal morphogenic signals.

Phytochrome is the best characterized of all plant photoreceptors. Reviews of its major photochemical properties can be found in several recent books (e.g., Kendrick and Kronenberg, 1986; Furuya, 1987; Attridge, 1990). Phytochromes are now known to be a family of proteins, all associated with a very similar (perhaps identical) chromophore (Sharrock and Quail, 1989; Furuya, 1989; Quail, 1991). Phytochrome molecules exist in two relatively stable forms, Pr and Pfr, with absorption maxima in the R and FR regions of the electromagnetic spectrum, respectively (Figure 1). Each form is converted into the other upon absorption of light. Because the absorption spectra of Pr and Pfr overlap to a large extent, a phytochrome population cannot be pushed to pure Pr or Pfr. Thus, under continuous illumination, a photoequilibrium (θ) between the two forms is reached that may be characterized by the proportion of phytochrome molecules that are in the Pfr form, (Pfr/P ratio). Values of θ may vary from 0.02 under monochromatic FR to 0.86 under monochromatic R (e.g., Mancinelli, 1988). Under natural polychromatic light, the Pfr/P ratio at θ is highly dependent on the ratio of R to FR photon fluxes in the incident radiation (R : FR ratio) (Smith and Holmes, 1977). Thus, θ values obtained by exposing purified phytochrome preparations to natural light typically vary from ca. 0.6 (direct sunlight; R : FR = 1.15) to 0.15 (light filtered through dense leaf canopy; R : FR = 0.2). The effects of light treatments on a number of physiological functions are well correlated with their effects on θ. Values of θ can be estimated analytically (Hartmann and Cohnen-Unser, 1972) from spectral irradiance data and published values of phytochrome photoconversion coefficients (e.g., Mancinelli, 1988). A similar approach can be used to calculate the rate of cycling between Pr and Pfr, which might modulate some responses to total irradiance (Section II,B,3), and to estimate how much Pfr is formed after pulse-light exposures too short to drive the pigment to photoequilibrium (see Cone and Kendrick, 1985; Scopel *et al.*, 1991). It should be noted here that physiological responses to different aspects of the light environment (e.g., total irradiance, R : FR ratio) might be mediated by different molecular species of phytochrome.

B. Plant Responses to Light Opportunities

1. Seed Germination   Successional environments are characterized by transient increases in light availability at the soil surface, which are caused by disturbances that eliminate or reduce the existing vegetation. Canopy gaps may be created by natural agents (i.e., fire, storms, or herbivores) or by man in managed ecosystems. Gaps may be of various sizes, from minute holes created by small herbivores foraging on vegetation to those involving hundreds of hectares caused by tillages in cultivated areas. Frequently, disturbance is followed by pulses of seed germination of colonizing species.

The environmental changes brought about by disturbances that are responsible for triggering seed germination are many and diverse. To a large extent the subject is reviewed by Vásquez-Yanes and Orozco-Segovia in Chapter 7 (this volume); therefore, coverage here will be selective and restricted to the perception of light opportunities by seeds in cultivated lands.

In arable lands, light at the soil surface level varies in a more or less cyclical fashion as a result of the management practices associated with the production of a particular crop, such as grazing, mowing, and soil cultivation. Several factors have been proposed to work as environmental signals for these events, including changes in soil temperature, soil atmosphere, the chemical composition of the soil solution, and light. The evidence for gap detection mechanisms based on these cues is briefly considered in the remainder of this section.

a. Light Opportunities Created by Canopy Removal   One of the major consequences of canopy removal is the modification of the thermal regime of the topsoil. In particular, the daily thermal amplitude (i.e., temperature$_{max}$-temperature$_{min}$) in the upper strata is much larger in the surface of bare soil than under a dense vegetation canopy. Seeds of weedy species usually show higher germination levels under alternating temperature regimes compared with constant temperatures (e.g., Aldrich 1984), which led Thompson *et al.* (1977) to propose increased thermal amplitude as a signal of gap formation. The best evidence for a gap-sensing mechanism based on soil temperature fluctuations was provided by Benech Arnold *et al.* (1988) working with Johnsongrass (*Sorghum halepense*). Germination of this weed was greater in experimental canopy openings than under undisturbed vegetation, and experimental manipulation of the soil temperature under the canopy to mimic the temperature fluctuations of adjacent bare plots led to germination equal to that in canopy openings.

In other cases, the environmental factor that informs seeds about the occurrence of openings in the leaf canopy appears to be light itself. Green leaves scatter light strongly and have well-defined absorption bands in the UV and visible wavelengths (i.e., <700 nm) and in the infrared (>1200 nm) region (e.g., Knipling, 1970). There is very little absorption between 700 and 1200 nm, so that most of the FR photons emerge from either side of the leaf as scattered radiation. This accounts for the low R : FR ratios that are observed in transmittance and reflectance spectra of green plant organs and in light measurements taken beneath vegetation canopies.

Classic studies on the photocontrol of seed germination (carried out mainly on freshly dispersed seeds or with seeds kept in dry storage for some time) have shown that, in order to be photostimulated, seeds must be exposed to light that establishes a relatively high level of the Pfr form of phytochrome (see references in Bewley and Black, 1982). Thus, the potential effect of light can be nullified if the Pfr content is immediately reduced to about 2% with a saturating pulse of FR radiation. Seeds exposed at the soil surface to light filtered through a leaf canopy (low R : FR ratio) would therefore be prevented from germinating. Opening the canopy increases the R : FR ratio and this may form enough Pfr in the seeds to promote germination (see Vásquez-Yanes and Orozco-Segovia, Chapter 7, this volume). The evidence for a gap-sensing mechanism based on perception of light quality changes in herbaceous plant communities under natural conditions is scant. Perhaps one of the adaptive drawbacks of a R : FR-driven dormancy mechanism is that seeds would need to be at or very near the soil surface in order to perceive changes in light environment, which would expose them to high risk of predation (e.g., Mittelbach and Gross, 1984; Scopel *et al.*, 1988; van Esso and Ghersa, 1989) and dessication.

In addition to temperature and light, there are other environmental factors associated with variations in aboveground biomass that, on theoretical grounds at least, might act as signals for seed germination. These include changes in NO_3^- concentration in the soil (Pons, 1989), organic acids produced by decaying vegetation (e.g., Simpson, 1990), or reduction in levels of allelopathic substances (e.g., Angevine and Chabot, 1979). However, the evidence for dormancy mechanisms utilizing these signals remains largely circumstantial.

b. Light Opportunities Preceded by Soil Disturbances   In agricultural settings the event that most frequently heralds a period of high light availability is disturbance of the topsoil by tillage operations. That seeds have the sensory machinery capable of detecting these events is fairly obvious from the resulting large populations of seedlings of weedy species (e.g., Chancellor, 1964; Roberts and Potter, 1980; Ballaré *et al.*, 1988b). For some species, germination occurs when seeds are carried back to the soil surface after being buried for some months, whereas germination of seed samples maintained at the soil surface or buried at constant depth is small or nil. Thus, in these species, germination seems to depend on at least two tillage episodes: one to bury the seeds and another to provide the necessary environmental signals to the (now sensitized) seeds. This suggests a remarkable degree of adaptation to the intensively disturbed agricultural environment (Soriano *et al.*, 1970).

Germination tests on exhumed seeds have demonstrated that the changes in dormancy status during burial can be cyclical, with dormancy reaching a minimum by the time of the year when germination usually occurs, and then increasing again in seeds that remain in the soil bank after that period (secondary dormancy) (reviews by Karssen, 1982; Baskin and Baskin, 1985). Laboratory tests with unearthed seed samples are not sufficient to establish which of the many alterations of the seed's microenvironment caused by soil disturbance are responsible for triggering germination. This is because (1) the expression of the dormancy status can be affected by sample manipulation and processing (e.g., desiccation, see Bouwmeester and Karssen, 1989) and (2) several environmental factors affected by tillages (e.g., light and temperature) can interact in the control of germination.

Seeds might detect soil disturbance by responding to soil temperature fluctuations. Since daily thermal amplitude diminishes rapidly with depth, seeds buried more than a few centimeters are subjected to very small temperature fluctuations in most soils, particularly at the beginning of the growing season. Higher thermal amplitudes may elicit germination of seeds moved from deep to shallow soil strata by tillage (see Koller, 1972; Ghersa *et al.*, 1992).

Light appears to play an important role in the detection of cultivation episodes by buried seeds of many species. Classic experiments by Sauer and Struik (1964) and Wesson and Wareing (1969) have shown that exclusion of light causes fewer seedlings to emerge from disturbed soil samples. Hartmann and Nezadal (1990) reported reduced weed abundance in fields cultivated during the night. Scopel *et al.* (1991) induced germination of buried seeds of *Datura ferox,* an annual weed of summer crops, by piping sunlight from the soil surface to the seeds without disturbing other microenvironmental factors. Germination of a fraction of the seed population (about 25%) was shown to be limited *only* by the lack of light. The experiment was carried out during the spring, when there were large daily temperature fluctuations even at 10 cm depth. When seeds were unearthed using light-tight equipment, exposed to sunlight, and placed again at the initial depth (simulating the effects of soil cultivation), germination exceeded 90% (Figure 2). The most significant aspect of these results is the observation that a short period of burial caused a dramatic (ca. 10,000-fold) increase in the sensitivity of *Datura ferox* seeds to light. In fact, after a few months of burial, the majority of the seed population could be induced to germinate in the field by irradiations equivalent to 0.1 to 10 ms of full sunlight (Figure 2). The amount of Pfr resulting from these short irradiations would be on the order of 0.0001 to 0.01%, that is, 2 to 4 orders of magnitude lower than the level established by a saturating exposure to pure FR (Scopel *et al.*, 1991)!

It is now well established that many plant responses to light can be elicited by irradiations that form only minute amounts of Pfr and, there-

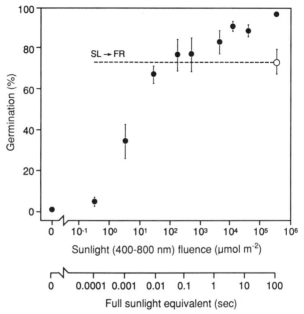

Figure 2   Effects of sunlight pulses on the germination of buried seeds of *Datura ferox* under field conditions. Seeds were unearthed near midday using light-tight equipment, exposed to sunlight, and buried again at − 7 cm. The open symbol indicates germination of seeds irradiated with FR after a saturating exposure to sunlight. (After Scopel *et al.,* 1991.)

fore, cannot be reversed by FR. This type of photoresponse has been termed very-low-fluence (VLF) response, to distinguish it from the classic, FR-reversible, low-fluence (LF) response (Mandoli and Briggs, 1981; Blaauw-Jensen, 1983; Kronenberg and Kendrick, 1986; VanDerWoude, 1989). Seeds of several species display VLF responses following artificial pretreatments. On the basis of fluence response curves obtained in laboratory and field conditions, Scopel *et al.* (1991) interpreted the sensitization observed in buried seeds as a *natural* transition from the LF to the VLF mode of phytochrome action. Taylorson (1972) and Baskin and Baskin (1979) also reported high light sensitivity in seeds retrieved from soil. Theoretical calculations of the amount of light that individual seeds would receive during soil tillage suggest that VLF responses are of great importance for the perception of light signals of soil cultivation (Scopel *et al.,* 1991; Scopel *et al.,* 1993).

The interactions between light and other microenvironmental factors have not been studied in the field. Temperature fluctuations might be the ultimate signal for germination induction even in light-requiring seed. In fact, a pool of phytochrome that is highly stable in the Pfr form appears to mediate promotion of germination by light, at least in some species (Casal *et al.*, 1991). Therefore, if the thermal regime is not appropriate immediately after light exposure, Pfr might be stored for a few hours (days?) and the seeds remain in a photostimulated state until germination is elicited by fluctuating temperatures.

2. Stem Phototropism and Leaf Movements   After germination, the transition to the photoautotrophic stage, involving leaf and plastid development and reduced axis elongation, is also regulated by light acting through photomorphogenic pigments. Once seedlings have emerged from the soil, they face a complex light environment that is characterized by extreme variations in intensity and direction of the incident rays. These variations are caused by changes of solar angle during the course of the day, changes in cloudiness, and light absorption and scattering by neighboring objects and plants. In this section, I briefly discuss how plants or individual organs detect the direction of illumination. Typical morphological responses to variations of light direction are phototropic movements of hypocotyls and young stems and modifications of leaf inclination and azimuth angles.

Phototropism is the bending of elongating shoots toward higher light in an uneven light field. This bending results from redistribution of growth, that is, inhibition on the illuminated side and stimulation on the shaded side, with little change in net growth (Rich *et al.*, 1987, and references therein; see also Briggs and Baskin, 1988; Iino, 1990). The subject has been reviewed (Firn, 1986; Iino, 1990) and only some general aspects will be considered here.

Most of our present understanding of phototropic mechanisms in higher plants comes from experiments with etiolated (dark-grown) seedlings exposed to unilateral light. These studies have shown that detection of the direction of illumination is based on fluence rate gradients that are created between the illuminated and dark sides of the phototropic organ (i.e., coleoptile tip or stem). These gradients occur because of internal light scattering (mainly at the air/cell-wall interfaces) and absorption (Seyfried and Fukshansky, 1983; Vogelmann and Haupt, 1985; Vogelmann, 1986). The B/UV-A region of the spectrum, sensed by one or more as yet unidentified specific receptors, is involved in signaling changes in light direction (e.g., Iino 1990). Most studies have indicated that phytochrome does not function as a detector of the direction of illumination in etiolated seedlings (see references in Iino, 1990), except under some special experimental conditions (see Iino *et al.*, 1984; Parker *et al.*, 1989). What is clear is that phytochrome is involved in accessory processes (e.g., sensitivity and responsivity adjustment) that might be essential for the phototropic machinery to operate under natural light conditions (see Woitzik and Mohr, 1988; Iino, 1990, and references therein).

In light-grown, green seedlings, phytochrome can participate in the detection of internal light gradients. Some early action spectra obtained using broad-band light sources show a peak in the R region in addition to the B/UV-A peak (Atkins, 1936; and references in Iino, 1990, and Ballaré *et al.*, 1992a). Iino (1990) suggested that the differential penetration of light of different wavelengths into green organs may allow phytochrome to function as a phototropic detector in plants receiving polychromatic radiation. Chlorophyll absorption causes an abrupt decline in the fluence rate of R light within the first millimeter of tissue, whereas the fluence rate of FR declines much more gradually (Seyfried and Fukshansky, 1983; Vogelmann and Björn, 1984; Vogelmann, 1986). Therefore, in unilateral light, a gradient of R : FR ratio and θ between the illuminated and shaded sides of the organ would be established. Since low θ's stimulate elongation (Section II,B,3), the shaded side would elongate faster than the illuminated one, causing the shoot to bend toward the lateral light source. Although consistent with the available data, we cannot yet conclude that this mechanism is actually important in phototropic responses in nature. Plants are seldom exposed to truly unilateral illumination. The "shaded" side of the stem may actually receive substantial amounts of diffuse radiation that, in open conditions, will be unaltered skylight, with even higher R : FR ratios than direct sunlight (Holmes and Smith, 1977a).

Recent results with cucumber (*Cucumis sativus*) seedlings provide some indication about the involvement of different photoreceptors in natural phototropism (Ballaré *et al.*, 1991b). Seedlings grown in a glasshouse with no supplemental lighting naturally bent along the hypocotyl toward the north (the experiments were carried out in the Southern Hemisphere). The degree of curvature in seedlings of a mutant deficient in phytochrome-B (phyB) (see Section II,B,3,a), was the same as in the near isogenic wild-type individuals. Furthermore, in both genotypes, bending was totally prevented by removing the blue (B) component from direct sunlight with an orange cellulose acetate filter. This suggests that the B wave band, acting through a sensor other than phyB, is essential for phototropic responses, at least in plants grown in sparse settings. The situation is different for plants growing in the vicinity of other vegetation, since light scattering by nearby plants can substantially increase the amount of FR that impinges laterally on vertically oriented shoots (see Figures 5 and 6, Section II,B,3,b). This creates R : FR gradients across stems that are related to the distribution of neighbors around the plant: the R : FR ratio is lower on the side that faces the canopy than on the side that faces open space. Evidence from field experiments demonstrates that, under these conditions, a phytochrome-dependent phototrophic mechanism, acting in addition to the B-driven system, does signal elongating organs to bend toward the gaps (Ballaré *et al.*, 1992a). These studies suggest that stem bending responses to directional light signals are an important component of the mechanisms that help the plant to forage for light and improve PAR capture in heterogeneous light environments.

Apart from being able to rapidly (within minutes or hours) change the direction of growth of young stems, many plants can adjust the spatial orientation of their leaves in response to directional light signals (reviews by Ehleringer and Forseth, 1989; Koller, 1990). Diaphototropic leaf movements reorient the lamina throughout the day so as to maintain it nearly normal to the direct sunlight vector (solar-tracking). Solar tracking by young leaves is analogous to phototropic stem movements in that reorientation depends on redistribution of growth between opposite sides of the petiole (Koller, 1990). In mature leaves, reorientation is based on turgor changes and reversible structural deformations of the pulvinus, located beween the lamina and the petiole (or at the base of each leaflet in species with compound leaves). Information about changes in the direction of illumination is acquired by leaves through a specific B-light receptor. The fascinating complexity of perception and response mechanisms was reviewed recently by Koller (1990).

Solar tracking maintains a high and relatively constant PAR interception during the day, increasing daily carbon gain in species with high photosynthetic capacity. It may also increase photosynthesis via increased leaf temperature in the morning and evening (Mooney and Ehleringer, 1978; Shackel and Hall, 1979). The adaptive value of solar tracking may be greatest in arid, high-light habitats with short growing seasons (e.g., Mooney and Ehleringer, 1978; Ehleringer, and Forseth, 1980, 1989). Solar tracking is also common in seedlings of pioneer species of secondary succession, and most likely contributes to increase plant growth rate during the early stages of canopy development.

3. Stem Elongation   Rapid responses to directional light signals are important to adjust the orientation of light-harvesting organs so as to increase PAR interception in temporally variable environments. In many situations, particularly in rapidly growing canopies formed by plants of similar stature, there may be little advantages to change orientation without rapid growth in height. Not surprisingly, a key feature of elongation growth in species of early succession is high responsivity to local canopy conditions.

Stem elongation is influenced by many factors, including those that affect overall plant growth (i.e., PAR, water and nutrient supply) and others with specific morphogenic activity. Some factors have more than one effect. Light, for instance, may stimulate stem elongation by promoting total growth but, under certain conditions, increased light intensity leads to inhibitory morphogenic effects. In the short term at least, shade-intolerant species with orthotropic stems typically respond to leaf shading or increased population density by accelerating internode elongation. Plants with diagravitropic stem systems may show the opposite reaction to shading (i.e., decreased elongation), and leaves are raised in the canopy by means of changes in shoot inclination (Section II,B,5) or increased petiole length (Solangaarachchi and Harper, 1987; Thompson and Harper, 1988; Méthy *et al.*, 1990).

a. Elongation Responses to Shading under Established Canopies   Several microenvironmental changes caused by the presence of a leaf cover may stimulate stem elongation under canopies. Stem elongation is highly sensitive to changes in air humidity, mechanical disturbances, total irradiance, and spectral light quality. Although the effects of air humidity (McIntyre and Boyer, 1983) and mechanical perturbations (e.g., movements caused by wind; Neel and Harris, 1971; Telewsky and Jaffe, 1986; Retuerto and Woodward, 1992; and refs. in Nobel, 1981) are well documented, little is known about the extent to which responses to these variables contribute to the putative effect of shading on elongation under field conditions.

Differences in light environment between open and leaf-covered sites do certainly influence elongation growth. The optical characteristics of leaves (Section II,B,1,a) create a light environment beneath canopies characterized by low levels of UV and visible wavelengths and reduced R : FR ratios (e.g., Vézina and Boulter, 1966; Holmes and Smith, 1977b). All of these factors are known to influence elongation, at least under controlled conditions. UV-B inhibits shoot elongation in many species. Evidence is accumulating to suggest that this is not merely a by-product of unspecific growth inhibition, but a true photomorphogenic response to UV-B (e.g., Steinmetz and Wellmann, 1986; Barnes *et al.*, 1990; Ballaré *et al.*, 1991a). Attenuation of UV-B by a leaf canopy would thus lead to increased elongation in sensitive species.

Another aspect of the light environment beneath leaf canopies that may stimulate elongation is the reduced fluence rate in the 320- to 800-nm wave band (i.e., "white" light). Increasing white-light fluence rate markedly inhibits hypocotyl elongation in young seedlings (grown in the light for a few hours) (e.g., Meijer, 1959). This effect is mediated by specific B/UV-A photoreceptor(s) (Gaba and Black, 1979; Thomas and Dickinson, 1979; Ritter *et al.*, 1981) and phytochrome(s) (e.g., Ritter *et al.*, 1981; Wall and Johnson, 1981; Holmes *et al.*, 1982; Gaba and Black, 1985; see also Cosgrove, 1986). Effects of white light irradiance on internode elongation of more mature, fully de-etiolated plants have frequently, but not always, been demonstrated (see references in Ballaré *et al.*, 1991c). Therefore there is some controversy about the significance of this variable in the control of stem elongation under leaf canopies (see Frankland, 1986; Child and Smith, 1987). Britz (1990) has recently shown that stem elongation in fully de-etiolated soybean plants can be substantially inhibited by increasing white-light fluence rate, and suggested that this response was, at least in part, mediated by a specific B light receptor.

The effect of reduced R : FR ratio in promoting stem elongation beneath plant canopies has been reviewed by Smith (1982, 1986) and Smith and Morgan (1983). The work of Smith and associates has established that: (1) the reduction of R : FR ratio beneath canopies is a function of the amount of leaf area above the point of consideration and is therefore closely correlated with the reduction of PAR (Holmes and Smith, 1977b); (2) changes in R : FR beneath canopies are within the range where variations of R : FR lead to relatively large changes in θ (Smith and Holmes, 1977; and (3) in controlled environments, under light levels comparable to those that can be found beneath dense canopies (ca. 10% of full sunlight), plants of shade-intolerant species respond to reduced R : FR with an increase of elongation rate, and the response is quantitatively related to the effect of the R : FR treatment on the photoequilibrium reached by samples of purified phytochrome (Morgan and Smith, 1978, 1979).

The importance of light signals (relative to other microenvironmental factors) in determining elongation responses to leaf shading has been tested recently by using photomorphogenic mutants. The *lh* mutant of cucumber (Adamse *et al.*, 1987; Kendrick and Nagatani, 1991) lacks a phytochrome-B (phyB) polypeptide (López-Juez *et al.*, 1992). After de-etiolation, *lh* seedlings have longer hypocotyls than the near isogenic wild type (WT), show reduced levels of spectrophotometrically detectable phytochrome (Adamse *et al.*, 1988; Peters *et al.*, 1991), and lack (or exhibit severely reduced) elongation responses to end-of-day (Adamse *et al.*, 1988; López-Juez *et al.*, 1990) or daytime (Ballaré *et al.*, 1991b, 1992a) R : FR treatments. The lack of phyB also appears to affect some of the responses to B light. Thus, B light, acting through a specific photoreceptor, inhibits hypocotyl elongation in de-etiolated WT seedlings but not in the *lh* mutant (Ballaré *et al.*, 1991b; see Mohr, 1986, and Gaba and Black, 1987, for discussions about interactions among photoreceptor systems). The *lh* mutant is therefore very useful as an experimental tool because, except for responses to UV-B radiation, it is unresponsive to the light signals that putatively mediate the effects of leaf shading on stem elongation. In a glasshouse experiment, transfer of WT seedlings from full sunlight to the understory of a dense canopy caused a 3.5-fold increase in the rate of hypocotyl elongation. This response to shading was totally absent in the *lh* mutant. Applications of gibberellin A_3 promoted elongation of *lh* seedlings in the understory, suggesting that the lack of response to shading was not a consequence of assimilate limitation, but a reflection of the inability of the mutant to sense variations in the B and R-FR environment caused by the leaf canopy. Qualitatively similar results were obtained in parallel field experiments during the summer at Buenos Aires. These studies suggest that mechanisms controlled by photomorphogenic signals and involving the activity of phyB play a major role in determining short-term elongation responses to shading (Ballaré *et al.*, 1991b).

One of the issues that received little attention in the foregoing discussion is the effect of the illumination history on the sensitivity of plants to light stimuli. The responses to shading in the cucumber experiments were measured 2 days after the beginning of treatments. In the long term (i.e., after 1 or 2 weeks), the promotive effect of shading on elongation of WT plants tended to disappear (Ballaré *et al.*, 1991b). This could indicate either that shaded plants could not elongate more rapidly because of assimilate limitations or that they changed their pattern of response to light signals after a prolonged period of shading. The latter does not appear to be an unreasonable possibility, since in cases of strongly asymmetric competition (i.e., seedlings versus established plants) the "escapist" response to shading is likely to be of limited value to the subordinated plants.

b. Elongation Responses during Canopy Development: Evidence for Early Detection of Neighboring Plants   Stem elongation responses of individuals already severely shaded by neighbors are unlikely to contribute to the success of pioneer species in secondary successions. Plants that emerge late in the season, when the aerial and belowground spaces have been already occupied by earlier individuals, tend to make little growth and usually experience high mortality (e.g., Black and Wilkinson, 1963; Weaver and Cavers, 1979; Gross, 1980; Peters, 1984; Ballaré *et al.*, 1987b; Panetta *et al.*, 1988; Scopel *et al.*, 1988). The sophisticated environmentally controlled dormancy mechanisms discussed in Section II,B,1 tend to result in a more or less synchronous germination of pioneer species at the beginning of periods of high light availability. Therefore, competition for light develops among individuals of relatively similar size, in canopies with little vertical organization and low species diversity (see Bazzaz, 1990). In this scenario, precise modulation of elongation growth is likely to be a major strategic priority, since slight differences in height may result in disproportionately large differences in PAR capture between neighboring plants (Grime and Jeffrey, 1965; Ballaré *et al.*, 1988a; see also Section III). There is now solid evidence that the control of stem elongation in even-height canopies entails a great deal of mechanistic sophistication, and that responses to the proximity of other plants can begin well before the leaves become severely shaded by neighboring individuals.

Figure 3 shows the elongation responses of seedlings of the annual weed *Datura ferox* placed in the center of even-height canopies of different densities, ranging from an average leaf area index (LAI) of about 0 (isolated plants) to a maximum of 1.5. Shading among neighboring seedlings was small in these experiments (Figure 3, bottom). However, in that range of LAIs, increasing neighborhood density led to a marked, rapid increase in stem elongation rate. In a related study (Ballaré *et al.*, 1987a), seedlings of white mustard (*Sinapis alba*) were grown on the northern (directly illuminated in the Southern Hemisphere) side of grass hedges, half of which had been bleached by herbicide treatment. Seedlings grown in front of green hedges developed longer internodes than those near bleached neighbors even though, in both cases, the seedlings were continuously exposed to unfiltered sunlight (Figure 4). What follows

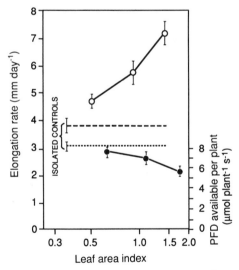

Figure 3 Elongation rates of the first internode of *Datura ferox* seedlings growing in even-height canopies of different densities under natural radiation. Also shown is the effect of density on PFD (photosynthetic photon flux density) interception of individual seedlings. The height of test plants was similar to that of plants forming the surrounding canopies. (After Ballaré *et al.*, 1990.)

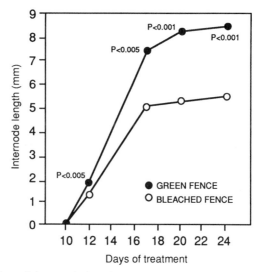

Figure 4 Effect of the proximity of a green grass canopy on the elongation of the first internode of *Sinapis alba* seedlings growing under full sunlight. Control seedlings were grown in front of canopies bleached with an application of paraquat. (After Ballaré *et al.*, 1987b.)

is a discussion of the role of light signals in these early elongation responses to the proximity of neighboring plants.

As pointed out in Section II,B,1,a, transmittance and reflectance spectra of green leaves are very similar and show a steep rise in the FR region. Ballaré *et al.* (1987a) hypothesized that the differential reflection of R and FR radiation from neighboring plants, acting through phytochrome, provides an environmental cue for the remote detection of these neighbors before shading occurs. Characterizations of the canopy light environment using conventional light sensors (e.g., Kasperbauer *et al.*, 1984) or cuvettes containing phytochrome preparations (Smith *et al.*, 1990) pointed to receive light back-scattered by green plants typically show low R : FR or Pfr/P ratios. However, to predict the effects of neighboring plants on the state of phytochrome in a particular organ of an intact plant is usually cumbersome. To a large extent this is because plants are much more complex structures than a flat light receiver. Individual organs are usually simultaneously exposed to light scattered by nearby objects and foliage and to direct sunlight. Ballaré *et al.* (1989) used a fiber optic probe implanted in different plant organs and attached to a spectroradiometer to investigate how the quality of radiation scattered within plant tissues was affected by the proximity of other plants. One limitation of the fiber optic approach is that we do not know the internal distribution of the phytochrome molecules involved in the control of stem elongation. However, it is a convenient way to visualize the integrated product of light beams entering the plant from all possible directions. The fiber optic studies showed that neighboring plants in uniform canopies would not cause a significant spectral change inside mostly horizontal leaves, unless they actually shade the leaves. This is due to the overwhelming contribution of direct sunlight to the internal light regime of leaves that are almost perpendicular to the direct light vector.

The situation is very different in the case of vertical stems or leaves. In these cases, light reflected back by neighboring plants may cause a substantial increase in the amount of FR light scattered inside the tissues, even if the LAI of the canopy is very low (Figure 5). Calculations showed that spectral changes of the magnitude shown in Figure 5 may cause a 30% decrease in the predicted value of θ inside the stem (Ballaré *et al.*, 1989; see also Mancinelli, 1991).

Stem internodes can respond with some degree of autonomy to local light-quality conditions (Garrison and Briggs, 1975; Morgan *et al.*, 1980; Casal and Smith, 1988a,b). Figure 6 shows the effects of increasing canopy density on the light environment of the stems in even-aged populations of *Sinapis alba* and *Datura ferox*. Measurements were taken with the

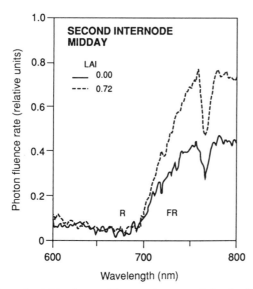

Figure 5 Effects of neighboring seedlings on the spectral distribution of the radiation scattered within the stems of *Datura ferox* seedlings in an even-height canopy of LAI = 0.72. Abbreviations: FR, far-red; R, red. (After Ballaré *et al.*, 1989.)

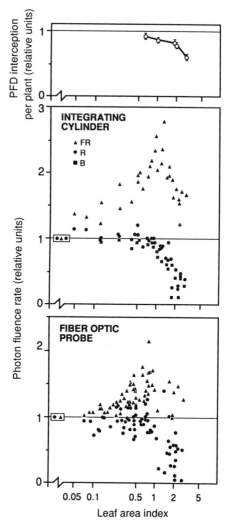

Figure 6   Effects of increasing density in even canopies of dicotyledonous seedlings on light interception by leaves (top) and the light climate of the stems. The integrating cylinder collects sidelight received by the stem surface; the fiber optic probe collects light scattered within the stem tissue. Abbreviations: B, blue; FR, far-red; PPFD, photosynthetic photon flux density; R, red. (After Ballaré *et al.,* 1991b.)

fiber optic probe or with an integrating cylinder (Ballaré *et al.,* 1987a) that accepts sidelight from all directions. At very low LAI (i.e., LAI between 0 and 1) the fluence rate of FR received by the stems increases substantially with an increase in canopy density while other wavelengths remain unaffected. The canopy works as a FR-reflecting light trap, and the R : FR ratio at the stem surface typically drops from ca. 0.9 to 0.4 (Ballaré *et al.*, 1987a). Further increases in LAI cause a strong reduction in the fluence rates of B, R, and FR radiation received by the stems. The most significant aspect is that all of these drastic changes in the stem's light environment take place well before shading at the leaf level becomes significant. In fact, measurements of light interception by leaves (Figure 6, top) show that in even-height canopies with a LAI of 2, neighbors have only a relatively modest effect on the light available for photosynthesis.

How do all these changes in the stems' light environment influence elongation rate and, hence, projection of new leaves into uncolonized space? In one series of field experiments (Ballaré *et al.*, 1987a), seedlings of three annual species from open habitats, *Sinapis alba*, *Datura ferox*, and *Chenopodium album*, were grown in front of selective-reflecting mirrors that provided additional FR, mimicking the presence of other seedlings (cf. Figures 5 and 6) without affecting PAR received by the leaves. The mirrors reduced the R : FR ratio received by the stems (measured with the integrating cylinder) from ca. 0.8 to 0.5 and the estimated θ inside the stem (calculated from spectral scans obtained with the fiber optic probe) from ca. 0.34 to 0.25 (see Ballaré *et al.*, 1989). These plants exposed to additional FR produced longer internodes than the controls grown either without mirrors or in front of mirrors that reflected R and FR light with similar efficiencies. Comparable results have been obtained in experiments where the additional FR treatment was applied only during the middle hours of the photoperiod (Ballaré *et al.*, 1991c).

In a series of glasshouse studies, 2-week-old seedlings of *Datura ferox* and *Sinapis alba* were placed for a short period of time in canopies formed by plants of similar height to the seedlings. Seedlings responded with a rapid increase in the rate of stem elongation that was related to the LAI of the surrounding canopy. This elongation response to the proximity of neighbors was much reduced when the internodes of test seedlings were covered by a FR-absorbing filter collar that maintained a high R : FR ratio at the stem level (Figure 7). These experiments demonstrated that in seedling canopies of very low LAI (i.e., LAI ≤ 1), where interference among neighbors for PAR capture is almost nil, the R : FR ratio perceived at the internode level plays a major role in driving adjustments of stem elongation rate to local density conditions (Ballaré *et al.*, 1990).

In canopies with LAI > 1, the efficiency of the FR-absorbing filters was generally reduced, even when they still maintained a relatively high R : FR (ca. 0.9) (Ballaré *et al.*, 1990). It appeared that some other factor promoted elongation growth in those populations. Light measurements presented in Figure 6 show that, when LAI reaches about 1, PAR interception by leaves continues to be high, but total irradiance at the stem level begins to decline markedly. Most experiments addressing the

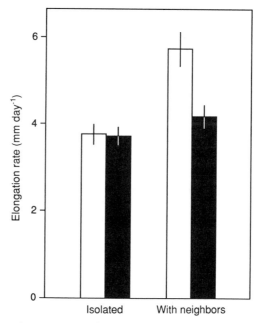

Figure 7 Elongation responses of *Datura ferox* first internodes when seedlings were placed in the center of an even-height canopy of LAI ~ 0.9 under natural radiation. During the experiment the internodes were surrounded by annular cuvettes containing distilled water (clear filter; open bars) or a $CuSO_4$ solution that absorbed FR radiation and maintained the R : FR ratio at ca. 1.1 (FR-absorbing filter; solid bars). (Adapted from Ballaré *et al.*, 1990.)

influence of total irradiance on elongation growth have been carried out using light treatments applied to the whole shoot (Section II,B,3,a), which is clearly not equivalent to the effects of neighboring plants in even-height canopies. When only the fluence rate of sidelight was reduced, by means of annular filters that did not affect PAR interception by leaves, a significant promotion of stem elongation could be demonstrated in seedlings grown under natural radiation (Ballaré *et al.*, 1991c). This response to fluence rate was not obviously affected by R : FR in the range of ratios typical of canopies of low LAI (i.e., $0.3 \leq$ R : FR ≤ 0.9). This suggests that both R : FR- and fluence-rate-dependent mechanisms modulate elongation responses to plant proximity when the canopy begins to close (Ballaré *et al.*, 1991c).

What photoreceptors are responsible for the perception of changes in fluence rate? Photosynthetic pigments are unlikely candidates. In fact, shading the stems with a pink collar that reduced PAR by filtering the green wave band (with minimal effects on spectral regions absorbed by phytochromes and B/UV-A receptors), did not result in increased elongation in *Sinapis alba* (Ballaré *et al.*, 1991c). Phytochrome(s) and B/UV-A sensors have been implicated to explain the fluence-rate-dependent photoinhibition of hypocotyl elongation in briefly de-etiolated seedlings (see Section II,B,3,a) and the same photoreceptors would seem to control elongation responses to stem shading in the case of more mature, fully de-etiolated plants (Ballaré *et al.*, 1991c).

4. Branching and Production of New Ramets   Most species rely on the production of new branches for colonization of open space in canopies, and it is well established that local conditions of crowding have a dramatic influence on branching (e.g., Langer, 1963; Harper, 1977). Multiple factors, including responses to resource levels and external morphogenic signals, are most likely involved in the control of branching and ramet production under field conditions. The possible involvement of morphogenic signals was first suggested by Kirby and Faris (1972) after observing in barley (*Hordeum vulgare*) crops that the effects of planting density on the fate of individual tiller buds appeared to be decided quite early during the transition from buds to tillers. In this section I briefly cover the effects of light on branching (see also Casal *et al.*, 1986; Hutchings and Slade, 1988; Sánchez *et al.*, 1993).

Reduced total irradiance has been shown to result in increased apical dominance and reduced branching in growth-cabinet (e.g., Mitchell, 1953), greenhouse (Slade and Hutchings 1987; Méthy *et al.*, 1990), and field experiments (Bubar and Morrison, 1984; Thompson and Harper, 1988). It is not clear whether this effect of shading is a consequence of assimilate limitations (which might act as an internal morphogenic signal) or a more direct response to light fluence rate mediated by specific photoreceptors. Frequently, the inhibitory effects of heavy natural or simulated leaf shading (i.e., low B and R, high FR) on branching are more severe than those caused by similar reductions of PAR obtained by means of neutral filters (e.g., Solangaarachchi and Harper, 1987; Méthy *et al.*, 1990). This is most likely due to the difference in R : FR ratio between shading treatments (see the following), but the role of direct effects of B and R fluence rates has not been investigated. Direct responses to UV-B fluence rate should also be considered in shading studies carried out in the field, because moderate levels of these wavelengths are known to promote tillering in some species (Barnes *et al.*, 1990).

Alterations of the ratios between wavelengths (i.e., R : FR ratio) influence apical dominance (Tucker and Mansfield, 1972) and are certainly involved in branching responses to the proximity of other plants. Deregibus et al. (1983) demonstrated that tillering in grasses is depressed by

FR light acting through phytochrome (see also Casal *et al.*, 1985; Kasper-bauer and Karlen, 1986; Casal, 1988). Deregibus *et al.* (1985) also showed that photomorphogenic mechanisms are involved in the control of tillering under field conditions. Working in a natural grassland, they were able to increase tillering rates by supplementing the radiation received at the base of *Paspalum dilatatum* and *Sporobolus indicus* plants with small amounts of R light from light-emitting diodes.

The interactions between light signals and resource levels in the control of branching and ramet production have not been thoroughly investigated in any system. The experiments of Casal *et al.* (1986) with *Paspalum dilatatum* show that in low-LAI grass canopies, where mutual shading is small, tillering responses to the proximity of other plants are most likely mediated by the altered R : FR balance. Grasses are very sensitive to small reductions in R : FR (Casal *et al.*, 1987), particularly if the treatments are applied during daytime (as opposed to end-of-day pulses) (Casal *et al.*, 1990). Therefore, the differential reflection of R and FR light by nearby leaves (e.g., Figure 5 and 6) might signal grass plants to reduce tillering rate early in canopy development (see Casal *et al.*, 1987). This is probably advantageous for the internal economy of the plant in environments where late tillers have little chance of contributing to reproduction. The mechanisms that control branch or tiller mortality in canopies are still not well understood. The popular belief that branch senescence is triggered by resource (light) starvation has been questioned by Lauer and Simmons (1989).

5. Spatial Orientation of Branches   The phototropic responses of leaves and stems considered in Section II,B,2 are examples of rapid, reversible movements that allow plants to rearrange their architecture in continuously changing light conditions. Plants can also change the orientation of whole branches in response to more permanent features of their light environments. Casal *et al.* (1990) found that the tillers of annual ryegrass (*Lolium multiflorum*) tended to adopt a more erect position in the canopy as plant density was increased, and there is some evidence that R : FR signals perceived by phytochrome activate mechanisms that control shoot inclination (e.g., Casal *et al.*, 1990; Aphalo *et al.*, 1991). Effects of total irradiance may also be involved in the responses of shoot inclination to canopy density. Grime *et al.* (1986) reported that neutral-shading treatments led to the production of more vertical shoots in several bryophyte species. Under natural conditions, these species tend to exploit the upper strata of pasture canopies. Montaldi and co-workers suggested that, in grasses with diagravitropic stems, shoot inclination angle is modulated by sucrose content (Montaldi, 1969; Willemoës *et al.*, 1988). They postulated that the effects of irradiance and R : FR ratio on inclination may be consequences of their influence on carbohydrate production and partitioning (e.g., Beltrano *et al.*, 1991).

An interesting example of how plants use directional light signals to adjust the orientation of branches in patchy canopies was provided by Novoplansky *et al.* (1990). They found that seedlings of *Portulaca oleracea*, a weed of intensively disturbed areas with a diagravitropic shoot system, tended to avoid growing toward their neighbors when colonizing bare soil. This response appeared to result from effects of neighbors on (1) the azimuthal orientation adopted by the main stem at the time it became horizontal and (2) the pattern of lateral shoot initiation. The authors implicated light signals in these responses after observing that seedlings also tended to grow away from green plastic objects that absorbed PAR and reduced the R : FR ratio. Furthermore, when seedlings were confronted with gray and green plastics that provided similar PAR transmission with different effects on the R : FR ratio, only a small proportion of the plants became recumbent toward the green (low R : FR) sector (Figure 8). Since the green plastic reflected more FR than the neutral gray, the orientation response observed in *Portulaca* might have been a consequence of the phytochrome-mediated phototropism discussed earlier (Section II,B,2; Ballaré *et al.*, 1992a).

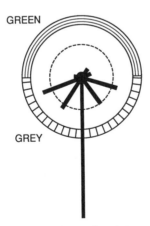

Figure 8 The effect of green filters that reduced the R : FR ratio of sidelight on the azimuthal orientation of main shoots of *Portulaca oleracea* seedlings grown in the field. The length of bars represents the relative frequency of plants with their stems developing in each of the different compass directions. Green and gray filters extended only 2 cm above the pot's rim and had similar PAR transmission. (After Novoplansky *et al.*, 1990.)

III. Competition for Light during Early Succession

In Section II, I have briefly covered some of the mechanisms whereby early successional plants perceive and react to changes in light conditions with emphasis on germination responses of arable weeds and morphological plasticity of vegetative shoots. The following presents an overview of how these mechanisms may contribute to increase species fitness in early secondary succession.

A. Acquisition of Information and the Race for Light

Secondary successions are generated by disturbances that increase light availability at the ground surface. Disturbance is followed by synchronous recolonization of the site by individuals of pioneer species (and cultivated plants in agricultural environments) that rapidly begin to compete with each other for access to light and other resources. We can visualize the process as a competitive race, where the success of an individual relative to its neighbors depends essentially on (1) its relative ability to capture light energy at the beginning of the period of interaction (which is mainly a function of its relative initial size) and (2) its relative ability to maintain high rates of light capture as the canopy develops.

Differences in initial size are largely due to differences in propagule (seed) size and emergence time (or emergence order). The production of large seeds with large embryos may have a number of associated costs (Harper *et al.*, 1970; Thompson, 1987). Thus small seeds appear to prevail among pioneer species in many different systems (Fenner, 1987). Given the importance of relative time of emergence in determining growth and fecundity (see references in Section II,B,3,b and Weiner, 1988), one can hardly dispute the notion that the ability of seeds to acquire information about the environment by the mechanisms discussed earlier (Section II,B,1) and relay this information to the systems that control dormancy is a key element of success in frequently disturbed ecosystems.

Maintaining high rates of light interception as the canopy develops is of primary importance in maintaining competitive advantage. The amount of leaf area generated per unit of dry matter produced, which is strongly influenced by light conditions (Blackman and Wilson, 1951; Evans and Hughes, 1961), is certainly a major determinant of the dynamics of light capture. Light interception per unit leaf area depends on the spatial arrangement of foliage in relation to the pattern of light availability. Since this pattern is characterized by major fluctuations imposed by abiotic factors and rapid growth of neighboring plants, a high degree of morphological plasticity, based on continuous acquisition and interpretation of environmental information, is most likely an essential requirement for species of early secondary succession. In Section II, I discussed how information acquired by B/UV-A receptors and phytochrome(s) is used by plants to actively orient leaves and shoots in patchy light environments and increase PAR capture.

Stem elongation is important to ensure that young leaves are situated in direct sunlight in even-height, rapidly growing canopies. Although stem internodes in herbaceous plants appear to be relatively cheap in terms of the opportunity value of the carbon invested in their construction (Ballaré *et al.*, 1991d), different types of indirect costs are most likely associated with the production of longer stems (e.g., reduced mechanical strength, increased exposure to herbivores). Therefore, for any individual in a population, there would be an optimum rate of stem elongation dictated by the rate of growth in height of the surrounding vegetation (see discussions by Givnish, 1982, and Waller, 1988). I have presented evidence in Section II,B,3,b that light signals perceived by phytochromes and B/UV-A sensors modulate the rate of stem elongation in even-height canopies of dicotyledonous seedlings, and that local light changes at the stem level can signal potential future variations of light interception by leaves. Morgan and Smith (1979) and Corré (1983) found that elongation responses to experimental changes in R : FR ratio are usually more dramatic in pioneer species than in species of shaded habitats. This result is consistent with the hypothesis that these responses are most valuable for plants that compete for access to light against individuals of similar morphology in relatively open canopies. Ballaré *et al.* (1988a) grew seedlings of the arable weed *Datura ferox* at three densities in the field and monitored the time course of stem elongation, leaf area expansion, and PAR interception per plant. After 2 weeks, the LAI of the high-density population was ~2 and shading among neighboring seedlings was still low. However, a significant effect of canopy density on seedling height was already evident, and seedlings transferred from low- to high-density canopies intercepted two-thirds less PAR than their (high-density) neighbors. These results suggest that a short lag in stem elongation would be enough to seriously compromise competitive ability in rapidly growing communities.

The foregoing discussion suggests that the systems that permit acquisition of information about present and future changes in PAR availability play a fundamental role as determinants of competitive success in early successional communities. Curiously, however, and with few exceptions (e.g., Grime *et al.*, 1986), the value of information-acquiring systems has been largely overlooked in experimental and theoretical approaches to the study of plant interactions. The likely reasons and potential consequences of this apparent neglect are discussed in the following section.

IV. Signals, "Decisions," and Models of Plant Competition

Two major avenues of research have contributed most to shape our current perception of the mechanisms of plant competition. Phenomenological studies, on the one hand, provided a wealth of information on the consequences of competition in relatively simple systems like monocultures and mixtures of cultivated species and arable weeds (reviews by Willey and Heat, 1969; White and Harper, 1970; Harper, 1977; Weiner, 1988; Firbank and Watkinson, 1990; Radosevich and Roush, 1990). These studies examined the effects of crowding on plant growth, morphology, and mortality. Physiological experiments, on the other hand, generated information on plant responses to light and other environmental resources that are presumably or certainly affected by changes in canopy density. These two disciplines grew up rather independently of each other and, traditionally, there has been very little exchange of ideas between groups working in descriptive population or community ecology and plant physiology. Physiological ecologists historically appeared to be more inclined to study plant responses to abiotic factors in extreme habitats (e.g., alpine areas and deserts; see Mooney, 1991) than to unravel physiological mechanisms of plant–plant interactions in more productive environments.

During the last few years, a number of mechanistic models of plant competition have been developed by ecologists and agronomists. These models vary greatly in complexity and purpose; the common denominator is that they are designed to predict the outcome of plant competition from assumptions about the way plants affect and respond to intermediary environmental resources (Goldberg, 1990). Models of this kind are proving to be useful in studying single-resource competition over short periods of time in cultivated plant associations (e.g., Rimmington, 1984), and in predicting results of competitive interactions among plants with contrasting life-forms or allocation patterns along gradients of resource supply or disturbance intensities (Tilman, 1988).

Most competition models assume that plants are "blind" organisms that grow over an arbitrarily defined period of time at a rate that is essentially proportional to the rate of resource supply. Most commonly plants are forced to co-occur, and the only possible way by which plants may influence each other's activities is by altering resource availability. In turn, the only process that is affected by a change in the rate of resource supply is total growth, since developmental, architectural, and biomass allocation patterns are most commonly rigid (i.e., genotype- or age-dependent); only in a few cases is some degree of phenotypic plasticity incorporated into the model by assuming that allocation among organs or parts depends on resource levels or ratios (e.g., Tilman, 1988). Strictly speaking these assumptions are false, or at best coarse simplifications of the physiology of real plants. Therefore, the analytic value of competition models based on these assumptions is limited when the goal is to pinpoint the specific physiological or genetic traits that determine competitive ability.

Previous sections (Section II,B,1) have presented an overview of the systems used by seeds to acquire information about the aerial environment (see also Vázquez-Yanes and Orozco-Segovia, Chapter 7, this volume). These systems, based on specific environmental signals, synchronize growth activity with periods of increased light availability. In highly disturbed habitats, with pulsed supplies of light and other resources, differences in ability to coordinate seed germination with the formation of natural or artificial gaps are likely to be far more important determinants of differences in competitive ability than variations in resource-assimilation physiology (e.g., photosynthetic capacity) or allocation patterns. Information-acquiring systems may help reduce the costs associated with the production of competitive offspring. Thus, weed species that evolved efficient mechanisms to detect light opportunities may afford to produce small seeds, which are cheap and suitable for dispersal but unlikely to generate seedlings with chances of success in occupied sites. Models or experimental designs that assume uniform germination across species or genotypes certainly miss a very important part of the competition story in early successional environments.

Our present understanding of green plant photobiology also highlights the significance of light as a vehicle of information that determines morphological development in plant canopies (Sections II,B,2 to II,B,5). The signals that plants use to monitor the status of the light environment are, by definition, directly associated or correlated with the amount of light available for photosynthesis. Why is it then necessary to recognize the information-acquiring systems as separate components in mechanistic models of plant competition? Conceptually, it is important to keep in mind that each individual plant in the canopy is influenced by its neighbors not only because they alter the availability of resources, but also because they modify factors of the environment that are specifically used by the plant to "decide" among alternative developmental programs. The use of specific cues to keep track of variations in light supply may offer several strategic advantages to the plant over direct monitoring of the resource, the most significant of which may be the ability to appropriately time the responses. In Section II,B,3,b, I discussed how perception of R : FR changes by individual internodes may allow rapid stem elongation responses that precede variations in light interception at leaf level. Thus, particularly where light availability is mainly a function of the size and growth rate of neighboring individuals, natural selection is likely to favor mechanisms of phenotypic response to environmental factors that are reliable and sensitive indicators of neighboring plant proximity. These mechanisms should, therefore, be considered as an integral part of the traits that determine competitive ability.

Practically, the distinction between plant responses to resources and signals may be of importance in applied fields like agronomy or forestry. Elongation growth (Section II,B,3), tillering (Section II,B,4), developmental timing (e.g., Mondal *et al.*, 1986), and carbon allocation to reproductive structures (Heindl and Brun, 1983) are all affected by canopy density in field crops, and different types of evidence have been presented showing that these responses are caused, at least in part, by photomorphogenic reactions to alterations of light climate caused by nearby plants (for further discussion, see Ballaré *et al.*, 1992b; Sánchez *et al.*, 1993). Some of the response mechanisms in modern crop species could have been left over from ancestral lines, where they were selected because they conferred competitive advantages under conditions that were different from the ones normally faced by crop plants in modern agricultural systems. Further, natural selection did most likely favor physiological responses that were beneficial to the individual plant, rather than to the whole population. If we accept these possibilities, we should accept that under certain combinations of planting density and environment, crop yields are likely to be limited by the presence (or lack) of particular physiological responses to specific proximity cues.

The design of crop ideotypes is usually based on identification of factors that limit commercial yield in a given cultural/environmental scenario and consists of deciding what morphological or physiological traits should be added or dropped from the crop plant in order to make it more efficient at converting resources into commercial yield. Until recently, the only way available for translating ideotypes into real plants was by repeated hybridization and selection. But with the arrival of molecular biological techniques, the opportunities for engineering plants with very specific physiological traits are increasing rapidly. Therefore, knowledge about the mechanistic details and genetic determinants of plant responses to their environment is now likely to be more important than in the past. Based on the limited information we have today, it would seem that manipulation of photomorphogenic behavior to induce or suppress developmental responses to signals of neighbors' proximity may be rewarding in commercial crops.

V. Summary

Under natural conditions, abiotic factors and the growth of vegetation determine dramatic variations in the flux of photosynthetically active radiation received at the ground surface. Light opportunities in heterogeneous environments are perceived by terrestrial plants through a great variety of specialized mechanisms. At the seed level, light (sensed by phytochrome) and the thermal regime of the soil are among the factors used to detect the onset of periods of high light availability. In the particular case of seeds of arable weeds, daily fluctuations of soil temperature and pulses of light operating through the very-low-fluence mode of phytochrome action appear to be very important to inform seeds about the occurrence of soil disturbances, which usually precede periods of abundant light. At the de-etiolated plant level, photoreceptors play a fundamental role in driving architectural plasticity and morphological responses to variations in the light environment. One or more specific photoreceptors for B light are essential for phototropic movements of leaves and stems, which, in many species, allow fine adjustment of leaf display to changes in the direction of incoming light. The involvement of phytochrome (most likely phyB) in the detection of light gradients leading to bending responses of stems in plant canopies has also been demonstrated. Precise control of stem elongation rate is most likely essential for plants that forage for light in plant communities formed by individuals of similar height. A number of recent experiments have indicated that phytochrome can detect FR-rich radiation scattered by neighboring plants and trigger changes in stem elongation rate before the plant is subjected to a reduction in light availability. In addition to this response to R : FR ratio, fluence rate signals, perceived by phytochrome(s) and B/UV-A sensors, influence stem elongation in even-height canopies. In the case of multilevel canopies, glasshouse and field experiments with photomorphogenic mutants strongly suggest that, at least in the short term, elongation responses to leaf shading depend to a large extent on processes in which phytochrome-B is directly or indirectly involved. Evidence has also been obtained over the last few years for the involvement of phytochrome in the control of branching and lateral spreading in clonal plants.

The ability to acquire information about present and future light conditions may be critical for plants of early successional communities, where dramatic changes of plant cover (and hence light availability) occur over relatively short periods of time because of disturbances and vegetation growth. Mechanistic models of plant competition usually fail to acknowledge that each plant in the community is affected by neighboring individuals not only because they modify the availability of energy and materials necessary for growth, but also because they affect factors of the environment that are specifically used by the plant to obtain information about neighbors' proximity. This information is used to make developmental "decisions," which may, in turn, have a dramatic impact on the resource-harvesting capacity of the plant and influence competitive ability. A better understanding of the mechanisms whereby plants acquire and "interpret" environmental information should provide new insights into the factors that determine differences between genotypes in the ability to compete for light. From a practical standpoint, this knowledge would be useful aid to the design of more productive crop genotypes.

Acknowledgments

Many of the ideas expressed in this chapter are the product of discussions with Pedro Aphalo, Jorge Casal, Claudio Ghersa, Steve Radosevich, Rodolfo Sánchez, and Ana Scopel. Les Fuchigami, Bob Pearcy, Mary Lynn Roush, Joe Zaerr, and an anonymous reviewer offered thoughtful suggestions on the various versions of the manuscript. Gretchen Bracher prepared the figures. I thank the Consejo Nacional de Investigaciones Científicas y Técnicas (CONICET, Argentina) and the University of Buenos Aires for supporting the research presented here. I am also indebted to CONICET and the Department of Forest Science for financing my stay at Oregon State University, where this review was written.

References

Adamse, P., Jaspers, P. A. M. P., Kendrick, R. E., and Koorneef, M. (1987). Photomorphogenic responses of a long hypocotyl mutant of *Cucumis sativus* L. *J. Plant Physiol.* **127,** 481–491.

Adamse, P., Jaspers, P. A. M. P., Bakker, J. A., Kendrick, R. E., and Koornneef, M. (1988). Photophysiology and phytochrome content of long-hypocotyl mutant and wild-type cucumber seedlings. *Plant Physiol.* **87,** 264–268.

Aldrich, R. J. (1984). "Weed-Crop Ecology: Principles in Weed Management." Breton Publishers, North Situate, Madison, WI.

Angevine, M. W., and Chabot, B. F. (1979). Seed germination syndromes in higher plants. *In* "Topics in Plant Population Biology" (O. T. Solbrig, S. Jain, G. B. Johnson, and P. H. Raven, eds), pp. 188–206. Columbia Univ. Press, New York.

Aphalo, P. J., Gibson, D. J., and Di Benedetto, A. H. (1991). Responses of growth, photosynthesis, and leaf conductance to white light irradiance and end-of-day red and far-red pulses in *Fuschsia magellanica* Lam. *New Phytol.* **117,** 461–471.

Atkins, G. A. (1936). The effect of pigment on phototropic response: A comparative study of reactions to monochromatic light. *Ann. Bot. (London)* [N.S.] **50,** 197–218.

Attridge, T. H. (1990). "Light and Plant Responses." Edward Arnold, London.

Ballaré, C. L., Sánchez, R. A., Scopel, A. L., Casal, J. J., and Ghersa, C. M. (1987a). Early detection of neighbour plants by phytochrome perception of spectral changes in reflected sunlight. *Plant, Cell Environ.* **10,** 551–557.

Ballaré, C. L., Scopel, A. L., Ghersa, C. M., and Sánchez, R. A. (1987b). The demography of *Datura ferox* (L.) in soybean crops. *Weed Res.* **27,** 91–102.

Ballaré, C. L., Sánchez, R. A., Scopel, A. L., and Ghersa, C. M. (1988a). Morphological responses of *Datura ferox* L. seedlings to the presence of neighbors. Their relationships with canopy microclimate. *Oecologia* **76,** 288–293.

Ballaré, C. L., Scopel, A. L., Ghersa, C. M., and Sánchez, R. A. (1988b). The fate of *Datura ferox* seeds in the soil as affected by cultivation, depth of burial and degree of maturity. *Ann. Appl. Biol.* **112,** 337–345.

Ballaré, C. L., Scopel, A. L., and Sánchez, R. A. (1989). Photomodulation of axis extension is sparse canopies. Role of the stem in the perception of light-quality signals of stand density. *Plant Physiol.* **89,** 1324–1330.

Ballaré, C. L., Scopel, A. L., and Sánchez, R. A. (1990). Far-red radiation reflected from adjacent leaves: An early signal of competition in plant canopies. *Science* **247,** 329–332.

Ballaré, C. L., Barnes, P. W., and Kendrick, R. E. (1991a). Photomorphogenic effects of UV-B radiation on hypocotyl elongation in wild-type and stable-phytochrome-deficient mutant seedlings of cucumber. *Physiol. Plant.* **82,** 652–658.

Ballaré, C. L., Casal, J. J., and Kendrick, R. E. (1991b). Responses of wild-type and *lh*-mutant seedlings of cucumber to natural and simulated shadelight. *Photochem. Photobiol.* **54,** 819–826.

Ballaré, C. L., Scopel, A. L., and Sánchez, R. A. (1991c). Photocontrol of stem elongation in plant neighbourhoods: Effects of photon fluence rate under natural conditions of radiation. *Plant, Cell Environ.* **14,** 57–65.

Ballaré, C. L., Scopel, A. L., and Sánchez, R. A. (1991d). On the opportunity cost of the photosynthate invested in stem elongation reactions mediated by phytochrome. *Oecologia* **86,** 561–567.

Ballaré, C. L., Scopel, A. L., Radosevich, S. R., and Kendrick, R. E. (1992a). Phytochrome mediated phototropism in de-etiolated seedlings. Occurrence and ecological significance. *Plant Physiol.* **100,** 170–177.

Ballaré, C. L., Scopel, A. L., Sánchez, R. A., and Radosevich, S. R. (1992b). Photomorphogenic processes in the agricultural environment. *Photochem. Photobiol.* **56,** 777–788.

Barnes, P. W., Flint, S. D., and Caldwell, M. M. (1990). Morphological responses of crop and weed species of different growth forms to ultraviolet-B radiation. *Am. J. Bot.* **77,** 1354–1360.

Baskin, J. M., and Baskin, C. C. (1979). Promotion of germination of *Stellaria media* by light from a green safe lamp. *New. Phytol.* **82,** 381–383.

Baskin, J. M., and Baskin, C. C. (1985). The annual dormancy cycle in buried weed seeds: A continuum. *BioScience* **35,** 492–498.

Bazzaz, F. A. (1990). Plant–plant interactions in successional environments. *In* "Perspectives on Plant Competition" (J. B. Grace and D. Tilman, eds.), pp. 239–263. Academic Press, San Diego.

Beltrano, J., Willemoës, J., Montaldi, E. R., and Barreiro, R. (1991). Photoassimilate partitioning modulated by phytochrome in Bermuda grass (*Cynodon dactylon* (L.) Pers.). *Plant Sci.* **73,** 19–22.

Benech Arnold, R. A., Ghersa, C. M., Sánchez, R. A., and García Fernández, A. E. (1988). The role of fluctuating temperatures in the germination and establishment of *Sorghum halepense* (L.) Pers. Regulation of germination under leaf canopies. *Func. Ecol.* **2,** 311–318.

Bewley, J. D., and Black, M. (1982). "Physiology and Biochemistry of Seeds in Relation to Germination," Vol. 2: Springer-Verlag, Berlin.

Blaauw-Jensen, G. (1983). Thoughts on the possible role of phytochrome destruction in phytochrome-mediated responses. *Plant, Cell Environ.* **6,** 173–179.

Black, J. N., and Wilkinson, G. N. (1963). The role of time of emergence in determining the growth of individual plants in swards of subterranean clover (*Trifolium subterraneum* L.). *Aust. J. Agric. Res.* **14,** 628–638.

Blackman, G. E., and Wilson, G. L. (1951). Physiological and ecological studies in the analysis of plant environment. VII. An analysis of the differential effects of light intensity on the net assimilation rate, leaf-area ratio, and relative growth rate of different species. *Ann. Bot. (London)* [N.S.] **15,** 374–408.

Bouwmeester, H. J., and Karssen, C. M. (1989). Environmental factors influencing the expression of dormancy patterns in weed seeds. *Ann. Bot. (London)* [N.S.] **63,** 113–120.

Briggs, W. R., and Baskin, T. I. (1988). Phototropism in higher plants—Controversies and caveats. *Bot. Acta* **101**, 133–139.

Britz, S. J. (1990). Photoregulation of root : shoot ratio in soybean seedlings. *Photochem. Photobiol.* **52**, 151–159.

Bubar, C. J., and Morrison, I. N. (1984). Growth responses of green and yellow foxtail (*Setaria viridis* and *S. lutescens*) to shade. *Weed Sci.* **32**, 774–780.

Casal, J. J. (1988). Light quality effects on the appearance of tillers of different order in wheat (*Triticum aestivum*). *Ann. Appl. Biol.* **112**, 167–173.

Casal, J. J., and Smith, H. (1988a). Persistent effects of changes in phytochrome status on internode growth in light-grown mustard: Occurrence, kinetics and locus of perception. *Planta* **175**, 214–220.

Casal, J. J., and Smith, H. (1988b). The loci of perception for phytochrome control of internode growth in light-grown mustard. Promotion by low phytochrome photoequilibria in the internodes is enhanced by blue light reaching the leaves. *Planta* **176**, 277–282.

Casal, J. J., Deregibus, V. A., and Sánchez, R. A. (1985). Variations in tiller dynamics and morphology in *Lolium multiflorum* Lam. vegetative and reproductive plants as affected by differences in red–far-red irradiation. *Ann. Bot. (London)* [N.S.] **56**, 553–559.

Casal, J. J., Sánchez, R. A., and Deregibus, V. A. (1986). Effects of plant density on tillering: The involvement of the R/FR and the proportion of radiation intercepted per plant. *Exp. Environ. Bot.* **26**, 365–371.

Casal, J. J., Sánchez, R. A., and Deregibus, V. A. (1987). Tillering responses of *Lolium multiflorum* plants to changes of red/far-red ratios typical of sparse canopies. *J. Exp. Bot.* **38**, 1432–1439.

Casal, J. J., Sánchez, R. A., and Gibson, D. (1990). The significance of changes in the red/far-red ratio, associated with either neighbour plants or twilight, for tillering in *Lolium multiflorum* Lam. *New Phytol.* **116**, 565–572.

Casal, J. J., Sánchez, R. A., Di Benedetto, A. H., and De Miguel, L. C. (1991). Light promotion of seed germination in *Datura ferox* is mediated by a highly stable pool of phytochrome. *Photochem. Photobiol.* **53**, 249–254.

Chancellor, R. J. (1964). Emergence of weed seedlings in the field and the effects of different frequencies of cultivation. *Proc. Br. Weed Control Conf.* **7**, 599–606. Br. Weed Control Council, London, UK.

Child, R., and Smith, H. (1987). Phytochrome action in light-grown mustard: Kinetics, fluence-rate compensation and ecological significance. *Planta* **172**, 219–229.

Chow, W. S., Goodchild, D. J., Miller, C., and Anderson, J. M. (1990). The influence of high levels of brief or prolonged supplementary far-red illumination during growth on the photosynthetic characteristics, composition and morphology of *Pisum sativum* chloroplast. *Plant, Cell Environ.* **13**, 135–145.

Cone, J. W., and Kendrick, R. E. (1985). Fluence-response curves and action spectra for promotion and inhibition of germination in wild-type and long-hypocotyl mutants of *Arabidopsis thaliana* L. *Planta* **163**, 43–54.

Corré, W. J. (1983). Growth and morphogenesis of sun and shade plants. II. The influence of light quality. *Acta Bot. Neerl.* **32**, 185–202.

Cosgrove, D. J. (1986). Photomodulation of growth. *In* "Photomorphogenesis in Plants." (R. E. Kendrick and G. H. M. Kronenberg, eds.), pp. 341–366. Martinus Nijhoff, Dordrecht, The Netherlands.

Deregibus, V. A., Sánchez, R. A., and Casal, J. J. (1983). Effects of light quality on tiller production in *Lolium* spp. *Plant Physiol.* **72**, 900–902.

Deregibus, V. A., Sánchez, R. A., Casal, J. J., and Trlica, M. J. (1985). Tillering responses to enrichment of red light beneath the canopy in a humid natural grassland. *J. Appl. Ecol.* **22**, 199–206.

Ehleringer, J. R., and Forseth, I. N. (1980). Solar tracking by plants. *Science* **210,** 1094–1098.

Ehleringer, J. R., and Forseth, I. N. (1989). Diurnal leaf movements and productivity in canopies. *In* "Plant Canopies: Their Growth, Form and Function" (G. Russell, B. Marshall, and P. G. Jarvis, eds.), pp. 129–142. Cambridge Univ. Press, Cambridge, UK.

Evans, G. C., and Hughes, A. P. (1961). Plant growth and the aerial environment. I. Effect of shading on *Impatiens parviflora. New Phytol.* **60,** 150–180.

Fenner, M. (1987). Seed characteristics in relation to succession. *In* "Colonization, Succession and Stability" (A. J. Gray, M. J. Crawley, and P. J. Edwards, eds.), pp. 103–115. Blackwell, Oxford.

Firbank, L. G., and Watkinson, A. R. (1990). On the effects of competition: From monocultures to mixtures. *In* "Perspectives on Plant Competition" (J. B. Grace and D. Tilman, eds.), pp. 165–192. Academic Press, San Diego.

Firn, R. D. (1986). Phototropism. *In* "Photomorphogenesis in Plants" (R. E. Kendrick and R. H. M. Kronenberg, eds.), pp. 369–389. Martinus Nijhoff, Dordrecht, The Netherlands.

Frankland, B. (1986). Perception of light quantity. *In* "Photomorphogenesis in Plants" (R. E. Kendrick and G. H. M. Kronenberg, eds.), pp. 219–235. Martinus Nijhoff, Dordrecht, The Netherlands.

Furuya, M., ed. (1987). "Phytochrome and Photoregulation in Plants." Academic Press, Tokyo.

Furuya, M. (1989). Molecular properties and biogenesis of phytochrome I and II. *Adv. Biophys.* **25,** 133–167.

Gaba, V., and Black, M. (1979). Two separate photoreceptors control hypocotyl growth in green seedlings. *Nature (London)* **278,** 51–54.

Gaba, V., and Black, M. (1985). Photocontrol of hypocotyl elongation in light-grown *Cucumis sativus* L.: Responses to phytochrome photostationary state and fluence rate. *Plant Physiol.* **79,** 1011–1014.

Gaba, V., and Black, M. (1987). Photoreceptor interactions in plant photomorphogenesis: The limits of the experimental techniques and their interpretations. *Photochem. Photobiol.* **45,** 151–156.

Garrison, R., and Briggs, W. R. (1975). The growth of internodes in *Helianthus* in response to far-red light. *Bot. Gaz. (Chicago)* **136,** 353–357.

Ghersa, C. M., Benech Arnold, R. L., and Martínez-Ghersa, M. A. (1992). The role of fluctuating temperatures in the germination and establishment of *Sorghum halepense* (L.) Pers. Regulation of germination at increasing depths. *Func. Ecol.* **6,** 460–468.

Givnish, T. J. (1982). On the adaptive significance of leaf height in forest herbs. *Am. Nat.* **120,** 353–381.

Goldberg, D. E. (1990). Components of resource competition in plant communities. *In* "Perspectives on Plant Competition" (J. B. Grace and D. Tilman, eds.), pp. 27–49. Academic Press, San Diego.

Grime, J. P., and Jeffrey, D. W. (1965). Seedling establishment in vertical gradients of sunlight. *J. Ecol.* **53,** 621–642.

Grime, J. P., Crick, J. C., and Rincon, J. E. (1986). The ecological significance of plasticity. Plasticity in plants. *Symp. Soc. Exp. Biol.* **40,** 5–29.

Gross, K. L. (1980). Colonization of *Verbascum thapsus* (mullein) of an old field in Michigan. Experiments on the effects of vegetation. *J. Ecol.* **68,** 919–927.

Harper, J. L. (1977). "Population Biology of Plants." Academic Press, New York.

Harper, J. L., Lovell, P. H., and Moore, K. G. (1970). The shapes and sizes of seeds. *Annu. Rev. Ecol. Syst.* **1,** 327–356.

Hartmann, K. M., and Cohnen-Unser, I. (1972). Analytical action spectroscopy with living systems: Photochemical aspects and attenuance. *Ber. Dtsch. Bot. Ges.* **85,** 481–551.

Hartmann, K. M., and Negadal, W. (1990). Photocontrol of weeds without herbicides. *Naturwissenschaften* **77**, 158–163.

Heindl, J. C., and Brun, W. A., (1983). Light and shade effects on abscission and ^{14}C-photoassimilate partitioning among reproductive structures in soybean. *Plant Physiol.* **73**, 434–439.

Holmes, M. G., and Smith, H. (1977a). The function of phytochrome in the natural environment. I. Characterization of daylight for studies in photomorphogenesis and photoperiodism. *Photochem. Photobiol.* **25**, 533–538.

Holmes, M. G., and Smith, H. (1977b). The function of phytochrome in the natural environment. II. The influence of vegetation canopies on the spectral energy distribution of natural daylight. *Photochem. Photobiol.* **25**, 539–545.

Holmes, M. G., Beggs, C. J., Jabben, M., and Schäfer, E. (1982). Hypocotyl growth in *Sinapis alba* L.: The roles of light quality and quantity. *Plant, Cell Environ.* **5**, 45–51.

Hutchings, M. J., and Slade, A. J. (1988). Morphological plasticity, foraging and integration in clonal perennial herbs. *In* "Plant Population Ecology" (A. J. Davy, M. J. Hutchings, and A. R. Watkinson, eds.), pp. 83–109. Blackwell, Oxford.

Iino, M. (1990). Phototropism: Mechanisms and ecological implications. *Plant, Cell Environ.* **13**, 633–650.

Iino, M., Briggs, W. R., and Schäfer, E. (1984). Phytochrome mediated phototropism in maize seedling shoots. *Planta* **160**, 41–51.

Karssen, C. M. (1982). Seasonal patterns of dormancy in weed seeds. *In* "The Physiology and Biochemistry of Seed Development, Dormancy and Germination" (A.A. Khan, ed.), pp. 243–269. Elsevier, Amsterdam.

Kasperbauer, M. J., and Karlen, D. L. (1986). Light-mediated bioregulation of tillering and photosynthate partitioning in wheat. *Physiol. Plant.* **66**, 159–163.

Kasperbauer, M. J., Hunt, P. G., and Sojka, R. E. (1984). Photosynthate partitioning and nodule formation in soybean plants that received red or far-red light at the end of the photosynthetic period. *Physiol. Plant.* **61**, 549–554.

Kendrick, R. E., and Kronenberg, G. H. M., eds. (1986). "Photomorphogenesis in Plants." Martinus Nijhoff, Dordrecht, The Netherlands.

Kendrick, R. E., and Nagatani, A. (1991). Phytochrome mutants. *Plant J.* **1**, 133–139.

Kirby, E. J. M., and Faris, D. G. (1972). The effect of plant density on tiller growth and morphology in barley. *J. Agric. Sci.* **78**, 281–288.

Knipling, E. (1970). Physical and physiological basis for the reflectance of visible and near-infrared radiation from vegetation. *Remote Sens. Environ.* **1**, 155–159.

Koller, D. (1972). Environmental control of seed germination. *In* "Seed Biology" (T. T. Kozlowski, ed.), Vol. 2, pp. 1–101. Academic Press, New York.

Koller, D. (1990). Light-driven leaf movements. *Plant, Cell Environ.* **13**, 615–632.

Kronenberg, G. H. M., and Kendrick, R. E. (1986). The physiology of action. *In* "Photomorphogenesis in Plants" (R. E. Kendrick and G. H. M. Kronenberg, eds.), pp. 99–114. Martinus Nijhoff, Dordrecht, The Netherlands.

Langer, R. H. M. (1963). Tillering in herbage grasses. *Herb. Abstr.* **33**, 141–148.

Lauer, J. G., and Simmons, S. R. (1989). Canopy light and tiller mortality in spring barley. *Crop Sci.* **29**, 420–424.

López-Juez, E., Buurmeijer, W. F., Heeringa, G. H., Kendrick, R. E., and Wesselius, J. C. (1990). Response of light-grown wild-type and long-hypocotyl mutant cucumber plants to end-of-day far-red light. *Photochem. Photobiol.* **58**, 143–150.

López-Juez, E., Nagatani, A., Tomizawa, K.-I., Deak, M., Kern, R., Kendrick, R. E., and Furuya, M. (1992). The cucumber long-hypocotyl mutant lacks a light-stable PHYB-like phytochrome. *Plant Cell* **4**, 241–251.

Mancinelli, A. L. (1988). Some thoughts about the use of predicted values of the state of phytochrome in plant photomorphogenesis research. *Plant, Cell Environ.* **11**, 429–439.

Mancinelli, A. L. (1991). Phytochrome-mediated detection of changes in reflected light. *Plant Physiol.* **91,** 144–151.

Mandoli, D. F., and Briggs, W. R. (1981). Phytochrome control of two low-irradiance responses in etiolated oat seedlings. *Plant Physiol.* **67,** 733–739.

McIntyre, G. I., and Boyer, J. S. (1983). The effect of humidity, root excision, and potassium supply on hypocotyl elongation in dark grown seedlings of *Helianthus annus. Can. J. Bot.* **62,** 420–428.

Meijer, G. (1959). The spectral dependence of flowering and elongation. *Acta Bot. Neerl.* **8,** 189–246.

Méthy, M., Alpert, P., and Roy, J. (1990). Effects of light quality and quantity on growth of the clonal plant *Eichhornia crassipes. Oecologia* **84,** 265–271.

Mitchell, K. J. (1953). Influence of light and temperature on the growth of ryegrass (*Lolium spp.*). I. Pattern of vegetative development. *Physiol. Plant.* **6,** 21–46.

Mittelbach, G. G., and Gross, K. L. (1984). Experimental studies of seed predation in old fields. *Oecologia* **65,** 7–13.

Mohr, H. (1986). Coaction between pigment systems. *In* "Photomorphogenesis in Plants" (R. E. Kendrick and G. H. M. Kronenberg, eds.), pp. 547–563. Martinus Nijhoff, Dordrecht, The Netherlands.

Mondal, M. F., Brewster, J. L., Morris, G. E. L., and Butler, H. A. (1986). Bulb development in onion (*Allium cepa* L.). III. Effects of the size of adjacent plants, shading by neutral and leaf filters, irrigation and nitrogen regime and the relationship between red: far-red spectral ratio in the canopy and leaf area index. *Ann. Bot. (London)* [N. S.] **58,** 207–219.

Montaldi, E. R. (1969). Gibberellin–sugar interaction regulating the growth habit of Bermudagrass (*Cynodon dactylon* (L.) Pers.). *Experientia* **25,** 91–92.

Mooney, H. A. (1991). Physiological plant ecology—Determinants of progress. *Funct. Ecol.* **5,** 127–135.

Mooney, H. A., and Ehleringer, J. R. (1978). The carbon gain benefits of solar tracking in a desert annual. *Plant, Cell Environ.* **1,** 307–311.

Morgan, D. C., and Smith, H. (1978). The relationship between phytochrome photoequilibrium and development in light grown *Chenopodium album* L. *Planta* **142,** 187–193.

Morgan, D. C., and Smith, H. (1979). A systematic relationship between phytochrome-controlled development and species habitat, for plant grown in simulated natural radiation. *Planta* **145,** 253–258.

Morgan, D. C., O'Brien, T., and Smith, H. (1980). Rapid photomodulation of stem extension in light-grown *Sinapis alba* L. Studies on kinetics, site of perception and photoreceptor. *Planta* **150,** 95–101.

Neel, P. L., and Harris, P. W. (1971). Motion-induced inhibition of elongation and induction of dormancy in *Liquidambar. Science* **173,** 58–59.

Nobel, P. S. (1981). Wind as an ecological factor. *In* "Encyclopedia of Plant Physiology" (O. L. Lange, P. S. Nobel, C. B. Osmond, and H. Ziegler, eds.), New Ser., Vol. 12A, pp. 475–500. Springer-Verlag, New York.

Novoplansky, A., Cohen, D., and Sachs, T. (1990). How portulaca seedlings avoid their neighbors. *Oecologia* **82,** 490–493.

Panetta, F. D., Gilbey, D. J., and D'Antuono, M. F. (1988). Survival and fecundity of wild radish (*Raphanus raphanistrum* L.) plants in relation to cropping, time emergence and chemical control. *Aust. J. Agric. Res.* **39,** 385–397.

Parker, K., Baskin, T. I., and Briggs, W. R. (1989). Evidence for a phytochrome-mediated phototropism in etiolated pea seedlings. *Plant Physiol.* **89,** 493–497.

Peters, J. L., Kendrick, R. E., and Mohr, H. (1991). Phytochrome content and hypocotyl growth of long-hypocotyl and wild type cucumber seedlings during de-etiolation. *J. Plant Physiol.* **137,** 291–296.

Peters, N. C. B. (1984). Time of onset of competition of various fractions of an *Avena fatua* L. population in spring barley. *Weed Res.* **24**, 305–315.

Pons, T. L. (1989). Breaking of seed dormancy by nitrate as a gap detection mechanism. *Ann. Bot. (London)* [N. S.] **63**, 139–143.

Quail, P. H. (1991). Phytochrome: A light-activated molecular switch that regulates plant gene expression. *Ann. Rev. Genet.* **26**, 389–409.

Radosevich, S. R., and Roush, M. L. (1990). The role of competition in agriculture. *In* "Perspectives on Plant Competition" (J. B. Grace and D. Tilman, *eds.*, pp. 341–363. Academic Press, San Diego.

Retuerto, R., and Woodward, F. I. (1992). Effects of windspeed on the growth and biomass allocation of white mustard *Sinapis alba* L. *Oecologia* **92**, 113–123.

Rich, T. C. G., Whitelam, G. C., and Smith, H. (1987). Analysis of growth rates during phototropism: Modifications by separate light-growth responses. *Plant, Cell Environ.* **10**, 303–311.

Rimmington, G. M. (1984). A model of the effect of interspecies competition for light on dry-matter production. *Aust. J. Plant Physiol.* **11**, 277–286.

Ritter, A., Wagner, E., and Holmes, M. G. (1981). Light quantity and quality interactions in the control of elongation growth in light-grown *Chenopodium rubrum* L. seedlings. *Planta* **153**, 556–560.

Roberts, H. A., and Potter, M. E. (1980). Emergence patters of weed seedlings in relation to cultivation and rainfall. *Weed Res.* **20**, 377–386.

Sánchez, R. A. Casal, J. J., Ballaré, C. L., and Scopel, A. L. (1993). Plant responses to canopy density mediated by photomorphogenic processes. *In* "International Crop Science I" (D. R. Buxton, ed.), Crop Science Society of America, Madison, Wisconsin. (In press).

Sauer, J., and Struik, G. (1964). A possible ecological relation between soil disturbance, light-flash, and seed germination. *Ecology* **45**, 884–886.

Scopel, A. L., Ballaré, C. L., and Ghersa, C. M. (1988). The role of seed reproduction in the population ecology of *Sorghum halepense* in maize crops. *J. Appl. Ecol.* **25**, 951–962.

Scopel, A. L., Ballaré, C. L., and Sánchez, R. A. (1991). Induction of extreme light sensitivity in buried weed seeds and its role in the perception of soil cultivations. *Plant, Cell Environ.* **14**, 501–508.

Scopel, A. L., Ballaré, C. L., and Radosevich, S. R. (1993). Photostimulation of seed germination during soil tillage. *New Phytol.* (in press).

Seyfried, M., and Fukshansky, L. (1983). Light gradients in plant tissue. *Appl. Opt.* **22**, 1402–1408.

Shackel, K. A., and Hall, A. E. (1979).Reversible leaflet movements in relation to drought adaptation of cowpeas, *Vigna unguiculata* (L.) Walp. *Aust. J. Plant Physiol.* **6**, 265–276.

Sharrock, R. A., and Quail, P. H. (1989). Novel phytochrome sequences in *Arabidopsis thaliana:* Structure, evolution, and differential expression of a plant regulatory photoreceptor family. *Genes Dev.* **3**, 1745–1757.

Simpson, G. M. (1990). "Seed Dormancy in Grasses." Cambridge Univ. Press, Cambridge, UK.

Slade, A. J., and Hutchings, M. J. (1987). The effect of light intensity on foraging in the clonal herb *Glenchoma hederacea. J. Ecol.* **75**, 639–650.

Smith, H. (1982). Light quality, photoperception, and plant strategy. *Annu. Rev. Plant Physiol.* **33**, 481–518.

Smith, H. (1986). The perception of light quality. *In* "Photomorphogenesis in Plants" (R. E. Kendrick and G. H. M. Kronenberg, eds.), pp. 187–217. Martinus Nijhoff, Dordrecht, The Netherlands.

Smith, H., and Holmes, M. G. (1977). The function of phytochrome in the natural environment. III. Measurement and calculation of phytochrome photoequilibria. *Photochem. Photobiol.* **25**, 547–550.

Smith, H., and Morgan, D. C. (1983). The function of phytochrome in nature. *In* "Encyclopedia of Plant Physiology" (W. Shropshire, Jr. and H. Mohr, eds.), New Ser., Vol. 16B, pp. 491–517. Springer-Verlag, New York.

Smith, H., Casal, J. J., and Jackson, G. M. (1990). Reflection signals and the perception by phytochrome of the proximity of neighbouring vegetation. *Plant, Cell Environ.* **13,** 73–78.

Solangaarachchi, S. M., and Harper, J. L. (1987). The effect of canopy filtered light on the growth of white clover *Trifolium repens. Oecologia* **72,** 372–376.

Soriano, A., De Eilberg, B. A., and Suero A. (1970). Effects of burial and changes of depth in the soil on seeds of *Datura ferox* L. *Weed Res.* **11,** 196–199.

Steinmetz, V., and Wellmann, E. (1986). The role of solar UV-B in growth regulation of cress (*Lepidium sativum* L.) seedlings. *Photochem. Photobiol.* **43,** 189–193.

Taylorson, R. B. (1972). Phytochrome controlled changes in dormancy and germination of buried weed seeds. *Weed Sci.* **20,** 417–422.

Telewsky, F. W., and Jaffe, M. J. (1986). Thigmomorphogenesis: Field and laboratory studies of *Abies fraseri* in response to wind or mechanical perturbation. *Physiol. Plant.* **66,** 211–218.

Thomas, B., and Dickinson, H. G. (1979). Evidence for two photoreceptors controlling growth in de-etiolated seedlings. *Planta* **146,** 545–550.

Thompson, K. (1987). Seeds and seed banks. *New Phytol.* **106,** Suppl., 23–24.

Thompson, K., Grime, J. P., and Mason, G. (1977). Seed germination in response to diurnal fluctuations of temperature. *Nature (London)* **267,** 147–148.

Thompson, L., and Harper, J. L. (1988). The effect of grasses on the quality of transmitted radiation and its influence on the growth of white clover *Trifolium repens. Oecologia* **75,** 343–347.

Tilman, D. (1988). "Plant Strategies and the Dynamics and Structure of Plant Communities." Princeton Univ. Press, Princeton, NJ.

Tucker, D. J., and Mansfield, T. A. (1972). Effects of light quality on apical dominance in *Xantium strumarium* and the associated changes in endogenous levels of abscisic acid. *Planta* **102,** 140–151.

VanDerWoude, W. J. (1989). Phytochrome and sensitization in germination control. (R. B. Taylorson, ed.) *NATO ASI Ser. A.,* **187,** 181–190.

van Esso, M. L., and Ghersa, C. M. (1989). Dynamics of *Sorghum halepense* seeds in the soil of an uncultivated field. *Can. J. Bot.* **67,** 940–944.

Vézina, P. E., and Boulter, D. W. K. (1966). The spectral composition of near ultraviolet and visible radiation beneath forest canopies. *Can. J. Bot.* **44,** 1267–1284.

Vogelmann, T. C. (1986). Light within the plant. *In* "Photomorphogenesis in Plants" (R. E. Kendrick and G. H. M. Kronenberg, eds.), pp. 307–337. Martinus Nijhoff, Dordrecht, The Netherlands.

Vogelmann, T. C., and Björn, L. O. (1984). Measurements of light gradients and spectral regime in plant tissue with a fiber optic probe. *Physiol. Plant.* **60,** 361–368.

Vogelmann, T. C., and Haupt, W. (1985). The blue light gradient in unilaterally irradiated maize coleoptiles: Measurements with a fiber optic probe. *Photochem. Photobiol.* **41,** 569–576.

Wall, J. K., and Johnson, C. B. (1981). Phytochrome action in light-grown plants: Influence of light quality and fluence rate on extension growth in *Sinapis alba* L. *Planta* **153,** 101–108.

Waller, D. M. (1988). Plant morphology and reproduction. *In* "Plant Reproductive Ecology: Patterns and Strategies" (J. Lovett Doust and L. Lovett, eds.), pp. 203–227. Oxford Univ. Press, New York.

Weaver, S. E., and Cavers, P. B. (1979). The effects of date of emergence and emergence order on seedling survival rates in *Rumex crispus* and *R. obtusifolius. Can. J. Bot.* **57,** 730–738.

Weiner, J. (1988). Variation in the performance of individuals in plant populations. *In* "Plant Population Ecology" (A. J. Davy, M. J. Hutchings, and A. R. Watkinson, eds.), pp. 59–81. Blackwell, Oxford.

Wesson, G., and Wareing, P. F. (1969). The role of light in the germination of naturally occurring populations of buried weed seeds. *J. Exp. Bot.* **20,** 402–413.

White, J., and Harper, J. L. (1970). Correlated changes in plant size and number in plant populations. *J. Ecol.* **58,** 467–485.

Willemoës, J. G., Beltrano, J., and Montaldi, E. R. (1988). Diagravitropic growth promoted by high sucrose contents in *Paspalum vaginatum* and its reversion by gibberellic acid. *Can. J. Bot.* **66,** 2035–2037.

Willey, R. W., and Heat, S. B. (1969). The quantitative relationships between plant population and crop yield. *Adv. Agron.* **21,** 281–321.

Woitzik, F., and Mohr, H. (1988). Control of hypocotyl phototropism by phytochrome in a dicotyledonous seedling (*Sesamum indicum* L.). *Plant, Cell Environ.* **11,** 653–661.

4

Canopy Gaps: Competitive Light Interception and Economic Space Filling—A Matter of Whole-Plant Allocation

Manfred Küppers

I. Introduction

Many studies have addressed the question of how physiological parameters either in leaves and shoots or in roots affect plant competitiveness. But one of the most striking characteristics of higher plants is their capacity to respond to the presence of neighbors by changing their pattern of morphological development. Despite this, only very few attempts have been made to understand the significance of physiological and morphological adaptations in their relative contribution to competitive ability. Except for cases of extremely harsh environments, physiological characters alone generally do not explain ecological success (Osmond *et al.*, 1987; Mooney and Chiariello, 1984; Bazzaz *et al.*, 1987; Chapin *et al.*, 1987; Pearcy *et al.*, 1987; Schulze and Chapin, 1987). For example, none of the leaf parameters such as photosynthetic characteristics, stomatal responses, annual carbon gain, or seasonal water losses explained successional position and competitive ability of woody species growing in an otherwise favorable environment (Küppers, 1984a,b,c). Instead, growth has often been best related to assimilate partitioning (Lambers, 1987; Körner, 1991) and competitive success has been clearly related to structural features (Küppers, 1985, 1989). The same conclusion can be drawn from the elegant studies of Givnish (1986), who assumed constant photosynthetic activity in forest understory herbs and was able to precisely relate optimum branching angle, twig and stem length, branching frequency, and stem bending from an analysis of cost–benefit relationships. Similarly, Field (1988) did not find any relationship between photosynthetic capacity and ecological success of tropical understory species. However, a combination of physiological and morphological parameters—including nitrogen-use efficiency, canopy depth, and leaf longevity—could explain quantitatively, but not qualitatively, the success of an understory generalist and a gap specialist in their respective habitats. Beyschlag *et al.* (1992) have shown that canopy structure is more important for competitiveness of roadside grasses than are their physiological properties. These examples demonstrate the need for more architectural and structural analyses with an emphasis on cost–benefit relationships (Bloom *et al.*, 1985).

However, advantageous structural characteristics must be considered in the context of the particular environment in question. An optimal biomechanical design depends strongly on the competitive situation and should not be taken as absolute as has been frequently done—although in excellent studies (e.g., Borchert and Tomlinson, 1984). In the following I will present examples of how both morphological and physiological characteristics must be considered together and that if only single parameters are considered this can lead to false predictions of competitive ability.

Space is an often neglected resource for all sessile organisms but may be essential in predicting community structure (Yodzis, 1978). Space, especially in canopy gaps, is characterized by gradients in light, nutrients, and plant-available water. Depending on gap size, environmental constraints on plants differ, resulting in shifts in optimum requirements of carbon gain versus costs of support. As a result of community-inherent processes and regular abiotic events, gaps and space will recur. Once gaps are occupied, plants differ in their frequencies and performance, which, in the long term, leads to a sorting-out of species. In this context I address the question: What is the role of carbon allocation in particular growth forms relative to the physiology of carbon acquisition in competitive success and filling space during succession? I will discuss this in two field studies conducted in intermediate to large gaps (for definition of gap size, see, e.g., Thompson *et al.*, 1988). The gaps are sufficiently large to allow at least microsuccessional cycles (Remmert, 1991). One case is from broad-leaved evergreen and the other from deciduous broad-leaved vegetation.

II. Carbon Allocation and Height Growth: An Example of Two Eucalypts

Two evergreen montane eucalypts from Australia exemplify how an understanding of plant growth, canopy development, and space occupation following disturbance requires an integrated view of leaf physiology and carbon allocation. Both species grow in adjacent populations and are of similar growth form but accomplish different functions during filling of open sites. *Eucalyptus pauciflora* Sieb. ex Spreng. ssp. *pauciflora* (= snow gum) is a tree (Figure 1B) dominating high mountain forests in the Brindabella Ranges and Australian alps of southeastern Australia (Slatyer and Morrow, 1977; Brooker and Kleinig, 1983). The second species, *E. delegatensis* R.T. Baker (= alpine ash, Figure 1A), also occurs in nearly monospecific stands in these same mountains but at lower elevations than the snow gum. Between 1000 to 1400 m elevation, both are found in adjacent populations separated by clearly defined borders often just 10 to 20 m wide that are not associated with landscape features. The reasons for this sharp boundary are still in question but appear to be partially linked with carbon economy, as I will show.

The two eucalypts have clearly different photosynthetic characteristics as determined by periodic measurement of leaf gas exchange in the field throughout the year (Küppers *et al.*, 1986, 1987, 1993). The maximum photosynthetic capacity exhibited during the season of snow gum was as much as twice that of alpine ash (Table I). As a consequence, the annual leaf carbon balance of fully developed leaves as determined by interpolation from measured diurnal courses of photosynthesis and from application of the model of Küppers and Schulze (1985) is 40% higher in snow gum. This photosynthetic potential was only occasionally realized in the field because of daily light fluctuations and other environmental conditions (Küppers *et al.*, 1986). Leaf age effects further reduced differences in annual carbon budgets (Table I). Because of these factors, there was only 25% more carbon per unit leaf area fixed by the average snow gum leaf as compared to the average mountain ash leaf even though there was 40% greater annual incident photon irradiance where the snow gum occurs due to topographic differences.

Figures 1C and D show the growth habit of the individual saplings studied, and from this it is evident that the plants allocate their assimilates differently to support growth. Both individuals have the same basal stem diameter and similar stem dry matter (Table I). But an 11-year-old alpine ash was 8.2 m tall whereas a 20-year-old snow gum was only 6 m tall despite its greater photosynthetic capacity and carbon balance. Part of the explanation may be found in differences in the ratio of leaf mass to

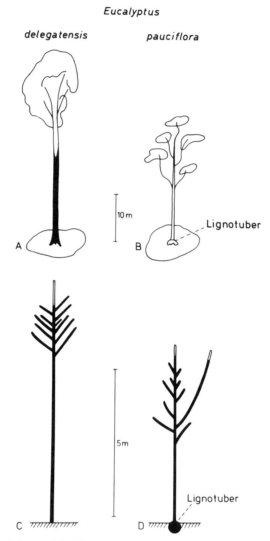

Figure 1   Typical growth habits and plant heights of (A) mature *Eucalyptus delegatensis* with a denser canopy and (B) *E. pauciflora* ssp. *pauciflora* with a more open crown. Individual saplings used for measurements are shown in (C) and (D), respectively, with maximal seasonal shoot elongation indicated as open sections. The generally more stunted growth in *E. pauciflora* is already indicated in this stage (D).

Table I Photosynthetic Characteristics, Annual Carbon Gain, and Canopy and Growth Parameters of Selected Mountainous Eucalypts at Natural Growth Conditions in the Brindabella Ranges near Canberra, Australia[a]

	Eucalyptus pauciflora Sieb. ex Spreng. ssp. *pauciflora*	*E. delegatensis* R. T. Baker
Highest observed photosynthetic capacity (μmol m^{-2} s^{-1})		
At ambient CO_2 partial pressure	26	13
At $c_i = 40$ Pa[b]	39	25
Incident annual photon irradiance (kmol m^{-2} yr^{-1})	8.4	4.7
Annual carbon gain (mol m^{-2} yr^{-1})		
Leaves fully expanded	83	67
0 year old	76	61
Senescing	80	64
Mean depending on leaf age distribution in canopy	79	63
Basal stem diameter (cm)	11.0	9.2
Stem dry matter (kg)	4.71	4.91
Height (m)	6.01	8.20
Plant age (yr)	20	11
Number of leaves in canopy	930	709
Total leaf area in canopy (m^2)	1.93	4.35
LMA (g m^{-2})	226	138
Mean leaf longevity (yr)	2.5	1.5
Leaf area index (individual plant) (m^2 m^{-2})	1.48	3.44
Leaf area index (mature stand) (m^2 m^{-2})	2	4
Insertion height of lowest branch (m)	2.8	5.3
No. of first-order branches shed	ca. 39	ca. 60
Height of crown (top to lowest leaves) (cm)	352	320
Estimated canopy volume of individual (m^3)	4.1	3.6
Leaf carbon balance in canopy (mol yr^{-1})	152	273
Maximal shoot elongation (cm yr^{-1})	27	70

[a] From Küppers *et al.* (1993) and unpublished data of B. I. L. Küppers, A. M. Wheeler, C. Godkins, and M. Küppers.

[b] c_i = intercellular CO_2 partial pressure.

area (LMA) and leaf longevity of the two species. The LMA was 39% lower in alpine ash (Table I), but these less costly leaves (in terms of dry matter) have a shorter life span. Despite this, the long-term costs of displaying a square meter of canopy foliage are similar in both species. Alpine ash regenerates its leaf mass within 1.5 years and snow gum every 2.5 years. Therefore, after 15 years, a snow gum invests at minimum (costs of leaf growth and maintenance respiration not included) 226 g × (15 years/2.5 years longevity) = 1356 g, which is almost the same as for alpine ash: 138 g × (15/1.5) = 1380 g. The importance of LMA (Lambers and Poorter, 1992) is evident when the same investment of the two species in leaf mass in a year's time is considered. From 500 g of foliage, alpine ash displays 3.62 m^2 of leaf area and snow gum only 2.21 m^2. Multiplying these areas with the annual carbon balance of 63 and 79 mol m^{-2} yr^{-1}, respectively, yields a 25% higher total carbon gain after one year (228 mol yr^{-1}) in alpine ash as compared to snow gum (175 mol yr^{-1}). As shown, long-term costs of having 1 m^2 of leaf area in the canopy are identical, so that the extra carbon gains by *E. delegatensis* clearly support faster growth and generate a compound-interest effect at the canopy level. A total of 600 g of leaf mass is found in this sapling of alpine ash as compared to 436 g in the snow gum sapling, which is twice as old. Thus, the much higher leaf area (Table I) easily compensates for a lower annual carbon balance per unit of leaf, so that the plant with lower photosynthetic capacity actually gains more carbon.

Mooney *et al.* (1978) have studied eucalypts of diverse growth forms from different moisture regimes. They concluded that those plants producing smaller individual leaf areas with greater LMA (typical for plants in dryer habitats at high irradiance levels) allocate more to roots and have lower growth rates. This is the case with snow gum, although for this species other important environmental parameters need to be considered that promote similar allocational patterns, that is, fire and frost disturbances (see the following). On the other hand, plants with thinner and larger leaves are found in moisture sites of lower light availability and have higher growth rates (as observed in alpine ash).

Eucalyptus pauciflora invests less in height gain and canopy leaf mass. Additionally, snow gums grow much thicker branches that insert much closer to the ground (Table I), resulting in a more stunted sapling, and they allocate carbon and nutrients into their lignotubers. Furthermore, they form a very thick bark, 25% of stem diameter at this age, as opposed to alpine ash with less than 1%. In general, the canopy of *E. pauciflora* is less compact (less leaf area per volume) than that of *E. delegatensis* (Table I). These differences must have ecological implications.

Grime (1979) and Caldwell (1987) pointed out that a highly competitive plant should be able to quickly establish a canopy and effectively achieve height and lateral spread. According to them, the competitive advantage should be mainly due to aboveground shading of competitors and a better ability to exploit water and nutrients below ground. One might hypothesize that a superior leaf photosynthetic performance should result in greater competitive ability. However, our results show that height gain and canopy development are not related to leaf photosynthetic performance but to specific allocation patterns. By investing less in height growth, snow gum can allocate more to resist freezing and fire. Its thick bark and lignotuber prevent it from being killed by fires (Gill and Ashton, 1968; Vines, 1968; Rundel, 1981) as well as by extreme frosts, both of which are typical for its habitat. Consequently, *E. pauciflora* is better able to survive where these events frequently recur. Snow gum also resprouts from its belowground lignotuber after wind-throws. Alpine ash does not occur in these sites, not because it is less competitive, but because of its inability to survive severe fires and frosts as a result of its thinner bark and lack of vegetative regeneration. *Eucalyptus pauciflora* regenerates in open forests and is not shade tolerant. If expanded leaves were bagged they were shed after 2 to 4 months (B. I. L. Küppers, unpublished observation of B. I. L. Küppers). In more moderate environments, gap filling by fast stem growth and canopy formation appears to be more successful. *Eucalyptus delegatensis* quickly overgrows and shades *E. pauciflora* because of its higher leaf area index (Table I), which is in agreement with Caldwell (1987).

Differences in shade tolerance in combination with different abilities to endure fires and temperature extremes likely explain the clear borders between populations of the two species, since fire lines and boundaries of frost hollows are often very distinct. Allelopathy, and fungi, may also be partially involved. However, it is unlikely that leachates from foliage litter are important since in stands of both species, leaves from either species can be found because of redistribution by wind. The species form almost monospecific stands over large areas, and their differential allocation patterns have impacts on landscape ecology, affecting runoff, fire frequency, and fire intensity.

III. Canopy Architecture in Space Exploitation during Secondary Succession

In the previous section I have shown that, despite generally similar branching characteristics among similar growth forms, differences in carbon allocation have important consequences for distribution in habitats with different disturbance frequencies and climatic conditions. It is well known that different growth forms replace each other during succession, starting, for example, from fallow land or gaps (Clements, 1916; Horn, 1974; Whittaker, 1975; Golley, 1977; Kempf and Pickett, 1981; Shugart, 1984; Küppers, 1989). If successful dominance in dissimilar habitats can be attributed to differences in structural allocation in similar growth forms as shown earlier for the two eucalypts, how much more might be expected in habitat adaptation among different growth forms? To study this, an easily accessible type of vegetation in early and intermediate states of forest succession was chosen—naturally developing hedgerows.

Unplanted hedgerows in central Europe typically arise from pioneer woody species, which grow, for example, on stone ridges built by farmers since the middle ages as they collected stones from their fields and placed them at field borders parallel to the contours. The hedgerows are at about 60-m intervals determined by the distance the farmers could easily throw stones. Community development is undisturbed in these hedgerows until they are cut back, which occurs at approximately 20- to 30-year intervals. This trimming resets the successional sequence. Without it, the hedgerow would continue to expand at its margins and would follow the general successional sequence to a tall forest characteristic of this region in Europe. After 15 to 30 years of growth, trees and treelike bushes form the upper canopy, while smaller bushes form the sides of the canopy (Küppers, 1984a,d, 1992a). Vines may form a veillike cover over the canopy sides (Wilmanns, 1983). Fallow land neighboring hedgerows is invaded by plants from the hedgerow. At this stage strong selective pressure from browsing may affect species composition. Within the hedgerow community, strong microclimatic gradients occur, especially with regard to light, air, and soil temperature (Küppers, 1984d), comparable to gradients at forest borders (Dierschke, 1974).

Although the successional sequence involves many species, six were chosen for detailed studies on the basis of their dominance or codominance of specific stages of the successional sequence. The early-successional species studied were: *Rubus corylifolius* agg. (Focke) Frid. and Gel., a blackberry species that spreads into fallow land by fast-growing runners (Rauh, 1950); *Prunus spinosa* L. var. *spinosa* (= blackthorn), a pioneer shrub that spreads by suckers but not at as fast a rate as does *R. corylifolius;* and *Crataegus* × *macrocarpa* Hegetschw. (a hawthorn hybrid), which establishes in early midsuccession and is a tall shrub (up to 8 m). The midsuccessional species chosen was the small tree (up to 12 m) *Acer campestre* L. ssp. *leiocarpum* (Opiz.) Tausch (= field maple), and the late midsuccessional species was *Prunus avium* L. var. *avium* (= wild cherry), a tree that may be found in tall forests (up to 25 m). The late-successional species investigated was *Fagus sylvatica* L.

(= European beech), a tree up to 40 m tall. Less intensive studies were also carried out on *Rosa canina* L. agg., a light-demanding pioneer, the early midsuccessional *Cornus sanguinea* L. (a dogwood species), and *Ribes uva-crispa* L. (= wild gooseberry), a species restricted to the shaded understory because it experiences unfavorable water relations in open sites (Küppers, 1984b). The pioneers and early midsuccessional species possess thorns (except *C. sanguinea*), suggesting that protection from herbivory is important for success in early stages of succession. Indeed, if exclosure fences remove browsing by deer, the later species such as *A. campestre* and *P. avium*, even occasionally *F. sylvatica*, can establish in the zone normally occupied only by pioneers. Consequently, it may be better to define the successional position of a species by its capacity for persistence or codominance at a particular stage rather than by the stage where reproduction is possible.

A. Role of Leaf Physiology and Allocation

Secondary succession involves replacement of growth forms that have been shown to differ not only in architecture but also in photosynthetic performance. However, contrary to a pattern that might be expected, photosynthetic capacity (A_{max}) in sun leaves declines from early to later stages. This pattern appears to be universal in secondary forest succession since it has been observed in temperate forests in North America (Bazzaz, 1979), central Europe (Küppers, 1984a, 1987), and northern Japan (Koike, 1987), as well as in tropical forests (Bazzaz and Pickett, 1980). For the hedgerow and forest species from central Europe, the differences were larger when expressed on the basis of leaf mass (Figure 2) than on leaf area (Table II). However, differences in seasonal carbon balances were greater on a leaf-area basis than on a leaf-mass basis (Table II). Thus, no matter whether the photosynthetic performances in sun leaves were expressed per unit area or per unit weight, they declined during succession. Lower, not higher, photosynthetic capacity or leaf carbon balance is, therefore, associated with a presumed greater competitive ability that is required for success later in succession. Such a relationship must reflect changes in the way resources are allocated or the effects of different resource availabilities in species adapted to different successional stages.

Foliage nitrogen (N) is one of the most important resources determining photosynthetic capacity. Nitrogen is required for biosynthesis of photosynthetic proteins, and much of the plant N capital is located in the leaves. Numerous studies have shown that photosynthetic capacity is highly correlated with leaf N (e.g., Field and Mooney, 1986; Küppers *et al.*, 1988; Evans, 1989). This is evident in the successional sequence where I found declining maximal N contents in parallel with declining

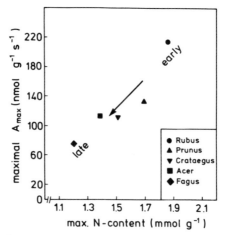

Figure 2 Maximal seasonal photosynthetic capacity (A_{max}) as related to maximal observed N content of sun leaves. Both decline in concert with successional position of species. The species are *Rubus corylifolius, Prunus spinosa, Crataegus* x *macrocarpa, Acer campestre,* and *Fagus sylvatica. A*$_{max}$ and N contents are not from the same leaves but from leaves in almost identical expositions. The relationship falls in the ranges given by Field and Mooney (1986) and Evans (1989).

maximum photosynthetic capacities of sun leaves. The decrease with successional stage may occur because N availability is often greater in earlier stages of succession (Chapin, 1983; Vitousek and Denslow, 1986). As succession proceeds, lower nitrogen availability may reflect increasing root competition as more vegetation occupies the site. Whereas total N uptake by the vegetation may increase in late succession, N acquisition by individual plants may decline (Chapin, 1983). There may also be changes in allocation involving trade-offs between thicker leaves, more N, and greater photosynthetic capacity per foliage area versus thinner leaves, which provide more foliage area with greater shading at perhaps little or no penalty for whole-plant carbon balance (Körner, 1991; Lambers and Poorter, 1992; Pearcy and Sims, Chapter 5, this volume). However, this trade-off does not seem to occur in the species I studied here since it is leaf N per foliage mass that changes most dramatically.

Table III shows seasonal balances of assimilate partitioning (Küppers, 1985) for a couple of species representing different growth forms. These partitioning patterns do not relate directly to standing dry matter since stems, for example, may contain large amounts of dead biomass and resulting root/shoot ratios would depend strongly on growth form and plant age (e.g., Bray, 1963; Mooney *et al.*, 1978; Kellomäki *et al.*, 1992). Here, only the partitioning of live material is considered. In most cases,

Table II Costs of Aboveground Volume Capture of Selected Woody Species from a Secondary Forest Succession in Central Europe, Their Photosynthetic Characteristics, Measured Seasonal Carbon Gain, and Certain Allometric Parameters[a]

	Rubus corylifolius	Prunus spinosa	Crataegus x macrocarpa	Acer campestre	Prunus avium	Fagus sylvatica
Successional position	Early pioneer	Pioneer	(Early) midsuccession	Midsuccession	(Late) midsuccession	Late succession
Mechanical defense	Spines	Thorns	Thorns	None	None	None
Growth form	Vinelike shrub	Shrub	Small tree	Small tree	Medium-sized tree	Tall tree
Photosynthetic capacity (μmol m^{-2} s^{-1}) at ambient CO_2 conc.	15	12	12	11	?	12
Seasonal carbon gain						
(mol m^{-2} yr^{-1})	33	31	43	27	?	18
(mol g^{-1} yr^{-1})	0.49	0.40	0.46	0.33	?	0.36
Aboveground primary production per total leaf dry matter (g g^{-1} yr^{-1})	2.4	12.6	12.1	9.7	?	2.4
Ratio (stem/twig) of measured seasonal carbon allocation (from Table III)	∞ (first year)	0.1	0.6	0.6	>1 (estimated)	~2 (estimated)
Costs for volume gain						
Twig level (kg m^{-3})						
of dry matter	n.d.[b]	26.3	12.8	7.5	?	1.8[c]
of nitrogen	n.d.	0.56	0.26	0.14	?	0.03
Canopy level (g m^{-3})						
of dry matter	n.d.	270	125	97	?	?
of nitrogen	n.d.	5.9	2.6	1.9	?	?
of assim. CO_2	n.d.	0.92	0.89	0.43	?	?
Leaf area density in newly gained space (m^2 m^{-3})	n.d.	0.58	0.51	0.72	?	0.70
Leaf area index (m^2 m^{-2})	3.3	3.8	8.5	8.9	?	6.9[d]

[a] After Küppers (1984c,d, 1985, 1987, 1992a,b).

[b] n.d. = not defined, since leaf areas are in more or less one plane.

[c] Recalculated after Schulze et al. (1986).

[d] After Schenk et al. (1989).

Table III Seasonal Balance of Assimilate Flows (% of Total Net Assimilate Gain by the Individual Canopy) in Woody Species of Different Growth Forms under Similar Climatic and Edaphic Conditions[a]

Growth form	Species	Age of investigated plant (yr)	Leaves	Twigs and branches	Stem	Flowers and fruits	Sum over shoot	Roots	Root–shoot ratio
Vinelike	Rubus corylifolius[b]	1	12	—[c]	17	—[c]	29	71	2.45
Shrub	Ribes uva-crispa[b]	9	8	46	12	4	70	30	0.43
	Prunus spinosa[b]	7	10	45	5	9	69	31	0.45
Small tree	Crataegus x macrocarpa[b]	12	10	35	20	6	71	29	0.41
	Crataegus x macrocarpa[b]	(13)	(11)	(22)	(3)	(38)	(74)	(26)	(0.35)
	Acer campestre[b]	13	14	34	21	—[d]	69	31	0.45
Tall tree	Larix decidua[e]	33	16	16	34	—[d]	66	34	0.52
	Larix leptolepis[e]	33	15	19	35	—[d]	69	31	0.45
	L. decidua x leptolepis[e]	33	15	16	38	3	72	28	0.39
	Picea abies[f] (healthy)	30	15	15	35	—[d]	65	35	0.54
	Picea abies[f] (declining)	30	10	13	58	—[d]	81	19	0.23
	Pinus sylvestris[g] (control)	20	16	12	12	—[d]	40	60	1.50
	P. sylvestris[g] (fertilized)	20	23	20	21	—[d]	64	37	0.58

[a] All species are from the same latitude in Germany (50°N at 11°E) except for *Pinus sylvestris*, which is from cooler habitats on sand in Sweden (60°49'N at 16°30'E); for details see references. Values for high fruit yield are in parentheses. These balances may strongly differ from dry matter partitioning, since they exclude accumulation of dead biomass (e.g., that contained in stems). "Roots" include all assimilate consumed below ground (in growth, respiration, exudates, and symbiosis) in roots, rhizomes, and belowground hypocotyls.

[b] After Küppers (1982, 1985, 1992a).

[c] Usually neither branching nor flowering in first year.

[d] No flowers or fruits.

[e] Recalculated after Matyssek (1985).

[f] Recalculated after Oren and Zimmermann (1989) and after R. Oren (personal communication).

[g] Recalculated after Linder and Axelsson (1982) and after S. Linder (personal communication).

more or less constant root/shoot ratios and allocation to foliage can be observed under similar climatic and edaphic conditions independent of plant age and growth form. Most plants from similar sites allocated about 30% of their net assimilate gain to belowground functions. According to McCree (1976, as quoted in Berry and Raison, 1982), 75% of the root allocation is used in dry matter increment with the remaining 25% going to either respiration or exudation from the roots. This is one-fourth of 30% (= 7%) of total net assimilate gain by the deciduous canopies, a figure very close to the estimates by Tranquillini (1959) for *Pinus cembra* and of similar magnitude to that reported by Oren and Zimmermann (1989) for *Picea abies*. The exception was *Rubus corylifolius*, which stores a large amount of carbohydrate below ground. If we assume that 30% of total net assimilate gain is allocated to roots, similar to the other species, 40% of this gain is stored below ground. This is sufficient stored carbon to regenerate runners completely without additional carbon gain.

The fairly constant root/shoot ratios and allocation to foliage suggest the importance of a close physiological balance between leaves, root, and shoot. The requirement for this balance may constrain plasticity independent of growth form. However, some plasticity is indicated by *Picea abies* in declining and healthy stands (Table III) and by *Pinus sylvestris* in stands with and without fertilizer treatments (Linder and Axelsson, 1982), the latter grown in a very different environment. A much greater plasticity is apparent in stem/twig ratios (Table II), which allows considerable variation in carbon allocation within canopies.

B. Role of Branching Patterns

Detailed examples of how certain combinations of branching patterns may determine growth form have been provided by Küppers (1985, 1989a). For example, a combination of epitony with basitony (Figure 3) does not permit growth of a tree and will always result in a shrub, whereas a combination of acrotony with hypotony or amphitony will result in a tree (Figure 3), although this combination does not determine the height of such a tree. These relationships hold for certain woody species from temperate and high-latitude areas but they do not necessarily apply in the tropics, where architectural development is often not interrupted by a regular period of dormancy and, thus, may be more complicated (Hallé *et al.*, 1978). Branching patterns are decisive whenever they constrain the degrees of freedom in architectural patterns, for example, by preventing growth of a distinct stem (epitony combined with basitony). In most cases, however, they permit considerable spatial plasticity including reiteration, thus enabling a plant to respond to changes in competitive pressure and other environmental factors (Hallé *et al.*, 1978). The architectural arrangement of leaves is largely independent of branching patterns. Most

	Rubus	Rosa	Prunus	Cornus	Crataegus	Ribes	Acer	Quercus	Fagus
Phyllotaxy	ALTERNATE	ALTERNATE	ALTERNATE	DECUSSATE	ALTERNATE	ALTERNATE	DECUSSATE	ALTERNATE	ALTERNATE
Branching pattern	SYMPODIAL	SYMPODIAL	SYMPODIAL	MONOPODIAL	SYMPODIAL	SYMPODIAL	MONOPODIAL	MONOPODIAL	MONOPODIAL
Longitudinal symmetry	BASITONIC	BASITONIC	MESOTONIC-ACROTONIC	MESOTONIC-ACROTONIC	MESOTONIC-ACROTONIC	WEAKLY BASITONIC	MESOTONIC-ACROTONIC	MESOTONIC-ACROTONIC	MESOTONIC-ACROTONIC
Growth form	VINE-LIKE SHRUB	SHRUB	TREE-LIKE SHRUB	TREE-LIKE SHRUB	SHRUB-LIKE TREE	SHRUB	SMALL TREE	TALL TREE	TALL TREE
Lateral symmetry	STRONGLY EPITONIC	STRONGLY EPITONIC	WEAKLY EPITONIC	WEAKLY EPITONIC	EPITONIC & HYPOTONIC	STRONGLY EPITONIC	STRONGLY HYPOTONIC	AMPHITONIC & HYPOTONIC	AMPHITONIC
Successional position	EARLY PIONEER	PIONEER	PIONEER	PIONEER & MID-SUCCESSIONAL	PIONEER & MID-SUCCESSIONAL	MID-SUCCESSION (UNDERSTORY PIONEER)	MID-SUCCESSION	LATE SUCCESSION (DRY OR MOIST SOILS)	LATE SUCCESSION

plants form more or less bell-shaped crowns when expanding without aboveground competition, but asymmetric canopies if competition is primarily from one side, a so-called riverbank effect (Hallé *et al.*, 1978; Fisher, 1986), and cylindrical crowns when competition is from all sides (e.g., Küppers, 1984d, 1987; Jurik, 1991; Weiner and Thomas, 1992). Costs to produce these canopies usually differ, depending on the specific branching patterns, and can be taken into account as costs to gain canopy space.

In aboveground competition, the lateral spread of a canopy appears to be initially very important in harvesting more light and shading com-

---

Figure 3 Phyllotaxy, branching patterns, symmetries of branching with resultant growth forms, and successional positions of selected woody species from central Europe. Species shown here are *Rubus corylifolius*, *Rosa canina* agg., *Prunus spinosa*, *Cornus sanguinea*, *Crataegus* x *macrocarpa*, *Ribes uva-crispa*, *Acer campestre*, *Quercus robur*, and *Fagus sylvatica*. Leaves and twigs of the plants shown here insert either alternately or decussately. A *monopodial* branch system results when the apical bud of the leading shoot remains active for several growth periods, whereas it is *sympodial* if the apical bud ceases its activity, for example, at the end of a vegetation period, after growth of a terminal inflorescence, or after growing an internode, so that an adjacent lateral or axillary bud continues shoot elongation. (A sympodium, of course, can be enforced in any typical monopodium when cutting the leading shoot.) *Longitudinal symmetry* refers to the lengths of current-year twigs (open sections) inserting on the previous season's main branch (solid sections). When the current year's twigs are longer the closer they are to the base of the main shoot, then branching is *basitonic*, and when they are longest in the middle or at the top, then it is *mesotonic* or *acrotonic*, respectively. Transitions often exist between mesotony and acrotony (acrotony being a special case of mesotony), as in *Acer campestre* and *Crataegus* x *macrocarpa*, although they do not occur regularly. *Lateral symmetry* considers the enhancement of growth of twigs from upper (*epitony*), lower (*hypotony*), or lateral surfaces (*amphitony*) of the main shoot. Branching symmetries are already indicated by the sizes of buds before bud break, and the growth activity of these meristems relative to the symmetry of axes determines the foregoing definitions (e.g., Troll, 1935; Rauh, 1950). Combinations of these symmetries determine the growth forms of many species. Strong basitony does not allow for growth of a single stem, whereas epitonic twigs overtop their main branch, again disabling the growth of a leading stem (e.g., *Rosa*), so that this combination must result in a shrub. On the other hand, acrotony as well as hypotony or amphitony promote growth of a leading shoot. From this a tree results, although height growth is an additional factor in determining a tree. In most species there are transitions between the strength of branching symmetries and their combinations. Therefore, it is possible to distinguish (1) a *shrub* without a single stem, (2) a *treelike shrub* with a very short single "stem" but otherwise shrub characteristics, (3) a *shrublike tree* with a short but distinct stem but partly epitonic branching, and (4) a *tree*. Relating these growth forms to position of species during secondary forest succession shows that, from pioneers to late-successional species, shrublike growth habits are increasingly replaced by treelike features. Many of the branching patterns are found in other species of the same genera outside Europe; for example, *Nothofagus* species from the Southern Hemisphere follow the same pattern as *Fagus sylvatica*. (From Küppers, 1989, with permission from Elsevier Science Publishers, Amsterdam and New York.)

petitors. When a single twig or a branch system grows into space, its axes and leaves occupy only a rather small volume. On the other hand, when several twigs in multiple layers cause the crown to expand, they effectively occupy much more volume. Although space is still available between individual branch systems, it is of lower quality for competitors since it is at least partially shaded. Obviously, a single branch has a different effect on space than a group of branches, and I have taken this into account when determining costs of twig and canopy volume gain.

Individual branch systems may differ in their bifurcation ratios and the distribution of lengths of lateral twigs. Therefore, species comparisons must be on the basis of representative model branches. Arbitrarily, I chose mean branch systems of 70-cm length (Figure 4) since these were long enough to represent more than two years of growth and generate the typical branching patterns characteristic of a species. Furthermore, species-specific branching patterns have been combined with their statistical mean bifurcation ratios, twig length distribution, and frequency of higher-order twigs. For such a model branch, its total leaf number, leaf area, and leaf and twig biomass follow from the data and procedure given by Küppers (1982, 1984c, 1985, respectively). Furthermore, it is assumed that leaves occupy a cylinder around each axis, the radius of which follows from the total length of a leaf and a mean inclination of 45° (Figure 4). The volume of lateral twigs that extends partially or totally into the space of neighboring twigs does not increase occupied volume.

The growth of twigs, of course, contributes to the overall crown volume gain that provides for the competitive occupation of space. The expansion zone is, sensu stricto, the outermost zone into which the outer twig

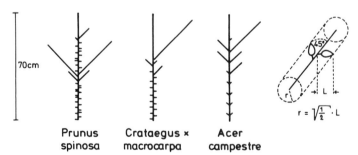

Prunus spinosa   Crataegus × macrocarpa   Acer campestre

$$r = \sqrt{\tfrac{1}{2}} \cdot L$$

Figure 4 Modeled mean branching patterns of branch systems of similar length. To define the volume occupied, an imaginary cylinder was positioned around the axis of every single twig, its radius being given by the leaf length. For simplicity, a similar leaf-to-twig angle was assumed in all cases. Twig volumes that overlap with the volume of another twig do not increase volume gain. (Redrawn from Küppers, 1985.)

axes grow (Figure 5). These current year axes are typically unbranched and have a negligible effect on space capture. However, they prepare for space occupation in the following season by the branching twigs that carry most of the leaf area. In the third season, space occupation appears to be almost complete with respect to an effective shading of any competitors. It is not necessarily complete with respect to a full leaf area packing, but, as discussed earlier, this is not necessary for preemption of space from competitors. Therefore, a period of three seasons was chosen to determine costs of space occupation (Figure 5). Measurements were made for each species studied in 20-m² monospecific areas of the canopy in the upper lateral portions where densities were typical. The expansion surfaces were defined by tips of the outer twigs (Figure 5). Thus, volume gain (as a mean over three seasons) can be calculated from annual radial expansion from the previous season's crown surface. Based on the dry matter of newly grown axes in this volume, volume gain per unit of dry matter invested can be deduced.

The species studied differ strongly with respect to volume gain per unit dry matter invested, either at the level of a representative model twig or at the level of the canopy (Table II). The pioneer *P. spinosa* and the early midsuccessional *C.* x *macrocarpa* produce many thorny twigs

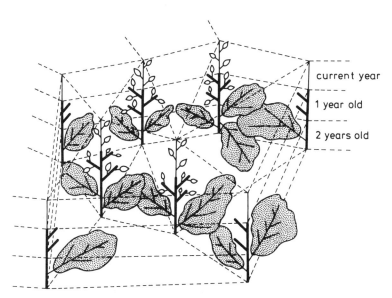

current year

1 year old

2 years old

Figure 5   Measurement of new volume capture at the canopy level after termination of elongation of the outermost shoots that grow into open space: A mean volume increase was calculated over three seasons (see the text) from space encompassed by these surfaces that were spread by bases of 2-year-old branches and tips of current year's shoots.

that deter browsing pressure. These costs, which may be as high as 50 to 65% of all assimilate invested into twigs, are clearly related to the fundamental branching patterns of these two species (Table II and Figure 3). Later successional species (*A. campestre, P. avium, F. sylvatica*) do not have such expenses of defense, so their volume gain should be less costly. However, they are forced to reach for more valuable space (valuable in terms of competition for light and lower browsing pressure) at greater height. Saving costs at the level of canopy volume gain, they lose pioneer characteristics in environments of high browsing pressure, but they have more assimilate available for stem and, thus, height growth. Interestingly, seedlings or saplings may strongly depend on the protection by pioneers that prevent access by browsers. They must also be shade tolerant (Schulze, 1972; Küppers, 1984a, 1987; Küppers and Schneider, 1993). Once they have reached a certain height, their crowns are safe from ground-dwelling browsers such as deer. Lower costs of space occupation permit them to grow a crown quickly and then outcompete shade-intolerant pioneers (Küppers, 1984a).

A second prerequisite to effectively cast shade in occupied space is the growth of sufficient leaf area. Despite lower costs, leaf area density as well as leaf area index increase from early- to late-successional species (Table II). Light acclimation of photosynthetic capacity (Björkman, 1981), which influences the distribution of N within the canopy (Hirose and Werger, 1987), may be important in this instance for efficiently maintaining a high leaf area density (Meister *et al.*, 1987; see also costs for volume gain of nitrogen in Table II). Leaf area density is determined here as leaf area in the newly occupied space (over the last three seasons), not as total canopy leaf area per total crown volume. The latter is actually greatest in the early pioneer *R. corylifolius* (Table IV), which occasionally may grow in dense, monospecific but low hedgerows (Küppers, 1982), if, for example, trimming or burning frequency is high. In the later successional species, leaf area density of the whole canopy, in fact, declines (Table IV). Height growth of whole crowns may take decades and is a competitive factor only in the long term. Consequently, whole-canopy costs are not equivalent to the costs of successful, rapid space occupation. Since competitive advantage in this circumstance means controlling valuable space by exploiting most of its resources before neighbors gain access to them (Louda and Renaud, 1991), it is the leaf area density in the newly occupied volume that is important.

C. Plant Forms of Woody Pioneers in Limited Light

During gap filling, competition for light and space increases and light gaps that are available become smaller but more frequent. The gaps become sufficiently small so that only parts of plants may be exposed to

Table IV Leaf Area Density of *Total* Crown of Selected Woody Species[a]

Species	Crown height (m)	Leaf area index $(m^2\ m^{-2})$	Leaf area density in *total* crown $(m^2\ m^{-3})$
Rubus corylifolius	1.0	3.29	3.29
Prunus spinosa	2.0	3.72	1.86
Crataegus x *macrocarpa*	4.5	8.15	1.81
Acer campestre	5.0	8.64	1.73
Fagus sylvatica	4–6[b]	6.3[c]	1.1–1.6
Ribes uva-crispa[d]	1.5	4.31	2.87

[a] Crown height is given for at least 95% of all leaves, and leaf area indices presented are from within this crown height. The remainder (less than 5%) of leaves are found along the stem and are not part of the more or less closed canopies. Recalculated after Küppers (1984d) and unpublished data.
[b] Estimated by eye.
[c] After Schenk *et al.* (1989).
[d] Pioneer in understory.

them. Under these conditions survival may strongly depend on the plant exposing sufficient leaf area in these gaps to counterbalance costs of maintenance of the whole plant. Once growth of a twig section in a given direction is complete, this vector cannot be reversed. Thus, the growth pattern of individual branches and twigs integrates and conserves competitive growth history. I give the following examples to illustrate this point.

An individual of the light-demanding pioneer *R. canina* (Figure 6E) grew long, orthotrophic shoots because the plant became completely shaded by its neighbors. Such "restoration" shoots (restoration in the sense of light capture) only branch near the apex and grow in length for only one season. They indicate light deprivation insofar as the dominant branching patterns of this species (a combination of epitony with basitony) in general do not support growth of a single trunk. As these restoration shoots were grown, I observed that another shoot section of well-developed twigs was sacrificed, most likely to reduce maintenance costs. Since it had been deeply shaded by neighboring shoots its leaves no longer contributed to carbon gain. In fact, for twigs of the light-demanding *P. spinosa* that were in a similar situation, negative daily carbon balances have been measured (Küppers, 1984a). Leaves in deep shade may significantly increase maintenance costs relative to total carbon gain in the understory while the plant as a whole merely succeeds in compensating by growth into "valuable" space with sufficient light. An unsuccessful attempt to reach for light results in short-lived twigs, loss

Cornus sanguinea

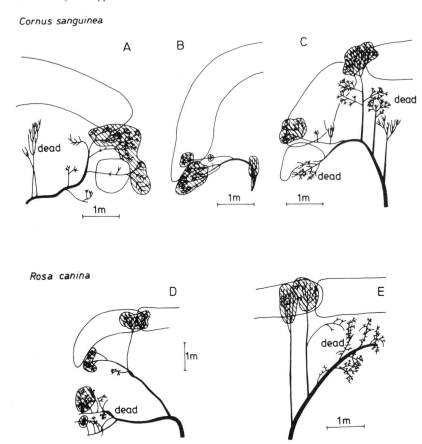

Figure 6   Growth habits of (A–C) the light-demanding dogwood *Cornus sanguinea* and (D,E) a species of *Rosa canina* agg. in the closed canopies of a hedgerow community. Open areas indicate dense canopies of neighboring plants. The distorted growth shapes indicate effects of aboveground competition. See text for further explanation. (Redrawn from Küppers, 1984d.)

of energy, and dry matter (Figures 6A and C orthotrophic shoots). The need to grow toward light may overcome apical dominance. In Figure 6B, an individual of *C. sanguinea* is shown that was growing from the dark interior of a hedgerow toward the canopy side. The highest light levels were found close to the ground. Basal buds were released from apical dominance and sprouted orthotrophic shoots, which in this situation was a waste of limited resources.

Responding at least in part to the spectral quality of the light environment (Ballaré *et al.*, 1990; Galinski, 1993; see also Chapter 3, this volume), length increment of twigs in the shaded portions of the canopy is larger than under unshaded conditions, as indicated by increased internode length (Table V). However, in certain situations this may not allow for very stable architectures as indicated by Figures 6A to E. Because of the distribution of tiny gaps in an otherwise closed canopy, individual canopy sections grew far separated from each other and were supported by long axis branching at critical shearing forces. Pathways of root–shoot communication and transport became very long and appear to be highly costly in relation to available energy. Most likely, this is another reason for gradual sorting out of these light-demanding species during succession.

D. Simulation of Woody Species Growth Forms from Measured Biological Data

We have seen that a possible mechanism of sorting out of early species with progression of secondary succession appears to be an increasing imbalance of costs to construct a canopy relative to the amount of valuable space acquired in this process. This conclusion can be drawn from the fact that these costs decline with progression of succession (Table II). However, this is only a correlation and it would seem nearly impossible to design an ecological experiment to test this interpretation. Growth of woody plants is slow relative to human lifetimes. Nevertheless, a certain test of this interpretation can be provided by computer modeling of the growth of woody species including both architectural development as well as biomass increments. For this modeling, the following primary features have been considered: (1) the annual carbon gain from measurements that determine the upper limit of biomass increments (Table II), (2) seasonal assimilate allocation (Table III), and (3) statistical branching patterns in space (Figure 3), and their dry matter increments (Küppers, 1984d, and unpublished data). In this model, the seasonal carbon gains of the different canopy layers that develop during growth are taken

Table V Lengths of Internodes (cm) in the Open and in the Understory of Selected Early and Midsuccessional Woody Species in Central Europe (Means and Standard Deviations)[a]

	Exposed	Understory
Prunus spinosa	1.14 ± 0.31	2.37 ± 0.79
Crataegus x *macrocarpa*	1.84 ± 0.41	2.86 ± 0.69
Acer campestre	4.06 ± 1.25	6.01 ± 2.18

[a] From Küppers (1984d).

into account. Also it assumes that each branch allocates an equal proportion of its available carbon gain to stem and roots. This approach differs from the many architectural models applying fractal analysis (e.g., Prusinkiewicz and Lindenmayer, 1990) insofar as it takes measured biological information into consideration.

Figure 7 shows the simulated architectural growth habit of the pioneer *P. spinosa* under light conditions not affected by neighboring plants. In the second year (Figure 7A) its typical branching patterns are already evident, alternate branching combined with meso- to acrotony. In the third year (Figure 7B) the role of epitony, allowing for upright shoots, is apparent. However, from this point, height gain is slow and even after 7 years (Figure 7F) the crown only gradually spreads out. After 9 years of simulated growth (not shown), height and space gain remain limited, but the canopy becomes dense, consisting of small, thorny twigs. This is close to what is observed in nature. *Prunus spinosa* grows much slower as compared to *A. campestre* (Figure 8), but many individual shoots, fre-

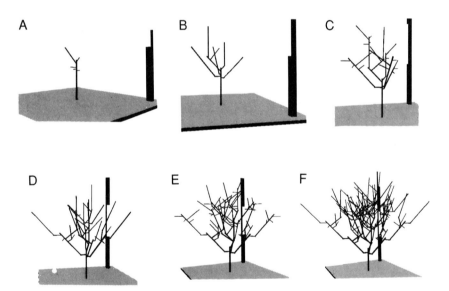

Figure 7 Simulation of architectural growth of the pioneer shrub *Prunus spinosa* from measured branching patterns, bifurcation ratios, annual carbon balance, and assimilate partitioning. The simulation starts at the end of the first year of growth with an unbranched shoot of 15 cm length. (A) 2 years of growth, (B) 3 years of growth, (C) 4 years of growth, (D) 5 years of growth, (E) 6 years of growth, and (F) 7 years of growth. Leaves are not shown. All major twigs and branches are projected into one plane. One segment of the scale equals 20 cm. See text for further details. (Unpublished data of R. List, F. Schröder, and M. Küppers.)

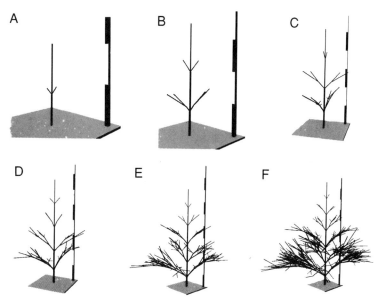

Figure 8 Simulation of architectural growth of the midsuccessional tree *Acer campestre* from measured branching patterns, bifurcation ratios, annual carbon balance, and assimilate partitioning. The simulation starts at the end of the first year of growth with an unbranched shoot of 15 cm length. (A) 2 years of growth, (B) 3 years of growth, (C) 4 years of growth, (D) 5 years of growth, (E) 6 years of growth, and (F) 7 years of growth. Leaves are not shown. All major twigs and branches are projected into one plane. One segment of the scale equals 20 cm. See text for further details. (Unpublished data of R. List, F. Schröder, and M. Küppers.)

quently connected by a widespread polycormic rooting system (Jakucs, 1969; Stephan and Stephan, 1971), gradually form a dense￼npenetrable, fencelike belt around the older parts of vegetation (compai ˙ Figures 9A and C). On the other hand, monopodial growth in *A. campɛ ʹre* (Figures 8A–F), and its acrotony (Figure 8B) and hypotony (Fiǧ re 8D) allow for a rapid height gain as well as a lateral expansion of the crown. Already after 5 years of simulated undisturbed growth, clear differences between these two species are evident. This simulation supports my interpretation of ecological success in these species based on field observations.

We have run the simulation up to 10 years. At this stage, both plants have entirely different growth habits but tend to fill remaining gaps within the canopy so that only shade-tolerant species are able to exist beneath their canopies. However, our simulation model needs further development. Our model for carbon gain is a simple one, assuming a linear relationship between total annual carbon gain and total annual

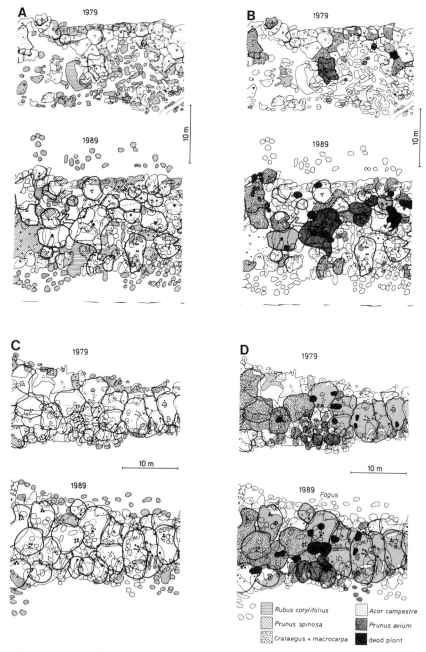

Figure 9 Maps of projected individual crowns of the species investigated in a hedgerow community in central Europe (compare with Küppers, 1985). Thick solid lines: dominant canopy; thin solid lines: subcanopy (if under thick solid line), otherwise also dominant, for example, in the open; broken lines: understory. (A) Early-successional stage on fallow

amount of incident light within a certain section of the canopy. This relationship is more likely than a light-saturation curve since sunshine hours and daylength increase carbon input linearly as a function of quantity of incident quanta. This is indicated for the hedgerow species from data of Küppers (1984a,c,d, and unpublished data), for *Fagus sylvatica* from data of Schulze (1970, 1972), and from a study of understory herbs of a beech forest by Kriebitzsch (1992). The section of the model for loss of twigs is entirely on the basis of carbon limitation for growth in portions of the canopy receiving low light. Actual data on frequency of twig loss need to be incorporated to improve this aspect of the model. Beyond a certain plant age, say of 8 years, such effects may gradually become more important. Nevertheless, this first simulation already clearly demonstrates that branching patterns may be decisive for growth form, canopy development, and ecological success of a species.

E. Vegetation Dynamics

A second test of my interpretation may be provided by predicting successional replacements from costs of aboveground space capture as given in Table II and then comparing them with field observations. *Fagus sylvatica* should replace *A. campestre* and *C.* x *macrocarpa*, and these species should in turn replace *P. spinosa* (and *R. corylifolius*). The distribution of individual crowns of the two pioneers at an early stage of succession (Figure 9A in 1979) demonstrates their dominance in the open. Sufficient space for further expansion is available between individual plants, but this space allows deer to penetrate and to exert browsing pressure. Midsuccessional species without mechanical protection are already present (Figure 9B in 1979) but occur in microsites that are naturally "fenced off" from browsing by pioneers. With further vegetation development, midsuccessional species gain in height, extend their canopies into more valuable space, and thus, gradually shade pioneers (Fig. 9B in 1989). Finally, in a mature midsuccessional stage (Figure 9C in 1979 and 1989), pioneers are entirely removed from the canopy and restricted to a fence-like belt around the hedgerow. In the midst of the hedgerow are primarily species without mechanical defense. At that stage, pioneers that grow in the immediate neighborhood of midsuccessional plants lose in competition because of their higher costs for crown expansion. A mature mid-

land in 1979 and in 1989: distribution of pioneers. (B) Early-successional stage on fallow land in 1979 and in 1989: distribution of midsuccessional species. (C) Midsuccessional stage in 1979 and in 1989: distribution of pioneers. (D) Midsuccessional stage in 1979 and in 1989: distribution of midsuccessional species. One single individual seedling of *Fagus sylvatica* was found in 1989 in the understory but disappeared by 1991.

successional community may be temporarily stable (compare Figure 9D in 1979 and in 1989), most likely because the shade tolerance of species and costs to gain canopy space are more balanced between mid- and late-successional species than between pioneers and midsuccessional plants. For example, the difference in these costs is generally smaller between *C*. x *macrocarpa* and *A. campestre* than between *P. spinosa* and either of these midsuccessional species (Table II). At this stage, community development depends mainly on replacement of dead individuals by suppressed individuals in the understory ("oskars"; Silvertown, 1987) that are recruited into the canopy. The major change over 10 years of development is an increased number of dead plants (Figure 9D, black areas). But in agreement with the lower costs to gain space (Table II), a slight increase of dominance of *A. campestre* and *P. avium* relative to *C*. x *macrocarpa* can be observed. However, to endure shade and finally replace overstory individuals, "oskars" have to be highly shade tolerant, especially in midsuccessional stages, which have extremely high leaf area indices. Broad-leaved canopies at this stage allow only 2 to 3% of the light to penetrate to the understory (Küppers, 1984d). Therefore, plants surviving in the understory must be able to utilize sparse lightflecks with high efficiency in order to gain carbon, irrespective of their potential shade tolerance (e.g., Woods and Turner, 1971; Küppers, 1984a; Tinoco-Ojanguren and Pearcy, 1992; Pearcy *et al.*, Chapter 6, this volume). We have started to investigate this aspect in a comparative study (Schneider *et al.*, 1993).

IV. General Discussion

I have provided two examples from broad-leaved vegetation on the role of leaf physiology relative to carbon allocation in light competition and filling of space. For both the deciduous or evergreen habits, the same result is apparent. Ecological success of the species is several steps removed from leaf physiology (Tables I and II). I will discuss this in relation to other topics that are relevant for vegetation processes such as the role of allocation of masses (flux balances) in plants, the role of allocation in space in specific architectural arrangements, and the role of aboveground competition.

The role of carbon acquisition by leaves can be compared with an engine supplying energy to drive a vehicle. The same engine can move several vehicles, for example, a plane, a ship, an automobile, or an escalator (analogous to different growth forms). Even the most efficient engine placed in an inefficient vehicle will fail to win a race, whereas an inefficient engine put into a highly efficient vehicle will do much better. "Engines" in plants, such as photosynthesis, are very similar among C_3 species. Photosynthesis is also fairly conservative in that when plants are under stress, cell division and elongation (Körner, 1991) and even shoots (Küppers *et al.*, 1985) are affected before any changes in photosynthesis are detectable. Those plants that are ultimately the stronger competitors in secondary forest succession have, in fact, lower photosynthetic capacity and annual carbon balances at the leaf level (Table II). Poorter (1989) investigated relationships between relative growth rate and net assimilation rate and concluded that LMA and leaf mass ratio are well correlated with relative growth rate whereas net assimilation is not. This clearly corresponds to the example of the two eucalypts presented here (Table I), but not to the deciduous hedgerow succession (see the following).

Allocation, which denotes (net) fluxes of assimilated resources among plant parts, is a plant-inherent program. It has mechanistic, biomass, and architectural components. Physiological supplies and demands have to be balanced, as expressed by the homeostatic root/shoot ratios that plants tend to maintain even after severe disturbances (Brouwer, 1983). For example, Chalmers and van den Ende (1975) observed a constant root/shoot ratio in *Prunus persica* L. from 4 to 16 years in age. This ratio did not change when the plants matured and produced fruit at an age of 6 years. In this case the dry matter ratio did not change which makes it likely that ratios of flux balances between roots and shoots as well as ratios of accumulation and storage stayed constant. This would be in agreement with my results obtained for *C.* x *macrocarpa* (Table III) with either a light or heavy fruit set. How strong this physiological balance is in similar environments follows from the generally very similar partitioning pattern between the leaves, the shoot as a whole, and the root system (Table III) that is more or less independent of growth form, plant size, plant age, and the evergreen or deciduous habit.

Rubus corylifolius is a special case since its belowground parts are loaded with starch in the autumn, allowing for quick regeneration of shoots growing away from the shade of overgrowing neighbors. This strategy is advantageous after catastrophic events such as fire and trimming, enabling faster regeneration than in other species without belowground storage. A similar strategy is found in snow gum as discussed earlier (Table I and Figure 1). Obviously, it is important to look at the timing of allocation to storage in relation to environmental events to understand its adaptive purpose. Since these carbohydrates can be fully remobilized they are not simply "accumulated" in the sense described by Chapin *et al.* (1990).

In all plants reported in Table III, about 10 to 15% of assimilate is partitioned into growth of new leaf mass. Obviously, plant growth cannot be predicted from this alone. For growth, it is essential to understand how a plant translates the carbohydrates into leaf area (Poorter, 1989). A lower LMA may support a compound-interest effect on carbon gain by the canopy as a whole. But these leaves frequently have shorter life spans. When considering costs and longevity, both snow gum and alpine ash have identical costs to maintain 1 m^2 leaf area over 15 years but display different foliage areas for a given new investment into leaf mass as discussed earlier. Kikuzawa (1991) has provided an interesting theoretical analysis of when a leaf should be replaced. Specific leaf weight may be less important in deciduous plants since no difference was found among the pioneer *Prunus spinosa*, midsuccessional *Crataegus* x *macrocarpa*, *Acer campestre* (90 to 110 g m^{-2} in sunleaves; Küppers, 1984a) and the late successional *Fagus sylvatica* (90 to 110 g m^{-2}; Stickan *et al.*, 1991). However, further studies appear to be essential.

Growth per se does not explain competitive success of species. I observed greatest aboveground primary production per total leaf dry matter in pioneers and lowest in strongly competitive late-successional species (Table II). To understand this it is important to examine how a plant invests its dry matter increments when in particular environmental situations.

Although many morphological features are species specific (e.g., branching patterns, flowers), the architectural arrangement of these features in space allows for considerable plasticity. Therefore, the way a plant develops its architecture appears to be one, if not the most decisive factor for competitive success. A successful sessile competitor should (1) be the first to arrive (Louda and Renaud, 1991), (2) be able to acquire the available resources, and (3) maintain occupation of the space as long as it contains useful resources. Branching patterns in space, as well as shoot elongation (Küppers, 1984d) and internode length (Table V), are clearly responsible for costs and speed of space occupation (Figures 4 and 5). This was tested by modeling architectural development of selected species (Figures 7 and 8) and predicting vegetation dynamics (Section IIIE, Figure 9). Maintaining space provides competitive advantage, but it certainly reduces the potential for quick structural adjustments. Growth along a certain vector is not easily shifted into another direction as indicated by the distorted plant forms in Figure 6. Therefore, gaps generated by death of individuals (Figure 9D) or other events are frequently not filled by established plants but by "oskars" from the understory. Suddenly released, they can grow very fast into canopy openings.

When examining competition among plants we have to take into consideration both spatial and temporal components. Future asymmetric growth in response to neighbors is sometimes already apparent in seedlings or saplings that are still growing at some distance from one another (compare Figures 9A and B, in both 1979 and 1989). Plants can recognize neighbors at a distance by sensing the red/far red (R : FR) ratios of the reflected light from the neighbors' foliage long before direct shading occurs (Ballaré, Chapter 3, this volume; Galinski, 1993; Schmitt and Wulff, 1993). R : FR-mediated changes in internode lengths (e.g., Ballaré *et al.*, 1990) appear to release competitive pressure by shade and help to reduce or delay self-thinning in dense stands (Schmitt and Wulff, 1993), such as a mature, midsuccessional hedgerow. Still the phytochrome mechanism does not explain how efficiently and rapidly a plant will develop its architecture in such situations.

V. Conclusions

Our simulations of plant growth and canopy expansion on the basis of biological field data provide evidence that branching patterns rather than leaf physiology explain the competitive roles played by the co-occurring species. Küppers (1984d, 1985) and Givnish (1986) came to a very similar conclusion; Field (1988, p. 348) stated that "photosynthetic characteristics are several steps removed from ecological success." The fairly conservative responses of carbon acquisition and even allocational fluxes among organs in different growth forms demonstrate the need for more integrative research. For future progress in understanding plant processes it appears to be essential to reduce attention to unnecessary details of leaf physiology, morphology, and partitioning that do not have significance at higher levels of integration. More research is needed on multiple factors to learn which salient features are essential in certain environmental situations and how frequently these factors control vegetation processes. For this purpose, developmental studies and the timing of space capture appear to be very important relative to the mechanisms of resource acquisition. Meristems partly determine branching and, thus, space capture and compound-interest effects on carbon gain. More comparative studies are essential, not just on growth and/or plant forms per se but on growth forms in a particular environmental context. Different environments through time, such as in forest succession, demand different optima for plant "behavior" that may strongly deviate from plant-inherent optima set by physiological features. We need to study at different levels of plant integration where plants have their greatest plasticity. For this purpose, plant histograms or data banks should be helpful. Nevertheless, the results shown and discussed here also make it likely that modeling effects of "global change" on vegetation may be entirely misleading if they are based only on physiological responses.

Acknowledgments

My special thanks go to Bob Pearcy for his continuous encouragement and his excellent comments when writing this chapter and for going through the pain of correcting the English. I also wish to thank both Bob Pearcy and Martyn Caldwell for their patience and general support here and in the past. The assistance of Mrs. Schäfer and Mrs. Heger (Darmstadt) with the illustrations is gratefully acknowledged.

References

Ballaré, C. L., Scopel, A. L., and Sanchez, R. A. (1990). Far-red radiation reflected from adjacent leaves: An early signal of competition in plant canopies. *Science* **247**, 329–332.

Bazzaz, F. A. (1979). The physiological ecology of plant succession. *Annu. Rev. Ecol. Syst.* **10**, 351–371.

Bazzaz, F. A., and Pickett, S. T. A. (1980). Physiological ecology of tropical succession: A comparative review. *Annu. Rev. Ecol. Syst.* **11**, 287–310.

Bazzaz, F. A., Chiariello, N. R., Coley, P. D., and Pitelka, L. F. (1987). Allocating resources to reproduction and defense. *BioScience* **37**, 58–67.

Berry, J. A., and Raison, J. K. (1982). Responses of macrophytes to temperature. *In* "Encyclopedia of Plant Physiology" (O. L. Lange, P. S. Nobel, C. B. Osmond, and H. Ziegler, eds.), New Ser., Vol. 12A, pp. 277–338. Springer Verlag, Berlin.

Beyschlag, W., Ryel, R. J., and Ullmann, I. (1992). Experimental and modelling studies of competition for light in roadside grasses. *Bot. Acta* **105**, 285–291.

Björkman, O. (1981). Responses to different quantum flux densities. *In* "Encyclopedia of Plant Physiology" (O. L. Lange, P. S. Nobel, C. B. Osmond, and H. Ziegler, eds.), New Ser., Vol. 12A, pp. 57–107. Springer Verlag, Berlin.

Bloom, A. J., Chapin, F. S., III, and Mooney, H. A. (1985). Resource limitations in plants—An economic analogy. *Annu. Rev. Ecol. Syst.* **16**, 363–392.

Borchert, R., and Tomlinson, P. B. (1984). Architecture and crown geometry in *Tabebuia rosea* (Bignoniaceae). *Am. J. Bot.* **71**, 958–969.

Bray, J. R. (1963). Root production and the estimation of net productivity. *Can. J. Bot.* **41**, 65–72.

Brooker, M. I. H., and Kleinig, D. A. (1983). "Field Guide to Eucalypts," Vol. 1. Inkata Press, Melbourne, Sydney.

Brouwer, R. (1983). Functional equilibrium: Sense or nonsense? *Neth. J. Agric. Sci.* **31**, 335–348.

Caldwell, M. M. (1987). Plant architecture and resource competition. *Ecol. Sud.* **61**, 164–179.

Chalmers, D. I., and van den Ende, B. (1975). Production of peach trees: Factors affecting dry-weight distribution during tree growth. *Ann. Bot.* **39**, 423–432.

Chapin, F. S., III (1983). Patterns of nutrient absorption and use by plants from natural and man-modified environments. *Ecol. Stud.* **44**, 175–187.

Chapin, F. S., III, Bloom, A. J., Field, C. B., and Waring, R. H. (1987). Plant responses to multiple environmental factors. *BioScience* **37**, 49–57.

Chapin, F. S., III, Schulze, E.-D., and Mooney, H. A. (1990). The ecology and economics of storage in plants. *Annu. Rev. Ecol. Syst.* **21**, 423–447.

Clements, F. E. (1916). Plant succession: An analysis of the development of vegetation. *Carnegie Inst. Washington Publ.* **242**.

Dierschke, H. (1974). "Saumgesellschaften im Vegetations- und Standortsgefälle an Waldrändern." Scripta Geobotanica, Göttingen, Germany.

Evans, J. R. (1989). Photosynthesis and nitrogen relationships in leaves of C_3 plants. *Oecologia* **78**, 9–19.

Field, C. B. (1988). On the role of photosynthetic responses in constraining the habitat distribution of rainforest plants. *Aust. J. Plant Physiol.* **15**, 343–358.

Field, C. B., and Mooney, H. A. (1986). The photosynthesis–nitrogen relationship in wild plants. *In* "On the Economy of Plant Form and Function" (T. J. Givnish, ed.), pp. 25–55. Cambridge Univ. Press, Cambridge, UK.

Fisher, J. B. (1986). Branching patterns and angles in trees. *In* "On the Economy of Plant Form and Function" (T. J. Givnish, ed.), pp. 493–523. Cambridge Univ. Press, Cambridge, UK.

Galinski, W. (1993). Non-random needle orientation in one-year-old Scots pine (*Pinus sylvestris* L.) seedlings when adjacent to vegetation unable to shade them. *Trees* (in press).

Gill, A. M., and Ashton, D. H. (1968). The role of bark type in relative tolerance to fire of three Central Victorian eucalypts. *Aust. J. Bot.* **16**, 491–498.

Givnish, T. J. (1986). Biomechanical constraints on crown geometry in forest herbs. *In* "On the Economy of Plant Form and Function" (T. J. Givnish, ed.), pp. 525–583. Cambridge Univ. Press, Cambridge, UK.

Golley, F. B. (1977). "Ecological Succession: Benchmark Papers in Ecology," Vol. 5. Dowden, Hutchinson & Ross, Stroudsburg, PA.

Grime, J. P. (1979). "Plant Strategies and Vegetation Processes." Wiley, Chichester.

Hallé, F., Oldeman, R. A. A., and Tomlinson, P. B. (1978). "Tropical Trees and Forests: An Architectural Analysis." Springer-Verlag, Berlin.

Hirose, T., and Werger, M. J. A. (1987). Nitrogen use efficiency in instantaneous and daily photosynthesis of leaves in the canopy of a *Solidago altissima* stand. *Physiol. Plant.* **70**, 215–222.

Horn, H. S. (1974). The ecology of secondary succession. *Annu. Rev. Ecol. Syst.* **5**, 25–37.

Jakucs, P. (1969). Die Sproβkolonien und ihre Bedeutung in der dynamischen Vegetationsentwicklung (Polycormonsukzession). *Acta Bot. Croat.* **28**, 161–170.

Jurik, T. W. (1991). Population distributions of plant size and light environment of giant ragweed (*Ambrosia trifida* L.) at three densities. *Oecologia* **87**, 539–550.

Kellomäki, S., Väisänen, H., Hänninen, H., Kolström, T., Lauhanen, R., Mattila, U., and Pajari, B. (1992). "Sima: A Model for Forest Succession Based on the Carbon and Nitrogen Cycles with Application to Silvicultural Management of the Forest Ecosystem." The University of Joensuu, Finland.

Kempf, J. S., and Pickett, S. T. A. (1981). The role of branch length and angle in branching pattern of forest shrubs along a successional gradient. *New Phytol.* **88**, 111–116.

Kikuzawa, K. (1991). A cost–benefit analysis of leaf habit and leaf longevity and their geographical pattern. *Am. Nat.* **138**, 1250–1263.

Koike, T. (1987). Photosynthesis and expansion in leaves of early, mid and late successional tree species, birch, ash and maple. *Photosynthetica* **21**, 503–508.

Körner, C. (1991). Some often overlooked plant characteristics as determinants of plant growth: A reconsideration. *Funct. Ecol.* **5**, 162–173.

Kriebitzsch, W.-U. (1992). Der CO_2- und H_2O-Gasaustausch von Pflanzen in der Krautschicht eines Kalkbuchenwaldes in Abhängigkeit von Standortsfaktoren. III. CO_2-Bilanzen und Netto-Primärproduktion. *Flora* (*Jena*) **187**, 135–158.

Küppers, M. (1982). Kohlenstoffhaushalt, Wasserhaushalt, Wachstum und Wuchsform von Holzgewächsen im Konkurrenzgefüge eines Heckenstandortes. Ph.D. Dissertation, University of Bayreuth, Germany.

Küppers, M. (1984a). Carbon relations and competition between woody species in a Central European hedgerow. I. Photosynthetic characteristics. *Oecologia* **64**, 332–343.

Küppers, M. (1984b). Carbon relations and competition between woody species in a Central

European hedgerow. II. Stomatal responses, water use, and hydraulic conductivity in the root/leaf pathway. *Oecologia* **64,** 344–354.

Küppers, M. (1984c). Carbon relations and competition between woody species in a Central European hedgerow. III. Carbon and water balance on the leaf level. *Oecologia* **65,** 94–100.

Küppers, M. (1984d). Kohlenstoffhaushalt, Wasserhaushalt, Wachstum und Wuchsform von Holzgewächsen im Konkurrenzgefüge eines Heckenstandortes. *In* "Die pflanzenökologische Bedeutung und Bewertung von Hecken" (E.-D. Schulze, A. Reif, and M. Küppers, eds.), Suppl. 3, Part 1, pp. 10–102. Akademie für Naturschutz und Landschaftspflege, Laufen/Salzach, Germany.

Küppers, M. (1985). Carbon relations and competition between woody species in a Central European hedgerow. IV. Growth form and partitioning. *Oecologia* **66,** 343–352.

Küppers, M. (1987). Hecken—Ein Modellfall für die Partnerschaft von Physiologie und Morphologie bei der pflanzlichen Produktion in Konkurrenzsituationen. *Naturwissenschaften* **74,** 536–547.

Küppers, M. (1989). Ecological significance of above-ground architectural patterns in woody plants: A question of cost–benefit relationships. *Trends Ecol. Evol.* **4,** 375–379.

Küppers, M. (1992a). Changes in plant ecophysiology across a Central European hedgerow ecotone. *Ecol. Stud.* **92,** 285–303.

Küppers, M. (1992b). Changes in resource-use efficiency in different woody growth forms during secondary forest succession in Central Europe. *In* "Responses of Forest Ecosystems to Environmental Changes" (A. Teller, P. Mathy, and J. N. R. Jeffers, eds.), pp. 628–630. Elsevier Applied Science, London.

Küppers, M., and Schneider, H. (1993). Leaf gas exchange of beech (*Fagus sylvatica* L.) seedlings in lightflecks: Effects of fleck length and leaf temperature in leaves grown in deep and partial shade. *Trees* **7,** 160–168.

Küppers, M., and Schulze, E.-D. (1985). An empirical model of net photosynthesis and leaf conductance for the simulation of diurnal courses of CO_2 and H_2O exchange. *Aust. J. Plant Physiol.* **12,** 513–526.

Küppers, M., Zech, W., Schulze, E.-D., and Beck, E. (1985). CO_2-Assimilation, Transpiration und Wachstum von *Pinus sylvestris* L. bei unterschiedlicher Magnesiumversorgung. *Forstwiss. Centralbl.* **104,** 23–36.

Küppers, M., Wheeler, A. M., Küppers, B. I. L., Kirschbaum, M. U. F., and Farquhar, G. D. (1986). Carbon fixation in eucalypts in the field—Analysis of diurnal variations in photosynthetic capacity. *Oecologia* **70,** 273–282.

Küppers, M., Swan, A. G., Tompkins, D., Gabriel, W. C. L., Küppers, B. I. L., and Linder, S. (1987). A field portable system for the measurement of gas exchange of leaves under natural and controlled conditions: Examples with field-grown *Eucalyptus pauciflora* Sieb. ex Spreng. ssp. *pauciflora, E. behriana* F. Muell. and *Pinus radiata* D. Don. *Plant, Cell Environ.* **10,** 425–435.

Küppers, M., Koch, G., and Mooney, H. A. (1988). Compensating effects to growth of changes in dry matter allocation in response to variation in photosynthetic characteristics induced by photoperiod, light and nitrogen. *Aust. J. Plant Physiol.* **15,** 287–298.

Küppers, M., Küppers, B. I. L., and Godkins, C. (1993). Kohlenstoff-Akquisition und Wachstumsstrategien zweier konkurrierender *Eucalyptus*—Arten auf einem hochmontanen Standort in Australien. *Verh. Ges. Oekol.* **22,** 423–430.

Lambers, H. (1987). Does variation in photosynthetic rate explain variation in growth rate and yield? *Neth. J. Agric. Sci.* **35,** 505–519.

Lambers, H., and Poorter, H. (1992). Inherent variation in growth rate between higher plants: A search for physiological causes and ecological consequences. *Adv. Ecol. Res.* **23,** 187–261.

Linder, S., and Axelsson, B. (1982). Changes in carbon uptake and allocation patterns as a result of irrigation and fertilization in a young *Pinus sylvestris* stand. *In* "Carbon Uptake and Allocation in Subalpine Ecosystems as a Key to Management" (R. H. Waring, ed.), Forest Research Laboratory, Oregon State University, Corvallis.

Louda, S. M., and Renaud, P. E. (1991). Site pre-emption: Competition between generations of plants. *Trends Ecol. Evol.* **6,** 107–108.

Matyssek, R. (1985). Der Kohlenstoff-, Wasser- und Nährstoffhaushalt der wechselgrünen und immergrünen Koniferen Lärche, Fichte, Kiefer. Ph.D. Dissertation, University of Bayreuth, Germany.

McCree, K. J. (1976). The role of dark respiration in the carbon economy of a plant. *In* "CO$_2$ Metabolism and Plant Productivity" (C. C. Black and R. H. Burris, eds.), pp. 177–184. University Park Press, Baltimore, MD.

Meister, H.-P., Caldwell, M. M., Tenhunen, J. D., and Lange, O. L. (1987). Ecological implications of sun/shade—Leaf differentiation in sclerophyllous canopies: Assessment by canopy modeling. *In* "Plant Response to Stress: Functional Analysis in Mediterranean Ecosystems" (J. D. Tenhunen, F. M. Catarino, O. L. Lange, and W. Oechel, eds.), pp. 401–411. Springer-Verlag, Berlin.

Mooney, H. A., and Chiariello, N. R. (1984). The study of plant function—The plant as a balanced system. *In* "Perspectives on Plant Population Ecology" (R. Dirzo and J. Sarukhan, eds.), pp. 305–323. Sinauer Assoc., Sunderland, MA.

Mooney, H. A., Ferrar, P. J., and Slatyer, R. O. (1978). Photosynthetic capacity and carbon allocation patterns in diverse growth forms of *Eucalyptus. Oecologia* **36,** 103–111.

Oren, R., and Zimmermann, R. (1989). CO$_2$ assimilation and the carbon balance of healthy and declining Norway spruce stands. *Ecol. Stud.* **77,** 352–369.

Osmond, C. B., Austin, M. P., Berry, J. A., Billings, W. D., Boyer, J. S., Dacey, J. W. H., Nobel, P. S., Smith, S. D., and Winner, W. E. (1987). Stress physiology and the distribution of plants. *BioScience* **37,** 38–48.

Pearcy, R. W., Björkman, O., Caldwell, M. M., Keeley, J. E., Monson, R. K., and Strain, B. R. (1987). Carbon gain by plants in natural environments. *BioScience* **37,** 21–29.

Poorter, H. (1989). Interspecific variation in relative growth rate: On ecological causes and physiological consequences. *In* "Causes and Consequences of Variation in Growth Rate and Productivity of Higher Plants" (H. Lambers, M. L. Cambridge, H. Konings, and T. L. Pons, eds.), pp. 45–68. SPB Academic Publishing, The Hague. The Netherlands.

Prusinkiewicz, P., and Lindenmayer, A. (1990). "The Algorithmic Beauty of Plants." Springer-Verlag, New York.

Rauh, W. (1950). "Morphologie der Nutzpflanzen." Quelle und Meyer, Heidelberg.

Remmert, H. (1991). The mosaic-cycle concept of ecosystems. *Ecol. Stud.* **85,** 168.

Rundel, P. W. (1981). Fire as an ecological factor. *In* "Encyclopedia of Plant Physiology" (O. L. Lange, P. S. Nobel, C. B. Osmond, and H. Ziegler, eds.), New Ser., Vol. 12A, pp. 501–538. Springer-Verlag, Berlin.

Schenk, J., Stickan, W., and Runge, M. (1989). Belaubungsverlauf und Blattmerkmale von Buchen unter dem Einfluss von Kalkung und Stickstoffdüngung. *Ber. Forschungszentrums Waldökosyst., Ser. A* **49,** 91–101.

Schmitt, J., and Wulff, R. D. (1993). Light spectral quality, phytochrome and plant competition. *Trends Ecol. Evol.* **8,** 47–51.

Schneider, H., Paliwal, K., and Küppers, M. (1993). Blattgasaustausch in Lichtflecken von Jungpflanzen unterschiedlicher sukzessionaler Stellung aus dem Unterwuchs eines mitteleuropäischen Buchenwaldes—eine analytische Grundlage für die Ellenbergschen Licht-Zeigerwerte? *Verh. Ges. Oekol.* **22,** 439–442.

Schulze, E.-D. (1970). Der CO$_2$-Gaswechsel der Buche (*Fagus sylvatic* L.) in Abhängigkeit von den Klimafaktoren im Freiland. *Flora (Jena)* **159,** 177–232.

Schulze, E.-D. (1972). Die Wirkung von Licht und Temperatur auf den CO_2-Gaswechsel verschiedener Lebensformen aus der Krautschicht eines montanen Buchenwaldes. *Oecologia,* **9,** 235–258.

Schulze, E.-D., and Chapin, F. S., III (1987). Plant specialization to environments of different resource availability. *Ecol. Stud.* **61,** 120–148.

Schulze, E.-D., Küppers, M., and Matyssek, R. (1986). The roles of carbon balance and branching pattern in the growth of woody species. *In* "On the Economy of Plant Form and Function" (T. J. Givnish, ed.), pp. 585–602. Cambridge Univ. Press, Cambridge, UK.

Shugart, H. H. (1984). "A Theory of Forest Dynamics: The Ecological Implications of Forest Succession Models." Springer-Verlag, Berlin.

Silvertown, J. (1987). "Introduction to Plant Population Ecology." Longman, London.

Slatyer, R. O., and Morrow, P. A. (1977). Altitudinal variation in the photosynthetic characteristics of snow gum, *Eucalyptus pauciflora* Sieb. ex Spreng. I. Seasonal changes under field conditions in the Snowy Mountains area of South-eastern Australia. *Aust. J. Bot.* **25,** 1–20.

Stephan, B., and Stephan, S. (1971). Die Vegetationsentwicklung im Naturschutzgebiet Stolzenburg und ihre Bedeutung für die Schutzmassnahmen. *Decheniana* **123,** 281–305.

Stickan, W., Schulte, M., Kakubari, Y., Niederstadt, F., Schenk, J., and Runge, M. (1991). Ökophysiologische und biometrische Untersuchungen in einem Buchenbestand (*Fagus sylvatica* L.) des Sollings als ein Beitrag zur Waldschadensforschung. *Ber. Forschungszen. Waldökosysteme, Ser. B.* **18,** 1–82.

Thompson, W. A., Stocker, G. C., and Kriedemann, P. E. (1988). Growth and photosynthetic response to light and nutrients of *Flindersia brayleyana* F. Muell., a rainforest tree with broad tolerance to sun and shade. *Aust. J. Plant Physiol.* **15,** 299–315.

Tinoco-Ojanguren, C., and Pearcy, R. W. (1992). Dynamic stomatal behavior and its role in carbon gain during lightflecks of a gap phase and an understory *Piper* species acclimated to high and low light. *Oecologia* **92,** 222–228.

Tranquillini, W. (1959). Die Stoffproduktion der Zirbe (*Pinus cembra* L.) an der Waldgrenze während eines Jahres. *Planta* **54,** 107–151.

Troll, W. (1935). "Vergleichende Morphologie der höheren Pflanzen," Vol. 1, Part I. Borntraeger, Berlin (reprint: Koeltz, Königstein, 1967).

Vines, R. G. (1968). Heat transfer through bark, and the resistance of trees to fire. *Aust. J. Bot.* **16,** 499–514.

Vitousek, P. M., and Denslow, J. S. (1986). Nitrogen and phosphorous availability in treefall gaps of a lowland tropical rainforest. *J. Ecol.* **74,** 1167–1178.

Weiner, J., and Thomas, F. C. (1992). Competition and allometry to three species of annual plants. *Ecology* **73,** 648–656.

Whittaker, R. H. (1975). Functional aspects of succession in deciduous forests. *In* "Sukzessionsforschung" (W. Schmidt, ed.), pp. 377–405. Cramer, Vaduz.

Wilmanns, O. (1983). Lianen in mitteleuropäischen Pflanzengesellschaften und ihre Einnischung. *Tuexenia* **3,** 343–358.

Woods, D. B., and Turner, N. C. (1971). Stomatal responses to changing light by forest tree species of varying shade tolerance. *New Phytol.* **70,** 77–84.

Yodzis, P. (1978). "Competition for Space and the Structure of Ecological Communities." Springer-Verlag, Berlin.

5

Photosynthetic Acclimation to Changing Light Environments: Scaling from the Leaf to the Whole Plant

Robert W. Pearcy and Daniel A. Sims

I. Introduction

A remarkable feature of the photosynthetic apparatus of plants is its adaptability to a wide range of light inputs. The leaves of a tree seedling at the bottom of a tropical forest may receive less than 1% of the light incident at the top. Yet, over its life span, the tree may need to cope with both conditions. Moreover, when a canopy gap forms, seedlings and understory plants suddenly receive irradiances equal to those in the canopy. The additional light will stimulate extra growth if the plant can utilize it and if new stresses such as photoinhibition of photosynthesis do not offset the extra potential carbon gain. Closure of the gap requires that these plants readjust to the lower available light. Survival in the shaded understory demands maximization of light capture for photosynthesis concomitant with minimization of losses of energy and carbon in respiration (Björkman, 1973). By contrast, leaves exposed to high light must be able to make efficient use of the available energy while avoiding the possibility of loss of photosynthesis because of photoinhibition or other environmental stresses. The capacity to accomplish these compromises is greatly influenced by changes in other environmental factors such as nutrient availability and temperature that often accompany changes in light availability.

Changes in the light environment experienced by forest plants during their lifetime may range from sunflecks lasting from seconds to minutes to more sustained changes occurring when gaps are formed or canopies develop. Because acclimation to a changed light environment involves

Copyright © 1994 by Academic Press, Inc.

changes in enzyme and pigment amounts, as well as leaf anatomy and resource allocation as new leaves are produced in the new environment, the time scale over which these processes can occur determines the types of light changes for which acclimation is important. Sunflecks or even the normal diurnal change in solar radiation occur on too fast a time scale for acclimation. Thus, regulatory mechanisms such as light activation of enzymes operating on time scales of minutes or less are of primary importance for these short-term light changes (see Pearcy *et al.*, Chapter 6, this volume). These regulatory mechanisms appear to function to maintain a metabolic balance at existing enzyme levels as the levels of external resources such as light and CO_2 change. By contrast, the acclimatory changes in the concentrations of enzymes or in leaf anatomy are a redeployment of internal resources (primarily nitrogen and carbon) in a way that either enhances assimilation or enhances the resistance to stress in the new environment. This redeployment appears to require at the minimum a few days (Chow and Anderson, 1987a) to, in some cases, several weeks (Bauer and Thöni, 1988). Thus, to be beneficial, redeployment should only occur in response to sustained changes lasting for periods longer than these response times.

In this chapter, we discuss the nature of acclimation to changing light environments and its role in adapting plants to patchy environments. We focus on studies of the tropical forest understory plant *Alocasia macrorrhiza* (L.) G. Don, which, despite its reputation as an obligate shade plant, has a substantial capacity for acclimation to high light (Chow *et al.*, 1988; Sims and Pearcy, 1989). Its simple growth form (rosette with 5 to 10 saggitate leaves supported on long petioles) facilitates studies of allocation and whole-plant carbon balance. We show that while acclimation has been extensively studied at the leaf and chloroplast level, leading to an elegant understanding of the functional consequences of the underlying changes, an ecological understanding of its role in adapting plants to patchy light environments requires that this information be put into a more whole-plant context. We use a cost–benefit approach (Jurik and Chabot, 1986; Givnish, 1988) to analyze the consequences of the changes in photosynthetic capacity in light of the resources required to bring about that change.

II. Photosynthetic Acclimation of Leaves to Light

Acclimation of leaf photosynthesis to different growth light regimes is best viewed in terms of the factors that increase the photosynthetic capacity measured at light saturation and those that determine photosynthetic rate at any given photon flux density (PFD). At low PFD, the rate of photosynthesis depends on the rate of light capture, and therefore on the absorptivity of the leaf for photosynthetically active radiation (PAR). Absorbed PAR drives the electron transport reactions that ultimately lead to the regeneration of the primary CO_2 acceptor, ribulose bisphosphate (RuBP) in the Calvin cycle. The capacity of the electron transport (ET) chain has almost no influence on the rate of ET at these low PFDs and thus the rate of RuBP regeneration is determined by the rate of PAR absorption. Similarly, the capacity of the primary carboxylation reaction, ribulose-1,5-bisphosphate carboxylase/oxygenase (Rubisco), has little influence, and the rate of carboxylation is limited by the supply of RuBP. At low PFD, therefore, there is little return on investment in increasing the capacity of the photosynthetic reactions and resources are better invested in increasing PAR absorption.

As the PFD increases, the capacity of the photosynthetic reactions becomes increasingly limiting to photosynthetic rate. These capacities must therefore be increased if high photosynthetic rates are to be achieved. Studies of von Caemmerer and Farquhar (1981, 1984) show that the ET and Rubisco capacities in leaves covary, increasing and decreasing in concert with PFD in the growth regime, or as environmental stresses (e.g., low nutrients) limit photosynthetic capacity. This is to be expected if resources are invested with high efficiency, since if one is limiting the excess capacity of the other could not be utilized. Indeed, photosynthesis often appears to operate at an intercellular CO_2 pressure where Rubisco and ET capacities are approximately colimiting (von Caemmerer and Farquhar, 1984; Farquhar and Sharkey, 1982). Because RuBP regeneration becomes more limiting as the intercellular CO_2 pressure (p_i) increases, the colimitation also depends on the stomatal conductance. Thus, for maximum efficiency it is also necessary for stomatal conductance to covary with the biochemical capacity for CO_2 fixation, as has been demonstrated in a wide variety of plant species (Wong *et al.,* 1979, 1985; Körner *et al.,* 1979). In high light, therefore, there should be increased investment in factors that determine the capacity of ET and carboxylation as well as a concomitant increase in stomatal conductance.

A. Factors Determining Light Harvesting by Sun and Shade Leaves

1. The Quantum Yield of Sun- and Shade-Acclimated Leaves   It is now well established that in the absence of stresses, sun and shade leaves do not differ in their quantum yield (Boardman, 1977; Björkman and Demmig, 1987). Quantum yields for O_2 evolution measured at saturating CO_2 for *Alocasia* declined only very slightly with an increase in the growth PFD (Figure 1C). These quantum yields approach the maximum theoretical value expected in the absence of a q cycle (Evans, 1988). Thus at low PFD, sun and shade leaves appear to use nearly all absorbed photons

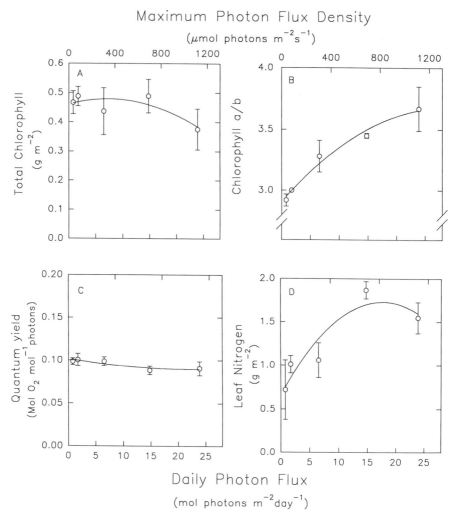

Figure 1 Relationships between the growth light environment and (A) total chlorophyll, (B) Chl a/b ratio, (C) quantum yield, and (D) leaf nitrogen content for *Alocasia macrorrhiza*. The maximum photon flux density and the daily integral of photon flux density are given on the top and bottom axes, respectively. Each point is the mean ± 1 S.E. for five leaves. (Redrawn from data in Sims and Pearcy, 1989.)

for photochemistry. Quantum yields of CO_2 uptake measured at saturating CO_2 and low O_2 are lower for reasons that are not understood (Björkman and Demmig, 1987). The difference may reflect the reduction state of the products being produced under the measurement conditions. Quantum yields for CO_2 uptake measured at normal CO_2 and O_2 pressures are still lower because of the oxygenation of RuBP and photorespiration and thus in C_3 plants are influenced by temperatures and intercellular CO_2 pressures.

2. Pigment Organization in Sun- and Shade-Acclimated Leaves   Shade leaves typically contain the same or even somewhat less chlorophyll (Chl) per unit area than sun leaves (Björkman, 1981). More Chl yields only diminishing returns. For example, Evans (1989a) found that doubling the Chl content of leaves from 0.3 to 0.6 mmol chl m^{-2} increased leaf absorptance from about 0.825 to 0.885, an increase of only 7%. On a per unit weight basis, however, Chl contents are higher in shade than in sun leaves, reflecting the general shift in allocation of carbon and N resources toward a function in light absorption. Chl itself contains only a small amount of N (4 mol N mol^{-1} Chl) but the associated chlorophyll proteins contain 25 to 70 mmol N mol^{-1} chl (Evans, 1989a). Thus, an increase in Chl has a substantial N cost.

For *Alocasia*, Chl per m^2 leaf area (Figure 1A) declined only slightly with increasing PFD during growth. However, Chl a/b ratios (Figure 1B) increased markedly as commonly observed in many species (Anderson, 1986). Decreased Chl a/b ratios in the shade result from greater investment in light-harvesting Chl a/b−protein complexes of PS II (LHCP-II), which contain most of the Chl b in the chloroplast (Anderson, 1986). The capacity to increase the investment in Chl in general and LHCP-II in particular appears to be greater in shade or tree species as compared to sun-adapted species (Chu and Anderson, 1984; Evans, 1989c). In shade-grown *Alocasia* leaves, 57% of the Chl was LHCP-II (Chu and Anderson, 1984), whereas in peas grown at low and high PFDs, LHCP-II accounted for 40 and 28% of the Chl, respectively (Leong and Anderson, 1984a). Since LHCP-II provides the adhesion between grana (Chow *et al.*, 1988), extensive grana stacking is a characteristic feature of shade plant chloroplasts (Chow *et al.*, 1988; Anderson *et al.*, 1973; Lichtenthaler *et al.*, 1981), but its function is unclear. The close stacking may allow greater Chl contents per unit chloroplast volume (Anderson, 1986) and was thought to enhance light absorption of shade light, which typically is depleted in wavelengths absorbed by PS II. However, careful calculations based on spectral and quantum yield measurements suggest no greater light capture efficiency at equal Chl concentrations of shade as compared to sun chloroplasts in shade light (Evans, 1986). This would appear to rule out the changes in Chl a/b as a factor in any spectral adaptation to the light quality of the understory. It is noteworthy that the nitrogen content per Chl in LHCP-II is only 43% of that of the other major Chl−protein complexes associated with the PS II reaction center (Evans, 1986). Thus, the increase in LHCP-II maintains light harvesting in preference to electron transport capacity at a minimum nitrogen cost.

B. Factors Determining the Photosynthetic Capacity of Sun- and Shade-Acclimated Leaves

1. Electron Transport and Carboxylation Capacities   The general relationships determining the photosynthetic capacity of sun and shade leaves briefly discussed in the foregoing have now been confirmed for many plant species (Björkman, 1981; Evans, 1987, 1989b; Walters and Field, 1987; Chow and Anderson, 1987b; Seemann *et al.*, 1987; Thompson *et al.*, 1988; Ferrar and Osmond, 1986). In this respect, *Alocasia* as a shade plant appears to be perhaps quantitatively but certainly not qualitatively different in its acclimation response from that of many crops and trees. Photosynthetic capacity reached a maximum at daily photon fluxes of 15 mol m^{-2} day^{-1} and then leveled off or even declined slightly at higher PFDs (Figure 2A). The lack of an increase in photosynthetic capacity at higher PFD was due primarily to decreased stomatal conductances (Figure 2E).

Acclimation of photosynthetic capacity in *Alocasia* to the different light environments was brought about by changes in the investment in electron transport capacity (V_{jmax}; Figure 2D) and carboxylation capacity (V_{cmax}; Figure 2C) per unit leaf area. V_{cmax} and V_{jmax} were determined by measuring the response of CO_2 assimilation to intercellular CO_2 pressure, followed by a least-squares fit of the model of Farquhar and von Caemmerer (1982) to estimate both parameters. Both V_{cmax} and V_{jmax} increased with increasing growth PFD, except at the highest PFD, where V_{jmax} decreased slightly. These increases are very similar to the increases in Rubisco activity and electron transport carrier concentrations measured by Chow *et al.* (1988) in *Alocasia* leaves grown at incident PFDs from 40 to 800 μmol m^{-2} s^{-1}. Thus, like other species investigated so far, changes in photosynthetic capacity are produced by a balanced and coordinated change in investments in Rubisco and ET carriers (Evans, 1988; Leong and Anderson, 1984a,b; Chow and Anderson, 1987a). The decrease in V_{jmax} in *Alocasia* observed at the highest PFD did not seem to be due to photoinhibition since quantum yield should have been even more sensitive than electron transport capacity but showed no effect (Figure 1). The reduction in V_{jmax} in the highest-light environment may have been due to the high light alone; transpiration and leaf temperatures were also higher in this environment. Leaf N per unit area (Figure 1D) also increased in parallel with the capacities of the photosynthetic reactions. This is because the increase in carboxylation and electron transport capacity requires a substantial investment of nitrogen (Evans, 1989a).

2. Leaf Structure   Since light is received on a per unit area basis it has been common to express leaf photosynthetic rates on this basis in acclimation studies. However, growth at increased PFD causes leaves to

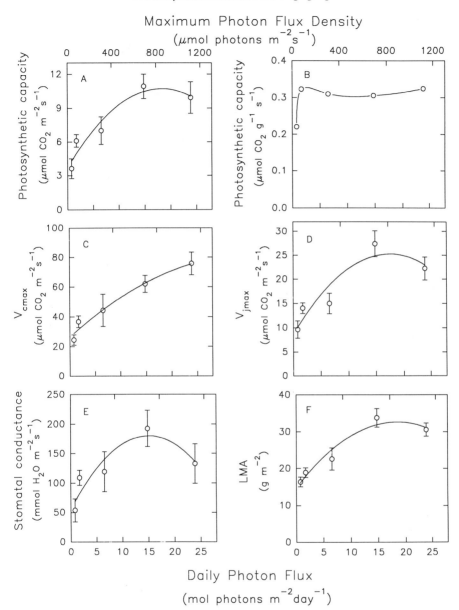

Figure 2 Relationships between the growth light environment and (A) photosynthetic capacity per unit leaf area, (B) photosynthetic capacity per unit weight, (C) carboxylation capacity (V_{cmax}), (D) electron transport capacity in CO_2 equivalents (V_{jmax}), (E) stomatal conductance, and (F) leaf mass per unit area (LMA) for *Alocasia macrorrhiza*. The maximum photon flux density and the daily integral of photon flux density are given on the top and bottom axes, respectively. Each point is the mean ± 1 S.E. for five leaves. (Redrawn from data in Sims and Pearcy, 1989.)

become thicker and to have an increased leaf mass per unit area (LMA), which is a major factor contributing to the increased photosynthetic capacity of *Alocasia* sun leaves (Sims and Pearcy, 1992). Consequently, assimilation rate per unit mass may reflect the return in terms of photosynthetic capacity of a given investment in leaf better than comparisons on a per unit area basis. For *Alocasia* leaves (Figure 2A), rates per unit leaf mass were almost independent of growth PFD except at the lowest PFD, where a decline was apparent. This is in marked contrast to the large changes per unit area. Changes in storage carbohydrate content, which can be up to 20% of leaf mass in high light (Björkman, 1981), can distort interpretations of photosynthetic rate per unit mass. However, careful measurements showed that for fully expanded leaves during their period of maximum photosynthetic capacity, photosynthetic rate per unit cell volume was constant but the increased cell volume per unit area in high-light leaves resulted in a higher photosynthetic capacity per unit area (Sims and Pearcy, 1992). Since chlorophyll content per unit area was constant, it decreased per unit mesophyll volume with increasing growth PFD. Thus, at the cellular level there was a shift in resources away from light harvesting, but this did not increase the photosynthetic capacity per unit cell volume.

The apparently rather fixed photosynthetic capacity per unit cell volume of *Alocasia* has significant implications for acclimation in this species. Fully developed leaves exhibit no ability for photosynthetic acclimation to either increases or decreases in PFD (Sims and Pearcy, 1991, 1992). It is possible to slightly increase the photosynthetic capacity per unit cell volume if the transfer is made just at completion of leaf expansion but before physiological development is complete. A consequence of this inflexible physiology after developmental maturity is that acclimation of *Alocasia* to a change in light environment requires production of new leaves.

Fully mature leaves in some other species can clearly exhibit substantial acclimation responses, suggesting changes in photosynthetic capacity per unit cell volume. In some experiments the changes in photosynthetic capacity have been related to increased or decreased amounts of Rubisco (Gauhl, 1979; Sebaa *et al.*, 1986; Chow and Anderson, 1987a; Davies *et al.*, 1986), depending on the direction of the transfer. Although anatomical examinations were not part of any of these studies, it seems unlikely that there could have been development of new cells that would account for the changes in photosynthetic capacity. Kamaluddin and Grace (1992), however, reported that acclimation of mature leaves of the tropical tree *Bischofia javanica* to high light, which resulted in increased photosynthetic capacity, was correlated with increases in cell volume per unit area. This was due to an increase in palisade mesophyll thickness. The acclimated leaves still differed from leaves fully developed in high light in that the former still had 30% lower photosynthetic capacities and maintained only one (albeit thicker) palisade cell layer whereas the latter had two layers. There is only one reported case of development of new cells in fully mature shade leaves after transfer to high light (Bauer and Thöni, 1988). Production of a new palisade layer was reported for *Hedera helix* following transfer to a light environment sufficiently bright to cause damage to the original palisade.

Further studies are needed to resolve the extent to which acclimation depends on anatomical changes vis-à-vis changes in photosynthetic capacity per unit cell volume. It is clear that some species such as *Alocasia* lack significant capacity to alter either photosynthetic capacity per unit cell volume or leaf anatomy in mature leaves. For these, acclimation of photosynthetic capacity to an increase in PFD is precluded, even though acclimatory changes in chloroplast ultrastructure and Chl a/b occur. Others (especially herbaceous but also a few woody species) can increase photosynthetic capacity of mature leaves in response to increased light, but the extent to which this is due to increased cell volume or to increased concentrations of photosynthetic enzymes per unit cell volume is unclear. Unfortunately, there are few studies that have addressed both anatomy and enzyme concentrations, so that the basis of any changes could be understood.

Leaf photosynthetic rates expressed on a cell volume basis are necessarily volume averaged, which may obscure significant differences among cells. Elegant experiments by Terashima and Inoye (1985) reveal a gradient in photosynthetic properties through a *Pelargonium* leaf reminiscent of sun/shade acclimation through a canopy (Osmond, 1989). If this also occurs in *Alocasia*, it implies that the leaf may be made up of individual cells differing in photosynthetic capacity depending on their position in the leaf but with allocation of resources among cells so that the same mean photosynthetic capacity per unit volume is maintained as extra cell layers are added. How such coordination is achieved is unknown.

III. Respiratory Costs of Sun- and Shade-Acclimated Leaves

The importance of respiration to sun and shade acclimation and carbon balance in high- and low-light environments has long been recognized (Grime, 1966; Björkman, 1973, 1981). Björkman (1973) proposed that shade adaptation was a matter of economics with the low photosynthetic capacities of shade leaves providing the benefit of minimizing maintenance costs and hence maximizing net photosynthesis in shaded environments. Indeed, carbon balance models of sun and shade leaves (Björkman, 1973; Givnish, 1988; Jurik and Chabot, 1986) typically have as their fundamental premise the economic trade-off between increased photosynthetic capacity and its consequences for photosynthetic rate in high light, and increased respiratory costs and its consequences for net photosynthesis in low light. However, there has been little systematic examination of the underlying causes for the differences in respiration rates between sun and shade leaves.

The relationship between respiration rate and photosynthetic capacity in *Alocasia* leaves was investigated by monitoring the time courses of dark respiration rate and photosynthetic capacity of mature leaves over a 28-day period following reciprocal transfer of plants between high- and low-light environments (Sims and Pearcy, 1991). Although sun leaves initially had much higher respiration rates than shade leaves, convergence occurred within a few days after transfer. Assimilation rates of sun leaves transferred to the shade declined slowly but they remained higher than those of shade leaves. After the initial period of rapid decrease, the measured respiration rates were always slightly greater in sun than in shade leaves, but the difference was so small that they were statistically indistinguishable. This small difference, however, may reflect the extra maintenance cost associated with the higher photosynthetic capacity of the sun leaves since, after the initial reduction, respiration rates of both sun and shade leaves were a remarkably constant 1.25% of photosynthetic capacity. If so, it implies that the additional maintenance costs of a higher photosynthetic capacity in the shade are small indeed.

The reasons for the high respiration rates in sun leaves may be the costs of carbohydrate processing or activity of alternative path respiration (Lambers, 1985; Azcon Bieto and Osmond, 1983; Azcon Bieto *et al.*, 1983) that may occur because of high carbohydrate concentrations. Both are related more to the high availability of light and therefore high daily photosynthesis rather than to high photosynthetic capacity. The costs of carbohydrate processing are more correctly a construction rather than a maintenance cost, and should be assigned to the sink where the construction is occurring. It may also be that the maintenance costs of leaves in the sun are higher because of a greater need for repair, or because protein turnover is higher in high light. Howeve these maintenance costs would apply to leaves in the sun but not tc igh-photosynthetic-capacity leaves in the shade. A lesson from these measurements is that, if taken too hastily, respiration rates measured for sun leaves grossly overestimate the maintenance cost of sun leaves in the shade. Consequently, carbon balance studies relying on respiration rates determined as part of a light–response curve probably overestimate the importance of maintenance respiration. As will be discussed in Section VI,A, construction costs are probably much more significant than maintenance costs to the return on investment of leaves in shade environments.

IV. Sensitivity to Photoinhibition

The problem immediately facing a shade plant that is suddenly exposed to bright light is the possibility of photoinhibition, and the constraint it may impose upon the capacity to acclimate and hence maximize the utilization of the increased light. The greater sensitivity to photoinhibition of shade plants as compared to sun plants has been widely documented (see reviews by Björkman, 1981; Anderson and Osmond, 1987). Photoinhibition is primarily evident as a reduction in the quantum yield, reflecting a change in the efficiency of light harvesting. Since it is widely believed that both the photodamage and dissipation of excess energy that contribute symptomatically to photoinhibition of photosynthesis primarily involve the PS II reaction center (Powles, 1984; Krause and Weiss, 1991), the reduction in light-saturated photosynthetic rate may be a secondary event.

Studies with *Alocasia* (Mulkey and Pearcy, 1992) reveal a strong interaction between acclimation and photoinhibition in this species. Sun leaves were much more capable of dissipating excess energy than shade leaves, which contributed to their greater resistance to photoinhibitory damage. This dissipation occurred in part because of the higher photosynthetic capacities of sun leaves but probably more importantly because of an increased capacity for thermal dissipation of excess energy (Demmig and Björkman, 1987; Demmig-Adams *et al.*, 1989). The greater capacity for nonradiative dissipation in sun leaves, as indicated by increased nonphotochemical fluorescence-quenching kinetics, may be related to the operation of the xanthophyll cycle in which depoxidation of violaxanthin to zeaxanthin within the thylakoid membrane promotes this capacity via an as yet unknown mechanism (Demmig *et al.*, 1987; Demmig-Adams *et al.*, 1989). When exposed to a 2-h high-light period (1900 μmol photons m^{-2} s^{-1}, 30°C leaf temperatures) designed to simulate the change in PFD following formation of a canopy gap, *Alocasia* sun leaves showed a more rapid development of nonphotochemical quenching and also a more rapid relaxation of quenching after the gap treatment than shade leaves. This is consistent with other measurements of a slower development of nonphotochemical quenching in shade leaves (Demmig and Björkman, 1987). Temperature also strongly interacted with light and acclimation status in determining the susceptibility to photoinhibition. When leaf temperatures were allowed to rise to 40°C during the gap treatment, as occurs for *Alocasia* leaves in natural gaps (R. W. Pearcy, unpublished observations), photoinhibition was increased and the recovery slowed. Whereas sun leaves recovered in 24 h from this treatment, shade leaves had not fully recovered after 5 days. The interactive effect of excessive light and high temperature was the important factor since no inhibition occurred at 40°C and a moderate but still saturating PFD (375 μmol photons m^{-2} s^{-1}) for the shade leaves.

With successive, daily gap treatments there was, however, evidence for an acclimation-based increase in resistance to photoinhibition in shade leaves even when leaf temperatures were allowed to rise to 40°C. The variable to maximum fluorescence ratio (F_v/F_m), which is highly correlated with quantum yield (Björkman and Demmig, 1987), recovered from an initial value on the first day of 0.52, indicating strong photoinhibition, to 0.75 by Day 10. The latter was only slightly lower than F_v/F_m ratios of shade leaves before the gap treatment (0.82). New leaves that expanded after the gap treatment commenced had F_v/F_m ratios of 0.82, a value indicating no inhibition. Light-saturated assimilation rates of leaves fully developed before the treatment commenced also recovered from the initial strong inhibition, but to a lesser extent than F_v/F_m. Leaves completing development after the gap treatment commenced exhibited substantial increases in photosynthetic capacity. This experiment shows that even though recovery from photoinhibition does not occur overnight following the first gap treatment, the effects of successive exposures are not additive. It can be postulated that acclimation of either the repair or protective mechanisms occurred, allowing both recovery and increased resistance to photoinhibition in later exposures. This was not related to photosynthetic capacity, which actually declined, but may be related to an increase in the pool of xanthophyll-cycle carotenoids. Sun leaves have been shown to contain about four times more xanthophyll-cycle carotenoid concentrations than shade leaves (Thayer and Björkman, 1990; Demmig-Adams and Adams, 1992a,b). Moreover, plants capable of recovery from photoinhibition after a sudden sustained increase in PFD also show increases in xanthophyll-cycle pigments that coincide with the recovery (Demmig-Adams et al., 1989b). The capacity to increase thermal dissipation in shade leaves may be important in the initial days after formation of a gap, especially in species with long-lived leaves. The resulting maintenance of function in these leaves, even if no increase in photosynthetic capacity occurs, is probably critical to the ability to respond to formation of a canopy gap.

V. Dynamics of Acclimation

To date, relatively little is known about the time course of the acclimation response to gap formation. Although the change in PFD is rapid and sustained, it clearly takes some time for an extant leaf to adjust to this change. The dynamics of this adjustment are determined by the capacity of these leaves to increase the concentrations of components that determine the photosynthetic capacity and pigment composition as well as the extent of any photoinhibition and subsequent recovery. Changes brought about by these processes are superimposed on the normal cycle of leaf development and aging so that the overall response is indeed very complex.

Only a few studies have actually examined the dynamics of acclimation following sustained increases or decreases in PFD. For leaves that are able to increase photosynthetic capacity when transferred to high PFD, the time required for completion of acclimation seems to range from about 4 to 14 days in herbs (Chow and Anderson, 1987a; Ferrar and Osmond, 1986; Gauhl, 1979) and 45 days in the evergreen species *Hedera helix* (Bauer and Thöni, 1988). The underlying acclimation involves coordinated changes in the whole system with each of the components, such as Rubisco activity, electron transport capacity, Cyt F content, and Chl a/b ratio, exhibiting similar time courses (Chow and Anderson, 1987a,b). Often some initial photoinhibition is observed followed by an increase in photosynthetic capacity. Conditions such as low nitrogen supply, which increase the susceptibility to photoinhibition (Osmond, 1983), increase the initial photoinhibition and delay or prevent any subsequent increase in photosynthetic capacity (Ferrar and Osmond, 1986). Low N clearly constrains the resources available for increasing protein concentrations and may also influence the balance between photoinhibitory damage and repair. Virus infection also may interfere with the acclimation response by intracellular competition for N (Osmond, 1990). Infected plants may function well in the shade but show symptoms and chronic photoinhibition in the sun.

The picture that emerges from the studies to date is that the dynamic response to an increase in PFD involves a complex interaction between photoinhibitory damage, repair, and photoprotection on one hand and acclimation of factors governing photosynthetic capacity on the other. Conditions that exacerbate photoinhibition, such as low nitrogen supply, high temperatures, or water stress (Gauhl, 1979; Björkman and Powles, 1984), shift the dynamic balance between these processes (Anderson and Osmond, 1987). If the limitations by other stresses are severe or the light increase is very large, then acclimation may be prevented and the net result is a sustained photoinhibition. Individual species differences also become important, since in plants such as *Alocasia* that are unable to increase photosynthetic capacity in fully developed leaves the response will only be an initial photoinhibition followed, if possible, by a recovery as the capacity for repair or photoprotection increases. The role of this dynamic balance in determining the differences between species that appear to differ in acclimation capacity, such as early- and late-successional species (Bazzaz and Carlson, 1982), needs attention.

At the whole-plant level, dynamic responses in individual leaves interact with the dynamic processes of leaf production and aging (Figures 3 and 4). Species such as *Alocasia* that lack the capacity to increase photosynthetic capacity of leaves once they are fully developed must necessarily depend on leaf production and turnover to bring about full acclimation. However, the acclimation response is rarely as large in fully developed leaves transferred between light environments as in leaves developing in contrasting light environments, so probably all species depend on leaf production and turnover for expression of their full acclimation

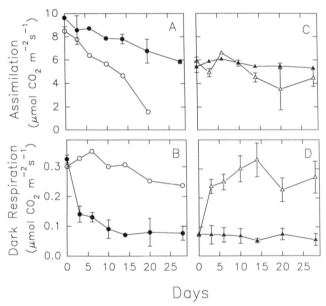

Figure 3 Effect of a transfer of *Alocasia macrorrhiza* plants from low to high or high to low PFD on light-saturated assimilation rate (A and C) and dark respiration rate (B and D). The leaves were fully developed at the time of transfer on Day 0. The left panels show the time courses for leaves transferred from high to low PFD (●) or remaining in high PFD (○). The right panels show time courses for leaves transferred from low to high PFD (△) or remaining in low PFD (▲). Each point is the mean ± 1 S.E. for three leaves.

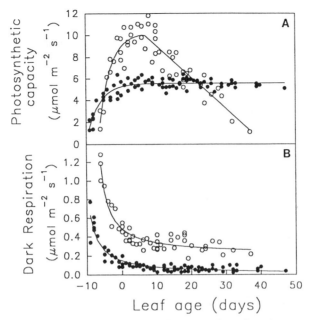

Figure 4   The dependence of (A) photosynthetic capacity and (B) dark respiration rate on leaf age for *Alocasia macrorrhiza* plants grown in sun (○) and shade (●) conditions. Leaf age is defined relative to the day that full expansion was reached (Day 0). (From Sims and Pearcy, 1991).

potential. Leaf longevity is influenced by the light environment, with shade leaves typically having longer leaf life spans than sun leaves on the same species (Chabot and Hicks, 1982). In *Alocasia,* sun leaves lived for about 40 days whereas shade leaves lived for 90+ days (Sims and Pearcy, 1992). In fact, the higher photosynthetic capacity of sun leaves as compared to shade leaves was sustained for only a short time. Photosynthetic capacity of sun leaves began to decline almost immediately after full expansion so that by 20 days and later, it was actually lower than in shade leaves. Sun plants, however, had about twice the rate of leaf production, partially compensating for the more rapid leaf senescence. Differences in leaf longevity and senescence patterns in the sun and shade greatly influence the return on investment in these different environments (Jurik and Chabot, 1986). There is no clear understanding of the interactions between the dynamics of leaf production and senescence and the dynamics of photosynthetic capacity and photoinhibition within the leaf. More studies involving a wider ecological range of species are needed.

VI. Costs and Benefits of Light Acclimation

A. Allocation of Resources in Sun- and Shade-Acclimated Plants

If a leaf possessing a particular suite of photosynthetic characteristics is to provide a benefit to a plant in a given environment, then clearly the carbon acquired by that leaf must exceed the costs associated with the leaf. Acclimation to a specific light environment should increase the amount that the benefits exceed the costs. Although this principle underlies much of the work on the ecological role of acclimation (Björkman, 1981; Givnish, 1988), there are few examples where an explicit accounting of the costs and benefits has been made. Most work has focused on nitrogen partitioning either between light harvesting and carbon metabolism enzymes (Evans, 1986, 1989a,b) or between leaves within a canopy (Field, 1988; Hirose and Werger, 1987). These studies reveal a tendency toward partitioning that in each instance maximizes carbon gain. Carbon costs themselves in sun and shade environments have ironically received less detailed attention even though they make up the majority of the costs of plant structure (Penning de Vries, 1974).

Figure 5 shows the distribution of N and the total construction costs (grams of glucose required per gram of leaf constructed) as measured using the calorimetric technique of Williams *et al.* (1987) for sun and shade *Alocasia* plants. Construction costs include the costs of the carbon skeletons, respiratory costs of biosynthesis, as well as the costs of nitrate and sulfate reduction required for producing the biomass. Estimates for light harvesting and the photosynthetic carbon reduction (PCR) cycle were based on the construction costs of protein and lipid (Penning de Vries, 1974). The sizes of the pies are proportional to the leaf area ratio of the plants, and therefore reflect the total amount of leaf area produced per unit of cost or per unit of N in the respective environment. On a per unit area basis, sun leaves were 80% more costly than shade leaves. Leaf lamina (major veins were lumped with petiole) were 50% of the cost and contained 68% of the total plant N in the shade. The proportions of total plant cost and N that were in leaves were each only 13% greater in shade than in sun leaves. Greater partitioning to leaves therefore accounted for much less of the increase in leaf area than the decrease in LMA that occurred within leaves. This result is consistent with findings for other shade-tolerant as well as shade-intolerant species (Hiroi and Monsi, 1963; G. C. Evans and Hughes, 1961; Poorter and Remkes, 1990).

The relative investments in the major components of the photosynthetic apparatus clearly shifted with acclimation to high light from light harvesting to ET and PCR enzymes as previously shown by Evans (1989c), but in total were a minority of the cost of leaf lamina in either environment. In the shade, ET and PCR enzymes accounted for 35% of the leaf

Shade Plant   Sun Plant

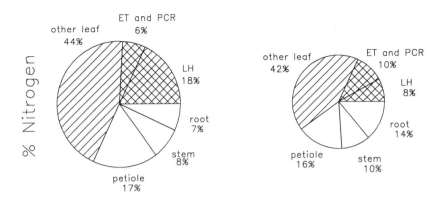

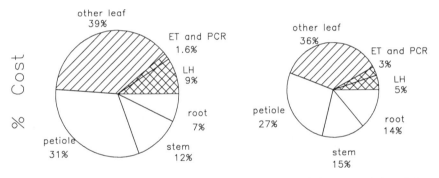

Figure 5   Pie charts showing the proportional allocation of nitrogen and the relative costs of plant components in sun- and shade-grown *Alocasia macrorrhiza* plants. The distribution of N among components of the photosynthetic apparatus in *Alocasia* leaves is based on data from Seemann *et al.* (1987) and Evans (1989c). The costs are based on the ratios of chlorphyll to protein in the LHCP complexes, and chlorophyll to lipid in thylakoid membranes (Anderson, 1986), and the average costs of lipids and proteins (Penning de Vries, 1974). The whole-plant costs are from Sims and Pearcy (1993).

N and 21% of the leaf costs. In the sun, the proportions decreased to 30% of the leaf N and 18% of the leaf costs. The majority of both leaf N and leaf cost was in the "other" category, which included all other cellular functions (including those of the chloroplast) and structural components, such as cell wall. *Alocasia* appears to invest less of its N in photosynthetic enzymes and LHCP than reported for typical sun plants (Evans and Seemann, 1989). Rubisco is only 9% of the total leaf N in sun *Alocasia* leaves whereas it can be 20% or more of the N in sun *Phaseolus vulgarus* leaves (Seemann *et al.*, 1987). Low investments in Rubisco, coupled with a low specific activity for this enzyme in *Alocasia* (Seemann, 1989), may be a major limitation on the maximum photosynthetic capacity expressed by this species.

Petiole investment per unit of leaf weight supported was greater in sun than in shade plants (Sims *et al.*, 1993), but since there was less allocation to leaves in sun plants the allocation to petioles was also somewhat less. Petiole investment scaled closely with the weight of the lamina supported when leaves of different sizes were compared, and therefore was consistent with the biomechanical requirement for support (Givnish, 1986; Wainwright *et al.*, 1976).

Allocation to roots in sun plants of *Alocasia* was found to be twice that of shade plants. A decrease in root/shoot ratio with acclimation to low light has been so widely reported that it may be considered to be almost a universal response (Blackman and Wilson, 1954; Björkman, 1981; Peace and Grubb, 1982; Rice and Bazzaz, 1989). Nutrients were well supplied to the *Alocasia* plants and hence were probably not limiting to growth, so differences in allocation to roots probably reflect the different transpirational demand. Daily transpiration on a leaf area basis was fourfold greater in sun than in shade plants because of their higher g_s and the higher leaf temperatures in the sun. However, on a weight basis, transpiration rates were only twice as high in sun than in shade plants because of the lower leaf area ratio of the former. Consequently, when viewed in terms of dry weight investment, the doubling of transpiration matched the twofold greater proportional allocation to roots in sun plants.

B. The Carbon Balance of Sun- and Shade-Acclimated Leaves

Given the construction costs (Figure 5), the respiration rates (Figure 3), and photosynthetic characteristics (Figure 2), a carbon balance can be calculated for sun and shade *Alocasia* leaves that yields an estimate of both the total cumulative carbon flux and the return on investment (Figure 6). We assumed that the construction costs remain fixed after leaf expansion for each leaf type but used the leaf aging patterns characteristic for each light environment (see Figure 4) to specify the change in photosynthetic capacity with time. We ignore in these calculations any possible effects of stress that might be expected to reduce carbon gain of shade leaves in high light.

As shown in Figure 6, the cumulative return on investment in low light was about 60% greater for shade-type than for sun-type leaves. Sun-type leaves in shade barely paid back their cumulative construction and maintenance costs by 100 days, whereas shade-type leaves earned a

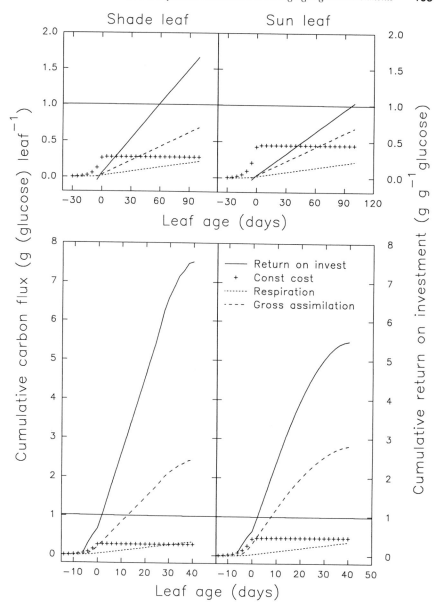

Figure 6 Simulations of the cumulative carbon fluxes in glucose equivalents [g (glucose) m^{-2} leaf] for gross assimilation, respiration, and construction cost for shade (left panels) and sun (right panels) leaves in shade (top: 0.5 mol photons m^{-2} day^{-1} and sun (bottom: 25 mol photons m^{-2} day^{-1}) light environments. The cumulative return on investment is shown by the solid line and the right axis. Leaf temperature was set to 25°C in all simulations. The horizontal line at a cumulative return on investment equal to one is the break-even point beyond which a net carbon profit is earned.

net profit after 60 days. In the shade, the rate of light absorption per unit area determines the photosynthetic rate, which was therefore identical for both leaf types. On a leaf weight or unit construction cost basis, however, sun leaves had a lower photosynthetic rate and consequently a lower return on investment in low light than did shade leaves. We assumed for the purposes of these calculations that the sun- and shade-type leaves had the same maintenance respiration rate (see Figure 3). The differences in return on investment evident in Figure 6 can therefore be attributed solely to the greater construction costs of sun leaves. If it is assumed that maintenance respiration rates differ between sun and shade plants in proportion to their investment in photosynthetic capacity (i.e., 1.25% of photosynthetic capacity as suggested for Figure 5), then the sun-leaf type would require a few days longer to reach the break-even point. However, the relative performances of sun- and shade-type leaves in the shade environment are still dominated by the effects of the different construction costs per unit area.

In the sun environment, carbon gain per unit area was higher for sun-type leaves (Figure 6, bottom). However, the higher construction costs per unit area of these leaves offset the higher carbon gain, so that the cumulative return on investment was initially the same as that of the shade-type leaves. As the leaves aged, the relative decrease in photosynthetic capacity was greater in the sun-type leaf. Consequently, the marginal return on investment, as reflected in the slope of the curve, declined more rapidly with age in the sun-type than in the shade-type leaf.

VII. Scaling Up the Consequences of Acclimation to Whole-Plant Performance

Scaling up the acclimation response to whole plants can be done either by direct measures or by a modeling approach. Measurements of whole-plant gas exchange of plants shifted to a new environment provide an overall estimate of performance (Rice and Bazzaz, 1989), but unfortunately do not allow the specific contribution of any particular characteristic to be determined. The modeling approach, on the other hand, can be utilized to examine the effects of particular characteristics (Jurik and Chabot, 1986). These approaches are complementary since the model predictions for whole-plant performance can be compared against the measured whole-plant performance as a validation. We have therefore undertaken both approaches.

Whole-plant net assimilation rates as a function of daily PFD for *Alocasia* sun and shade plants are shown in Figure 7. Measurements of whole-plant gas exchange over 24-h periods were made beginning just

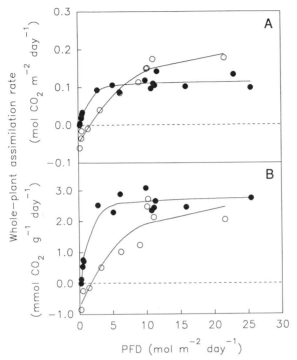

Figure 7 Dependence of whole-plant CO_2 exchange on the daily PFD for sun- (○) and shade-grown (●) plants. Each point represents a measurement of daily CO_2 exchange (roots and shoots combined) for a single plant transferred from the respective growth environment the prior evening. Different light environments were created by placing layers of shadecloth and cheesecloth over the whole-plant assimilation chambers located in the greenhouse.

after sunset on plants enclosed in a 30-liter chamber. The chamber and gas exchange system was located in the greenhouse and different daily PFDs were obtained by adding layers of shadecloth or cheesecloth. Plants were transferred from sun or shade growing conditions in the same greenhouse. Each data point in Figure 7 is the net daily gas exchange of an individual plant. Whole-plant assimilation rates per unit leaf area (Figure 7A) were higher in shade plants at a low daily PFD and in sun plants at a high daily PFD. On a per unit weight basis (Figure 7B), shade plants performed better than sun plants in low PFD because of a higher ratio of leaf area to total plant weight (leaf area ratio;LAR). Differences in respiration contributed to the differences in net photosynthetic performance at low PFD but, as discussed earlier, these were primarily due to the high carbohydrate status of sun plants at the time of transfer.

Thus, in agreement with the results for single leaves (Figure 6), sun plants had no advantage in terms of carbon gain per unit of investment. Shade plants performed better in low PFD because of their greater LAR. In high PFD their greater LAR also more than compensated for the low photosynthetic capacities of shade leaves.

We used an empirical model of *Alocasia* photosynthesis, partitioning, and relative growth rate (RGR) to further examine the consequences of acclimation, and specifically the contribution of individual phenotypic characteristics, for whole-plant performance in different light environments. A full description of this model is beyond the scope of this chapter and the reader is referred to Sims *et al.* (1993) for details. It suffices to say here that the model uses empirically derived relationships between light environment and acclimation determined for *Alocasia* to first predict daily carbon gain and then to partition this to support growth of leaves, support structures, and roots. LMA is a central driving variable since it depends strongly on the light environment (Figure 2F) and, in turn, is a strong predictor of the maximum photosynthetic capacity achieved by *Alocasia* leaves in different environments. The effects of development and aging were also described by empirical relationships that caused light-saturated assimilation rate to increase asymptotically during expansion and then decrease as leaf life span was approached. These simple relationships closely simulated the variations in photosynthetic capacity with leaf age (Figure 4). The effects of PFD and temperature on the photosynthetic rate were described by empirical relationships. Leaf temperature was predicted from the energy balance and stomatal conductance from the empirical relationship of Ball *et al.* (1986) as parameterized for *Alocasia* by Kirschbaum *et al.* (1988). Leaf boundary layer conductance was predicted from leaf dimensions and wind speed. Using these relationships it was possible to predict the photosynthesis and transpiration rate of the leaves in the canopy that in total simulated the actual responses observed in the whole-plant chamber.

Once a daily pool of carbon was acquired in the model, it was partitioned to achieve balance between leaves, support, and roots. Partitioning to roots was based on the allometric relationship between root biomass and daily transpiration. Once sufficient carbon was allocated to roots to meet the transpiration demand as set by the environment, leaf area, and stomatal conductance, the remainder was allocated to construction of new leaf and petiole. Partitioning of carbon between petiole and leaf was based on the measured allometric relationships. New leaves were produced in the model at set intervals depending on the light environment, with the weight of the leaf determined by the pool of available carbon. Thus leaf size generally scaled with plant size, in agreement with observations of young *Alocasia* plants. Clearly these relationships say nothing about the actual mechanisms underlying partitioning and their control but they did indeed allow a good simulation of the relative growth rates achieved by *Alocasia* in high and low light.

In real plants, LAR, photosynthetic capacity, maintenance respiration rates, and partitioning are, of course, all linked during acclimation. The model, however, is not so constrained and therefore could be used to calculate the sensitivity of RGR to both the combined (acclimation) and the separate effects (LAR, photosynthetic capacity, maintenance respiration rate) of each. The results of this sensitivity analysis are shown in Figure 8. Photosynthetic capacity per unit area greatly affected carbon gain and consequently growth in high light, but, as expected, little affected growth and carbon gain when light limited photosynthesis. Maintenance respiration rates had a greater effect on growth in shade, as opposed to sun, environments, whereas LAR has a similar effect in both environments. Maintenance respiration is proportional to biomass but not to daily carbon gain, so its impact will be greater when total daily carbon gain per unit of biomass is small. In contrast, a change in LAR, with the rate of carbon gain per unit area held constant, affects the ratio of investment to return. Doubling the LAR will double the return on investment, whatever the net daily carbon gain.

The acclimation response shows the effect of linkage between pheno-

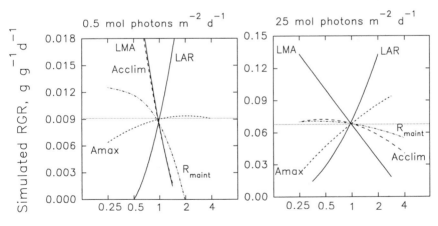

Figure 8 Sensitivity of modeled relative growth rate (RGR) to variation in photosynthetic capacity (A_{max}), maintenance respiration rate (R_{maint}), leaf mass per unit leaf area (LMA), and leaf area ratio (LAR) in sun and shade light environments. Values of one on the abscissa correspond to the values of the measured parameters for plants acclimated to the respective light environment. The acclimation response is the combined effects of modeled changes in LMA, A_{max} and R_{maint}.

typic characteristics observed in sun and shade plants. In this simulation, LMA was varied, but since photosynthetic capacity was a function of LMA, it also varied. LMA also determined how much leaf area was produced (and with what stomatal conductance), which in turn caused variation in partitioning to roots and support. In effect, the simulations changed the relationship between LMA and growth light environment, but all the consequences of a particular LMA for photosynthetic capacity, partitioning, and maintenance remained unchanged. It should be emphasized that although these empirical relationships between LMA and photosynthetic capacity and stomatal conductance describe the response of *Alocasia*, it would not necessarily hold for other species. In particular, species in which it is possible to have acclimation of mature leaves and hence different relationships between photosynthetic capacity and LMA would require a different approach. In these species, there may be more of an apparent advantage to acclimation in high PFD.

In low PFD (0.5 mol photons $m^{-2} day^{-1}$), where photosynthetic capacity had little influence on carbon gain, the combined acclimation response was nearly identical to that observed for LMA alone. A decrease in LMA (and changes in the associated parameters) beyond that actually observed would increase RGR. There are presumably limits on how thin leaves can be, which, along with other limits imposed by partitioning and support, constrain acclimation. These constraints are of course not part of the model. In high PFD (25 mol photons $m^{-2} day^{-1}$), a different picture emerges. Here, a decrease toward a shade phenology caused almost no increase in RGR whereas an increase caused a moderate decline in RGR. Changes in LMA and hence leaf area were almost exactly offset by changes in photosynthetic capacity.

VIII. Discussion

Our studies on acclimation in *Alocasia* show that when acclimation is considered on scales ranging from single leaves to whole plants, different perspectives develop in different parts of this scale. At the single-leaf level, and when instantaneous photosynthetic rates per unit leaf area are considered, the substantial changes in photosynthetic capacity and their underlying biochemical basis stand out. Indeed, much of the attention given to acclimation has focused on the underlying basis of these changes in photosynthetic capacity, which is now known to result from highly coordinated changes at the biochemical level as well as from changes in leaf structure. However, our work shows that while acclimation provided clear benefits in low PFD, the higher photosynthetic rates per unit leaf area of sun-acclimated leaves of *Alocasia* did not equate with a greater return on investment in high PFD. Similarly, when acclimation was considered at the whole-plant level, it provided a clear benefit in the shade environment but no direct benefit in terms of whole-plant carbon gain in the high-light environment. Other studies have also shown that the relative growth rate or whole-plant photosynthetic rate in high PFD of shade-acclimated plants can, in the short term at least, actually be higher than those of sun-acclimated plants (Hughs, 1966; Rice and Bazzaz, 1989).

A picture of lessened significance of the changes in photosynthetic rate per unit area also emerges when the relative costs incurred to change different components are considered. Overall, the direct construction costs of the photosynthetic apparatus are a small fraction (5–9%) of the total plant costs. Therefore, a doubling would be possible with relatively minor effects elsewhere. However, as shown by Evans (1989a), relatively little benefit but a substantial cost would accrue to increasing chlorophyll content in either the sun or shade. The costs of electron transport components and PCR-cycle enzymes are an even smaller fraction of the total and, if possible, a doubling of these components could lead to substantial increases in photosynthetic capacity. The maintenance costs of increased photosynthetic capacity may also be small. Costs in terms of nitrogen are clearly a larger fraction of the total plant costs but even for these, the direct costs of reduction and metabolism do not translate into high overall construction costs for increased photosynthetic capacity. More significant than either the direct costs of increasing photosynthetic capacity, which occurs largely through increased investment of nitrogen, or those of maintenance of this capacity, might be the higher transpiration rates and higher nitrogen demand, which would require a higher allocation to roots, and thus less resources for production of leaf area.

The importance of LAR in the acclimation response of *Alocasia* to different light environments is clearly evident. Increased LAR is far more significant than increased Chl per unit leaf area in increasing the total light absorbed. Moreover, the simulations showed that increased LAR significantly enhanced growth in the shade but also effectively compensated for the lower photosynthetic capacity when shade plants were measured in high PFD. The single most important factor in LAR of *Alocasia* was the trade-off between the production of low LMA and consequently greater leaf area as compared to high LMA and lesser leaf area per plant. The importance of LAR in relative growth rate has been emphasized in comparisons of fast- and slow-growing plants (Poorter *et al.*, 1990; Poorter and Remkes, 1990; Körner, 1991).

If the high LAR of shade plants significantly improves their performance in the shade but is also a benefit in terms of growth and whole-plant photosynthesis in the sun, then should not sun plants be like shade plants? The answer to this question must lie not in the photosynthetic behavior itself, but possibly in the relationships between LAR and resource demands and between LMA and stress resistance. Thicker leaves and hence a lower LAR may confer resistances to stresses that may be more prevalent in high light. The experiments were done under PFDs that saturated photosynthesis but not excessively. PFDs and leaf temperatures in gaps would typically reach higher values. Higher photosynthetic capacities may confer greater resistance to photoinhibition, which is exacerbated by high leaf temperatures in gaps. Higher transpiration is required for maintaining favorable leaf temperatures, but since A_{max} depends on LMA, the increased g_s at low LMA would result in a much lower water use efficiency, and clearly require greater investment in roots. Thus, a primary function of acclimation to high-light environments may be increased resistance to stresses usually associated with this environment.

The sensitivity arguments presuppose that A_{max}, LMA, and LAR are independent. Clearly LMA and LAR are inversely related and A_{max} is directly dependent on LMA. Similarly, partitioning appears to depend on the demand for resources, which is in turn a function of A_{max} and LMA. In reality, therefore, a change in one parameter constrains all others. An increase in leaf photosynthetic capacity during acclimation to high PFD necessitates a decrease in LAR, partially or even wholly offsetting the carbon gain benefit. The linkage may be particularly strong in species such as *Alocasia*, which have relatively little flexibility in adjusting the photosynthetic capacity per unit cell volume. In species with more of a capacity for acclimation of photosynthetic capacity at a cellular level, the linkage may be weaker. Nevertheless, even in these plants, changes in root to shoot ratio and ultimately developmental changes in LMA under constant PFD will act in a similar manner as in *Alocasia*.

References

Anderson, J. M. (1986). Photoregulation of the composition, function and structure of thylakoid membranes. *Annu. Rev. Plant Physiol.* **37,** 93–136.

Anderson, J. M., and Osmond, C. B. (1987). Shade–sun responses: Compromises between acclimation and photoinhibition. *In* "Photoinhibition" (D. J. Kyle, C. B. Osmond, and C. J. Arntzen, eds.), pp. 1–38. Elsevier, Amsterdam.

Anderson, J. M., Goodchild, D. J., and Boardman, N. K. (1973). Composition of the photosystems and chloroplast structure in extreme shade plants. *Biochim. Biophys. Acta* **325,** 573–585.

Azcon-Bieto, J., and Osmond, C. B. (1983). Relationship between photosynthesis and respiration. The effect of carbohydrate status on the rate of CO_2 production by respiration in darkened and illuminated wheat leaves. *Plant Physiol.* **71,** 574–581.

Azcon-Bieto, J., Lambers, H., and Day, D. A. (1983). Effect of photosynthesis and carbohydrate status on respiratory rates and the involvement of the alternative pathway in leaf respiration. *Plant Physiol* **71**, 574–581.

Ball, J. T., Woodrow, I. E., and Berry, J. A. (1986). A model predicting stomatal conductance and its contribution to the control of photosynthesis under different environmental conditions. *Prog. Photosynth. Res.* **4**, 221–224.

Bauer, H., and Thöni, W. (1988). Photosynthetic light acclimation in fully developed leaves of the juvenile and adult life cycle phases of *Hedera helix*. *Physiol. Plant.* **73**, 31–37.

Bazzaz, F. A., and Carlson, R. W. (1982). Photosynthetic acclimation to variability in the light environment of early and late successional plants. *Oecologia* **54**, 313–316.

Björkman, O. (1973). Comparative studies of photosynthesis in higher plants. *Curr. Top. Photobiol., Photochem. Photophysiol.* **8**, 1–63.

Björkman, O. (1981). Responses to different quantum flux densities. *In* "Encyclopedia of Plant Physiology" (O. L. Lange, P. S. Nobel, C. B. Osmond, and H. Ziegler, eds.), New Ser., Vol. 12A, pp. 57–107. Springer-Verlag, Berlin.

Björkman, O., and Demmig, B. (1987). Photon yield of O_2 evolution and chlorophyll fluorescence characteristics at 77K among vascular plants of diverse origins. *Planta* **170**, 489–504.

Björkman, O., and Powles, S. B. (1984). Inhibition of photosynthetic reactions under water stress: Interaction with light level. *Planta* **161**, 490–504.

Blackman, G. E., and Wilson, G. L. (1954). Physiological and ecological studies in the analysis of plant environment. IX. Adaptive changes in the vegetative growth and development of *Helianthus annuus* induced by an alteration in light level. *Ann. Bot. (London)* [N.S.] **18**, 71–94.

Boardman, N. K. (1977). Comparative photosynthesis of sun and shade plants. *Annu. Rev. Plant Physiol.* **28**, 355–377.

Chabot, B. F., and Hicks, D. J. (1982). The ecology of leaf life spans. *Annu. Rev. Ecol. Syst.* **13**, 229–259.

Chow, W. S., and Anderson, J. M. (1987a). Photosynthetic responses of *Pisum sativum* to an increase in irradiance during growth. I. Photosynthetic activities. *Aust. J. Plant Physiol.* **14**, 1–8.

Chow, W. S., and Anderson, J. M. (1987b). Photosynthetic responses of *Pisum sativum* to an increase in irradiance during growth. II. Thylakoid membrane components. *Aust. J. Plant Physiol.* **14**, 9–19.

Chow, W. S., Qian, L., Goodchild, D. J., and Anderson, J. M. (1988). Photosynthetic acclimation of *Alocasia macrorrhiza* (L.) G. Don to growth irradiance: Structure, function and composition of chloroplasts. *Aust. J. Plant Physiol.* **15**, 107–122.

Chu, Z.-X., and Anderson, J. M. (1984). Modulation of the light harvesting assemblies in chloroplasts of a shade plant, *Alocasia macrorrhiza*. *Photobiochem. Photobiophys.* **8**, 1–10.

Davies, E. C., Chow, W. S., LeFay, J. M., and Jordan, B. (1986). Acclimation of tomato leaves to changes in light intensity: Effects on the function of the thylakoid membrane. *J. Exp. Bot.* **37**, 211–220.

Demmig, B., and Björkman, O. (1987). Comparison of the effect of excessive light on chlorophyll fluorescence characteristics at 77K among vascular plants of diverse origins. *Planta* **171**, 171–184.

Demmig, B., Winter, K., Kruger, A., and Czygan, F.-C. (1987). Photoinhibition and zeaxanthin formation in intact leaves. A possible role of the xanthophyll cycle in the dissipation of excess light energy. *Plant Physiol.* **84**, 218–224.

Demmig-Adams, B., and Adams, W. W. (1992a). Photoprotection and other responses of plants to high light stress. *Annu. Rev. Plant Physiol. Plant Mol. Biol.* **43**, 599–626.

Demmig-Adams, B., and Adams, W. W. (1992b). Carotenoid composition in sun and shade leaves of plants with different life forms. *Plant, Cell Environ.* **15**, 411–419.

Demmig-Adams, B., Winter, K., Winkelmann, E., Kruger, A., and Czygan, F.-C. (1989). Photosynthetic characteristics and the ratios of chlorophyll, β-carotene and the components of the xanthophyll cycle upon sudden increase in growth light regime in several plant species. *Bot. Acta* **102**, 319–325.

Evans, G. C., and Hughes, A. P. (1961). Plant growth and the areal environment. I. Effects of artificial shading on *Impatiens parviflora*. *New Phytol.* **60**, 150–180.

Evans, J. R. (1986). A quantitative analysis of light distribution between the two photosystems, considering variation in both the relative amounts of the chlorophyll–protein complexes and the spectral quality of light. *Photobiochem. Photobiophys.* **10**, 135–147.

Evans, J. R. (1987). The relationship between electron transport components and photosynthetic capacity in pea leaves grown at different irradiances. *Aust. J. Plant Physiol.* **14**, 157–170.

Evans, J. R. (1988). Acclimation by the thylakoid membranes to growth irradiance and the partitioning of nitrogen between soluble and thylakoid proteins. *Aust. J. Plant Physiol.* **15**, 93–106.

Evans, J. R. (1989a). Photosynthesis—The dependence on nitrogen partitioning. *In* "Causes and Consequences of Variation in Growth Rate and Productivity of Higher Plants" (H. Lambers, M. L. Cambridge, H. Konings, and T. L. Pons, eds.), pp. 159–174. SPB Academic Publishing, The Hague, The Netherlands.

Evans, J. R. (1989b). Partitioning of nitrogen between and within leaves grown under different irradiances. *Aust. J. Plant Physiol.* **16**, 533–548.

Evans, J. R. (1989c). Photosynthesis and nitrogen relationships in leaves of C$_3$ plants. *Oecologia* **78**, 9–19.

Evans, J. R., and Seemann, J. R. (1989). The allocation of protein nitrogen in the photosynthetic apparatus: Costs, consequences and control. *In* "Photosynthesis" (W. R. Briggs, ed.), pp. 183–205. Alan R. Liss, New York.

Farquhar, G. D., and Sharkey, T. D. (1982). Stomatal conductance and photosynthesis. *Annu. Rev. Plant Physiol.* **33**, 317–345.

Farquhar, G. D., and von Caemmerer, S. (1982). Modelling of photosynthetic response to environmental conditions. *In* "Encyclopedia of Plant Physiology" (O. L. Lange, P. S. Nobel, C. B. Osmond, and H. Ziegler, eds.), New Ser., Vol. 12B, pp. 549–588. Springer-Verlag, Berlin.

Ferrar, P. J., and Osmond, C. B. (1986). Nitrogen supply as a factor influencing photoinhibition and photosynthetic acclimation after transfer of shade-grown *Solanum dulcamara* to bright light. *Planta* **168**, 563–570.

Field, C. B. (1988). On the role of photosynthetic responses in constraining the habitat distribution of rainforest plants. *Aust. J. Plant Physiol.* **15**, 343–358.

Gauhl, E. (1979). Sun and shade ecotypes of *Solanum dulcamara* L.: Photosynthetic light dependence characteristics in relation to mild water stress. *Oecologia* **39**, 61–70.

Givnish, T. J. (1986). Biomechanical constraints on crown geometry in forest herbs. *In* "On the Economy of Plant Form and Function" (T. J. Givnish, ed.), pp. 525–584. Cambridge Univ. Press, Cambridge, UK.

Givnish, T. J. (1988). Adaptation to the sun and shade: A whole-plant perspective. *Aust. J. Plant Physiol.* **15**, 63–92.

Grime, J. P. (1966). Shade avoidance and shade tolerance in flowering plants. *In* "Light as an Ecological Factor" (G. C. Evans, R. Bainbridge, and O. Rackham, eds.), pp. 187–207. Blackwell, Oxford.

Hiroi, T., and Monsi, M. (1963). Physiological and ecological analysis of shade tolerance of plants. 3. Effect of shading on growth attributes of *Helianthus annuus*. *Bot. Mag.* **77**, 121–129.

Hirose, T., and Werger, M. A. (1987). Maximizing daily canopy photosynthesis with respect to the leaf nitrogen pattern in the canopy. *Oecologia* **72**, 520–526.

Hughs, A. P. (1966). The importance of light compared with other factors affecting plant growth. *In* "Light as an Ecological Factor" (G. C. Evans, R. Bainbridge, and O. Rackham, eds.), pp. 121–145. Blackwell, Oxford.

Jurik, T. W., and Chabot, B. F. (1986). Leaf dynamics and profitability in wild strawberries. *Oecologia* **69**, 296–304.

Kamaluddin, M., and Grace, J. (1992). Photoinhibition and light acclimation in seedlings of *Bischofia javanica*, a tropical forest tree from Asia. *Ann. Bot. (London)* [N.S.] **69**, 47–52.

Kirschbaum, M. U. F., Gross, L. J., and Pearcy, R. W. (1988). Observed and modelled stomatal responses to dynamic light environments in the shade plant *Alocasia macrorrhiza*. *Plant, Cell Environ.* **11**, 111–121.

Körner, C. (1991). Some often overlooked plant characteristics as determinants of plant growth—A reconsideration. *Funct. Ecol.* **5**, 162–173.

Körner, C., Scheel, J. A., and Bauer, H. (1979). Maximum leaf diffusive conductance in vascular plants. *Photosynthetica* **13**, 45–82.

Krause, G. H., and Weiss, E. (1991). Chlorophyll fluorescence and photosynthesis: The basics. *Annu. Rev. Plant Physiol. Plant Mol. Biol.* **42**, 313–349.

Lambers, H. (1985). Respiration in intact plants and tissues: Its regulation and dependence on environmental factors, metabolism and invaded organisms. *In* "Encyclopedia of Plant Physiology" (R. Douce and D. A. Day, eds.), New Ser., Vol. 18, pp. 418–473. Springer-Verlag, Berlin.

Leong, T.-Y., and Anderson, J. M. (1984a). Adaptation of the thylakoid membranes of pea chloroplasts to light intensities. I. Study on the distribution of chlorophyll protein complexes. *Photosynth. Res.* **5**, 105–115.

Leong, T.-Y., and Anderson, J. M. (1984b). Adaptation of the thylakoid membranes of pea chloroplasts to light intensities. II. Regulation of electron transport capacities, electron carriers, coupling factor CF, activity and rates of photosynthesis. *Photosynth. Res.* **5**, 117–128.

Lichtenthaler, H. K., Buschmann, C., Doll, M., Fietz, H.-J., Bach, T., Kozel, U., Meier, D., and Rahmsdorf, U. (1981). Photosynthetic activity, chloroplast ultrastructure, and leaf characteristics of high-light and low-light plants and of sun and shade leaves. *Photosynth. Res.* **2**, 115–141.

Mulkey, S. S., and Pearcy, R. W. (1992). Interactions between acclimation and photoinhibition of photosynthesis of a tropical forest understory herb, *Alocasia macrorrhiza* (L.) G. Don, during simulated canopy gap formation. *Funct. Ecol.* **6**, 719–729.

Osmond, C. B. (1983). Interactions between irradiance, nitrogen nutrition, and water stress in the sun–shade responses of *Solanum dulcamara*. *Oecologia* **57**, 316–321.

Osmond, C. B. (1990). Photosynthesis from the molecule to the biosphere: A challenge for integration. *In* "Photosynthesis" (W. R. Briggs, ed.), pp. 5–17. Alan R. Liss, New York.

Peace, W. J. H., and Grubb, P. J. (1982). Interaction of light and mineral nutrient supply in the growth of *Impatiens parviflora*. *New Phytol.* **90**, 127–150.

Penning de Vries, F. W. T. (1974). Use of assimilates in higher plants. *In* "Photosynthesis and Productivity in Different Environments" (J. Cooper, ed.), pp. 459–480. Cambridge Univ. Press, London.

Poorter, H., and Remkes, C. (1990). Leaf area ratio and net assimilation rate of 24 wild species differing in relative growth rate. *Oecologia* **83**, 553–559.

Poorter, H., Remkes, C., and Lambers, H. (1990). Carbon and nitrogen economy of 24 wild species differing in relative growth rate. *Plant Physiol.* **94**, 621–627.

Powles, S. B. (1984). Photoinhibition of photosynthesis induced by visible light. *Annu. Rev. Plant Physiol.* **35**, 15–44.

Rice, S. A., and Bazzaz, F. A. (1989). Growth consequences of plasticity of plant traits in response to light conditions. *Oecologia* **78**, 508–512.

Sebaa, E. D., Prioul, J. L., and Brangeon, J. (1986). Acclimation of adult *Lolium multiflorum* leaves to changes in irradiance: Effect on leaf photosynthesis and chloroplast ultrastructure. *J. Plant Physiol.* **127**, 431–441.

Seemann, J. R. (1989). Light adaptation/acclimation of photosynthesis and the regulation of ribulose-1,5-bisphosphate carboxylase activity in sun and shade plants. *Plant Physiol.* **91**, 379–386.

Seemann, J. R., Sharkey, T. D., Wang, J. L., and Osmond, C. B. (1987). Environmental effects on photosynthesis, nitrogen-use efficiency, and metabolite pools in leaves of sun and shade plants. *Plant Physiol.* **84**, 796–802.

Sims, D. A., and Pearcy, R. W. (1989). Photosynthetic characteristics of a tropical forest understory herb, *Alocasia macrorrhiza*, and a related crop species, *Colocasia esculenta* grown in contrasting light environments. *Oecologia* **79**, 53–59.

Sims, D. A., and Pearcy, R. W. (1991). Photosynthesis and respiration in *Alocasia macrorrhiza* following transfers to high and low light. *Oecologia* **86**, 447–453.

Sims, D. A., and Pearcy, R. W. (1992). Response of leaf anatomy and photosynthetic capacity in *Alocasia macrorrhiza* (Aeraceae) to a transfer from low to high light. *Am. J. Bot.* **79**, 449–455.

Sims, D. A., Gebauer, R., and Pearcy, R. W. (1993). A model of growth and allocation in *Alocasia macrorrhiza* taking into account photosynthetic acclimation. Submitted.

Terashima, I., and Inoye, Y. (1985). Vertical gradient in photosynthetic properties of spinach chloroplasts. *Plant Cell Physiol.* **26**, 781–785.

Thayer, S. S., and Björkman, O. (1990). Leaf xanthophyll content and composition in sun and shade determined by HPLC. *Photosynth. Res.* **23**, 331–343.

Thompson, W. A., Stocker, G. C., and Kriedemann, P. E. (1988). Growth and photosynthetic response to light and nutrients of *Flindersia brayleyana* F. Muell., a rainforest tree with Broad tolerance to sun and shade. *Aust. J. Plant Physiol.* **15**, 299–315.

von Caemmerer, S., and Farquhar, G. D. (1981). Some relationships between the biochemistry of photosynthesis and the gas exchange of leaves. *Planta* **153**, 376–387.

von Caemmerer, S., and Farquhar, G. D. (1984). Effects of partial defoliation, changes of irradiance during growth, short-term water stress and growth at enhanced p(CO$_2$) on the photosynthetic capacity of leaves of *Phaseolus vulgaris* L. *Planta* **160**, 320–329.

Wainwright, S. A., Biggs, W. D., Currey, J. D., and Gosline, J. M. (1976). "Principles of Mechanical Design in Organisms." Princeton Univ. Press, Princeton, NJ.

Walters, M. B., and Field, C. B. (1987). Photosynthetic light acclimation in two rainforest *Piper* species with different ecological amplitudes. *Oecologia* **72**, 449–456.

Williams, K., Percival, F., Merino, J., and Mooney, H. A. (1987). Estimation of tissue construction cost from heat of combustion and organic nitrogen content. *Plant, Cell Environ.* **10**, 725–734.

Wong, S. C., Cowan, I. R., and Farquhar, G. D. (1979). Stomatal conductance correlates with photosynthetic capacity. *Nature (London)* **282**, 424–426.

Wong, S., Cowan, I. R., and Farquhar, G. D. (1985). Leaf conductance in relation to rate of CO$_2$ assimilation. I. Influence of nitrogen nutrition, phosphorous nutrition, photon flux density and ambient partial pressure of CO$_2$ during ontogeny. *Plant Physiol.* **78**, 821–825.

6

Photosynthetic Utilization of Sunflecks: A Temporally Patchy Resource on a Time Scale of Seconds to Minutes

Robert W. Pearcy, Robin L. Chazdon, Louis J. Gross, and Keith A. Mott

I. Introduction

Leaves within plant canopies or in the understory are subject to rapid fluctuations in light because of the occurrence of sunflecks that cause the photon flux density (PFD) available for photosynthesis to change by as much as 20-fold within a second. Sunflecks are often clustered into periods of when they occur in rapid succession separated by periods of little or no sunfleck activity. In addition to this temporal patchiness, sunflecks are also responsible for much of the spatial patchiness in light availability that has been shown to contribute to spatial variation in growth rates of a number of species (Pearcy, 1983; Oberbauer *et al.*, 1988). Within canopies, sunfleck characteristics also vary along the vertical gradient, adding further to the complexity of the light environment.

Sunflecks are usually received by a leaf in a forest understory for less than 10% of the day, but contribute from as little as 10 to over 70% of daily PFD (Chazdon, 1988). Although sunflecks received early attention as a potentially important factor in the carbon balance of understory plants (Lundegarth, 1921; Evans, 1939), there has been relatively little experimental work designed to examine their actual importance to photosynthetic carbon gain or the mechanisms governing their utilization. This is largely because, until recently, gas-exchange systems responded too slowly to resolve the photosynthetic response to sunflecks. Recent measurements show that utilization of sunflecks can account for a large fraction of the photosynthesis in understory leaves (e.g., Pfitsch and

Pearcy, 1989b). Sunfleck frequency and duration are important determinants of their utilization. Thus, the temporal pattern of sunfleck occurrence as influenced by canopy structure must be considered. In steady-state photosynthesis, the amounts of photosynthetic enzymes, the equilibrium stomatal conductance, and the rate of supply of light are the primary factors determining the rate of leaf photosynthesis. For sunflecks, the dynamic responses of photosynthesis as determined by stomatal movements, light regulation of enzymes, and the accompanying dynamics of metabolite pool sizes are also important. Each of these is characterized by different time constants governing the change in response to a change in PFD and to the temporal pattern of sunflecks.

In this chapter we first briefly review the characteristics of sunflecks relevant to their photosynthetic utilization and then discuss the mechanisms governing their use. We then present results from a model simulating the response to sunflecks that allows determination of how each dynamic component contributes to limiting or enhancing photosynthetic utilization of sunflecks.

II. Sunflecks as a Patchy Resource in Space and Time

A. What Is a Sunfleck?

A sunfleck is created by penetration of the direct solar beam through a canopy gap to a deeper layer, creating a patch of much higher irradiance levels than received by the surrounding area. The nature of direct-beam light penetration through canopies, approaches to quantifying it, and spatial and temporal patterns in relation to canopy structure in various communities are discussed in Baldocchi and Collineau (Chapter 2, this volume). Thus, in this chapter we only briefly review the features of sunflecks that are directly relevant to their utilization in photosynthesis.

As pointed out by Chazdon (1988), a sunfleck is hard to quantify. Sunflecks under tall canopies have a substantial penumbra or partial shadow, causing their boundary from the surrounding diffuse light to be much more indistinct than those under short canopies. The rotation of the earth creates slowly moving sunflecks, whereas leaf fluttering and the waving of plants in the wind create many rapid excursions of PFD on leaves below. Should each of these shorter excursions be called a sunfleck, or should we consider them part of a larger sunfleck defined by the gap? And what should we consider to be the size of a sunfleck given that its boundaries may be fuzzy?

In ecophysiological studies, sunflecks have typically been defined by the excursion of the PFD above some threshold just above the shade light PFD or based on the photosynthetic response of the plants studied

(Pearcy, 1983; Chazdon and Fetcher, 1984; Pearcy *et al.*, 1990; Tang *et al.*, 1988). Once a sunfleck event is determined, it can be further classified in terms of the maximum PFD, the integrated PFD, the sunfleck duration, or time from the previous sunfleck (Pearcy, 1983; Pearcy *et al.*, 1990). Unfortunately, the number of sunfleck events and the duration (or size) of sunflecks depend on the threshold chosen (see Baldocchi and Collineau, Chapter 2, this volume). The same threshold cannot be applied in crops and forest understories or even at different heights in a forest canopy because of differences in levels of diffuse radiation. A second, related problem is that sunflecks should be distinguished from larger sun patches or gap light environments, as these distinctions are both physiologically and ecologically relevant. Chazdon and Field (1987) found no acclimation in photosynthetic capacity within understory microsites that differed primarily in the daily PFD supplied by sunflecks, whereas acclimation was observed among microsites varying in daily PFD within plant canopies in a large gap. Acclimation of either the responses to sunflecks or photosynthetic capacity does not seem to occur in response to different sunfleck frequencies and durations but instead responds to the daily total PFD (Sims and Pearcy, 1993). The longer durations of direct PFD in sun patches or gaps cause temperature and water status changes and other environmental effects that may not accompany most sunflecks (Bazzaz and Wayne, this volume).

Smith *et al.* (1989) have attempted to develop a more rational and objective method of defining sunflecks in terms of gap geometry. In their scheme, sunflecks are those excursions of PFD created by gaps with a gap diameter ratio (GDR = G/D, where G = gap diameter and D = gap height) ≤ 0.01. By contrast, GDRs of 0.01 to 0.05 and >0.05 delineated "sun patches" and "sun gaps," respectively. Sunflecks are often dominated by penumbral effects so that even in the center, the PFD is less than full, direct-beam values. The durations delineating sun gaps, sun patches, and sunflecks depend on canopy height; for sunflecks it will be less than 10 min, even under a tall canopy. Penumbra are much wider under tall than short canopies and therefore sunflecks under these canopies are much larger. "Sun patches" and "sun gaps" are each longer and receive full, direct-beam PFD at the center, but differ in the proportion of the total area occupied by penumbra. These definitions have the advantage of being objective but do ignore canopy movement effects and are thus difficult to apply to real light changes under canopies. Moreover, the canopy creating the gap edges is often at different heights on different sides. The study of Smith *et al.* (1989) does highlight the contribution of penumbra and its potentially different role in sunflecks, sun patches, and sun gaps under contrasting canopy heights. Since penumbral effects result in a spreading of the PFD, yielding a larger area of elevated PFD but lower maximum PFDs (Miller and Norman, 1971a,b), a consequence may be more efficient utilization of the available light (Oker-Blom, 1984; Myneni *et al.*, 1986).

At present there is still no completely satisfactory system for classifying sunflecks. New approaches using wavelet analysis (see Baldocchi and Collineau, Chapter 2, this volume) offer hope that objective methods can be developed for detecting and classifying sunfleck events. However, these methods have not yet been tested under different canopy conditions in the field.

B. How Do Sunflecks Contribute to the Light Environment in and under Canopies?

Most of the understanding of the nature and contribution of sunflecks comes from direct measurement rather than theoretical derivations from canopy structure. Chazdon (1988) has summarized the studies in which the contribution of sunflecks to forest understory light environments was estimated. These ranged from 10 to 78% of the daily PFD. Some of the variation is clearly due to differences among forest types and weather conditions at the time of measurement, but measurements in different microsites within a forest on a single day yielded a similar range of variation (R. W. Pearcy and C. B. Field, unpublished results from a Queensland rain forest). Only two studies have been done within crop or shorter herbaceous canopies; these show a somewhat higher mean contribution of sunflecks, and also a high spatial variability (Pearcy *et al.*, 1990; Tang *et al.*, 1988). Diffuse, shade light PFD exhibits much less spatial or temporal variation than direct PFD in forest understories. Consequently, the major source of variability in the light available for photosynthesis in these understories is the incidence of sunflecks.

The characteristics of sunflecks depend on canopy attributes such as the height, the flexibility of canopy elements, and leaf area distribution, as well as weather conditions. In forests most sunflecks are brief, but most of the daily PFD is contributed in a few long sunflecks (Pearcy, 1983; Chazdon and Fetcher, 1984; Pfitsch and Pearcy, 1989b; Chazdon, 1988; Gildner and Larson, 1992). For example, the measurements in a Queensland rain forest, which were made with photosensors attached to a Campbell Scientific 21X micrologger, revealed that 60% of the sunflecks recorded were less than 2 s in duration, but the 2% that exceeded 1 min duration contributed 80% of the PFD (Figure 1). Most sunflecks were below the full solar beam PFD received above the canopy because of penumbral effects. Only the few longest sunflecks approached direct-beam PFDs. Microsites varied in the number of sunflecks received, which ranged from 20 to 200. The contribution of sunflecks to the total daily PFD was highly correlated with the total sunfleck duration.

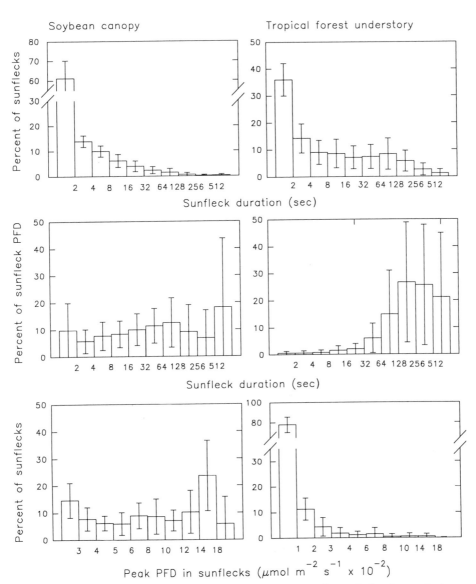

Figure 1 Frequency histograms for the maximum PFD in sunflecks, the sunfleck durations, and the fraction of total sunfleck PFD contributed by sunflecks of different durations in a tropical forest understory near Atherton, Queensland, Australia (right side) and a soybean canopy grown in Davis, California (left side).

Sunflecks under this canopy were clustered into periods with a high frequency of occurrence separated by periods with few or none. These results seem broadly representative of tall forest canopies (Chazdon, 1988; Pfitsch and Pearcy, 1989b). Sunflecks under a shorter forest canopy in Hawaii were brighter and more frequent, probably because of the windy conditions. Similarly, aspen canopies were characterized by many short sunflecks because of the flexible petioles and leaf flutter even at low wind speeds (Roden and Pearcy, 1993).

The few studies under crop or grassland canopies reveal a greater frequency of sunflecks with shorter average durations than found in forests. Leaves within a soybean canopy received 50 to 1700 (mean = 425) sunflecks per day. Short sunflecks within this canopy made a much larger contribution to the total PFD than they did in the Queensland rain forest understory (Figure 1). Sunflecks greater than 0.8 s in duration typically reached the full direct-beam PFD, which is consistent with the less significant role of penumbral effects in a short canopy. Wind increased the number of sunflecks and shortened their mean duration.

III. Responses of Leaf Gas Exchange to Sunflecks

Photosynthetic responses to sunflecks are complex because several components with very different time constants are involved. Leaves have the potential to respond rapidly to a sunfleck, but the extent to which this potential is realized is determined by factors that change on a much longer time scale than most sunflecks. It has long been known that, after an extended period in the shade or darkness, the photosynthetic apparatus exhibits an induction requirement in which the photosynthetic rate rises slowly when the PFD is suddenly increased (Osterhout and Hass, 1919). Typically, assimilation exhibits an initial rapid increase (within 5 s) up to a rate considerably below the light-saturated rate (Figure 2). This is followed by a second, slow increase for 10 to 30 min before the light-saturated assimilation rate is reached. The second increase may be either first-order and asymptotic with the light-saturated rate or it may be sigmoidal with an initial lag phase. In contrast, leaves preexposed to saturating PFDs so that full induction is achieved and then shaded for a minute or two exhibit no induction requirement (Figure 2). They reach their light-saturated assimilation rate within a few seconds.

Following long periods in low light, the initial rapid increase at the beginning of induction may only be from 10 to 30% of the final light-saturated photosynthetic rate. Chazdon and Pearcy (1986a) termed the rate measured 60 s after the light increase expressed relative to the light-saturated rate of assimilation the induction state (IS_{60}). This induction

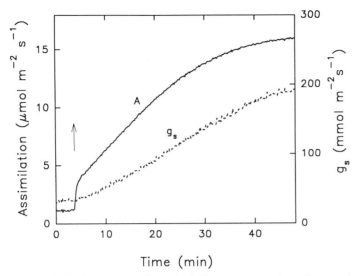

Figure 2 An induction response for assimilation and stomatal conductance (g_s) for a soybean leaf. The PFD was increased from 55 to 1650 μmol photons m^{-2} s^{-1} at the time indicated by the arrow.

state is a measure of the capacity of a leaf to respond immediately to an increase in PFD. Pons *et al.* (1992) measured the induction state at both 5 and 60 s after the light increase. The induction state after long shade periods may initially limit the photosynthetic rate achieved during a sunfleck by 80 to 90%. This limitation is gradually removed as induction proceeds.

Under some conditions, an intermediate phase in the induction response is evident. Immediately after the initial rise, a second phase occurs, which is characterized by an initial short shoulder or plateau for 3–5 s followed by an increase to a second plateau 60 to 120 s later (Figure 3). Then the assimilation increases further and more gradually up to the light-saturated assimilation rate. This intermediate phase has been termed the "fast-induction phase" to separate it from the following slower increase, finally leading to steady-state assimilation. This fast-induction phase is not very prominent in leaves that have been shaded for long periods since limitations by factors responsible for the slow phase are so great. It is, however, quite significant after 5–10 min of low light, because the half-time for development of the fast-induction phase is only about 300 s whereas it is 10–15 min for the slow phase. How much the fast-induction phase may limit sunfleck use probably varies among species. It is a less prominent feature of induction in the tropical forest understory herb *Alocasia macrorrhiza* than in the redwood forest understory herb

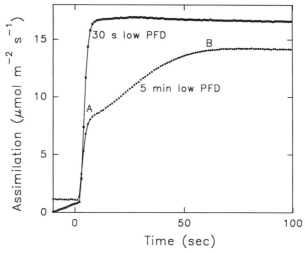

Figure 3 Responses of assimilation of an initially fully induced soybean leaf to an increase in PFD from 54 to 1650 μmol photons m^{-2} s^{-1} after 30 s or 5 min of low PFD. The response from A to B is the fast-induction component. The difference in the curves at B is the loss of induction due to the decrease in Rubisco activity and g_s during the 5-min low-light period. An additional 10 min was required for assimilation to increase back to the fully induced rate.

Adenocaulon bicolor (Pfitsch and Pearcy, 1989a). Sun-grown *Piper auritum* leaves exhibit a large fast-induction phase whereas shade-grown leaves do not (Tinoco-Ojanguran and Pearcy, 1993b). Küppers and Schneider (1993) also found a much greater limitation in the fast phase for partial-shade versus full-shade beech leaves. Gildner and Larson (1992) found that induction was very rapid in the desiccation-tolerant fern *Polypodium virginianum* and posed little limitation to utilization of sunflecks under field conditions. The mechanistic basis for the wide variation in induction responses is not understood.

   Pearcy *et al.* (1985) reported that continuous illumination is not required for induction since it continued for brief (120 s low-light) periods between lightflecks[1] in a series. Thus, the assimilation rate achieved was higher in each successive lightfleck. Induction occurring in response to a series of lightflecks has now been confirmed for a number of species of diverse types (e.g., shade-adapted tree seedlings and soybeans), indicating that it probably is a feature for all plants (Chazdon and Pearcy, 1986a; Pearcy, 1988, 1990; Pearcy and Seemann, 1990). Induction occurring in

---

[1] We use the term lightfleck to indicate a simulated sunfleck created with a lamp and shutter system.

response to the early lightflecks enhances the use of later lightflecks. Such an enhancement will hold as long as lightflecks are reasonably frequent so that the loss of induction in the intervening low-light period is not too great. Under natural canopy conditions, where typically periods of high sunfleck activity are separated by periods with few or no sunflecks, the induction state of the leaf is probably continuously changing.

The actual transient response of a leaf to a lightfleck is determined by a complex interaction between induction state, which sets the maximum photosynthetic rate possible in the lightfleck, and the transient increases and decreases in assimilation rate itself occurring in response to the light changes. Representative responses of a shade-adapted *Alocasia macrorrhiza* leaf to lightflecks of different durations and at different induction states are shown in Figure 4. Only a low maximum photosynthetic rate was achieved in an uninduced leaf during a 20-s lightfleck, but assimilation continued for a long period after the lightfleck. After the leaf was induced, the increase in assimilation is much faster, approaching the light-saturated assimilation rate within a few seconds. Similarly, after the lightfleck the decrease in assimilation was also faster than in the uninduced leaf. Nevertheless, substantial postillumination CO_2 fixation was still evident from the slower decrease than increase in assimilation rate. The response to a 1-min lightfleck differed in that a substantial postillumination burst of CO_2 developed because of the buildup of photorespiratory metabolites during this long lightfleck

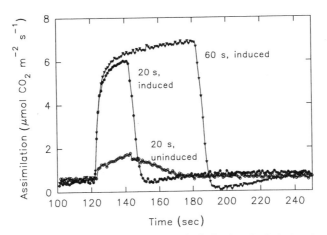

Figure 4 Responses of photosynthetic CO_2 assimilation by a leaf of *Alocasia macrorrhiza* to lightflecks (500 μmol photons m^{-2} s^{-1}) of different duration and given when the leaf was either fully induced or uninduced. The shade light before and after the lightfleck was 10 μmol photons m^{-2} s^{-1}. (From Chazdon and Pearcy, 1991).

(Vines *et al.*, 1982). This is explained by a requirement for these pools to build to high levels before significant rates of photorespiratory CO_2 release occurs, and by slow time constants for the photorespiratory pathway.

Postillumination CO_2 fixation evident in Figure 4 has been shown to contribute significantly to the total carbon gain during sunflecks. It is difficult to separate postillumination CO_2 fixation from CO_2 assimilation during the lightfleck proper because the instrument response times are sufficiently slow to confound them. Thus, a more indirect approach must be used in which the integrated carbon gain that is due to the lightfleck is compared to an "expected" carbon gain assuming an instantaneous response to a light change (Chazdon and Pearcy, 1986b). The ratio of these two gives a lightfleck use efficiency (LUE). The expected carbon gain is equivalent to a square-wave response where the rate increases instantaneously at the beginning of the lightfleck up to the light-saturated rate and then decreases instantaneously back to the steady-state rate in shade at the end of the lightfleck. In other words, it is as if there was no dynamic (time-dependent) response in photosynthesis.

Calculation of LUE for a number of species shows that it can be considerably greater than 100%, indicating a substantial contribution of postillumination CO_2 fixation, for short (1–10 s) lightflecks (Chazdon and Pearcy, 1986b; Kirschbaum and Pearcy, 1988c; Pons and Pearcy, 1992). LUE has been shown to decrease markedly with increasing lightfleck duration so that for lightflecks longer than 20 to 40 s it is often less than 100% (Figure 5). The decreasing contribution of postillumination CO_2 fixation in longer lightflecks is only relative and not absolute, since 4–5 s of lightfleck are sufficient to result in the maximum postillumination CO_2 fixation, which then remains relatively constant as lightfleck duration increases. Therefore, in longer lightflecks, postlightfleck CO_2 fixation is simply a smaller proportion of the total assimilation. Low induction state also limits the LUE, in this case primarily by limiting the assimilation rate during the lightfleck proper. There appears to be relatively little reduction in the amount of postillumination CO_2 fixation by induction state.

IV. Physiological and Biochemical Mechanisms Governing the Use of Sunflecks

A. Photosynthetic Induction

As discussed earlier, induction response in leaves involves a gradual removal of limitations that develop when a leaf has been in low light for an extended period. Although it is possible to measure increases in

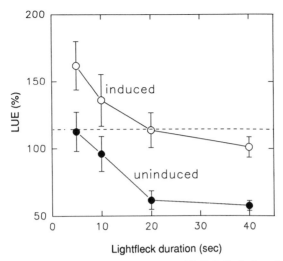

Figure 5 Decrease in lightfleck use efficiency with lightfleck duration for *Alocasia macrorrhiza* leaves before (solid circles) and after induction (open circles). (Redrawn from data in Chazdon and Pearcy, 1986b.)

activity of light-regulated enzymes and increases in metabolite pool sizes that correlate with the increase in assimilation during induction, an understanding of the role, if any, of these changes still depends on gas-exchange analysis. The foundation for this analysis is the idea that stomatal conductance regulates the supply of CO_2 to the mesophyll whereas the biochemical capacity for CO_2 fixation sets the demand for CO_2 (Raschke, 1979; Farquhar and Sharkey, 1982). Both change during induction but can assume different levels of significance, depending on the initial conditions and the time following the light increase. The biochemical capacity is given by the dependence of assimilation on intercellular CO_2 pressure (the A/p_i curve). A change in assimilation during induction that is due to stomata alone will move A and p_i along this curve, whereas a change in assimilation that is due to a change in biochemical capacity will cause a change in the A/p_i curve itself. Although heterogeneous (patchy) stomatal distributions can affect the interpretation of A/p_i curves, this phenomenon occurs only very rarely during light transients in well-watered leaves at moderate to high humidities (I. E. Woodrow and K. A. Mott, unpublished observations). The possibility that it is significant under low-humidity conditions or when stresses are present needs investigation (Tinoco-Ojanguren and Pearcy, 1993b).

1. The Role of Leaf Biochemistry Evidence for the importance of a biochemical limitation has been found in A/p_i curves constructed at different times during induction (Chazdon and Pearcy, 1986a) or by normalizing gas-exchange rates to a constant p_i so that the change in assimilation reflects only the change in the slope of the A/p_i curve (Woodrow and Mott, 1989). In both cases, an increase in the biochemical capacity for assimilation during the first 5 to 10 min of induction was apparent. After the first minute or so the further increase in biochemical capacity has been shown to correspond closely to the increase in Rubisco activity in the leaf (Seemann *et al.*, 1988; Woodrow and Mott, 1989). Both exhibit first-order increases with time constants in the range of 3–5 min. It has also been shown in these studies that the loss of induction in leaves that become shaded corresponds closely to the loss of Rubisco activity due to down-regulation in the shade. The loss of induction state and the down-regulation of Rubisco are both much slower than the increases during induction, having time constants on the order of 15 to 20 min (Seemann *et al.*, 1988; Woodrow and Mott, 1989). The close correspondence between the dynamics of Rubisco and of the mesophyll conductance is strong evidence for the important role of light regulation of this enzyme in limiting assimilation during induction (Pearcy, 1988).

The light regulation of Rubisco activity has been shown to be a complex process involving both activators and inhibitors (Portis, 1992). Native Rubisco is not catalytically competent, and the active form of the enzyme is produced by covalent addition of CO_2 and Mg^{2+} (carbamylation). This process is light dependent with the percentage of the carbamylated enzyme at steady state being roughly proportional to the photosynthetic rate (Woodrow and Berry, 1988). In many species, carbamylation and decarbamylation appear to be the primary mechanisms underlying light-induced changes in Rubisco activity (Kobza and Seemann, 1988; Seemann *et al.*, 1990). In others, however, a tight-binding inhibitor, carboxyarabinitol-1-phosphate (CA1P), is synthesized in low light or darkness and may function to further regulate the activity of Rubisco. CA1P is known to be catabolized in the light by a phosphatase (Salvucci *et al.*, 1988), but its synthesis in the dark is not well understood. Other phosphorylated compounds such as sugar-phosphates are also known to bind to Rubisco (Portis, 1992) but their role in light regulation of Rubisco activity has not been studied. Plants that differ in the mechanism of Rubisco regulation do not seem to differ significantly with respect to rates of up- or down-regulation of this enzyme (Kobza and Seemann, 1988). Evidence suggests that p_i may also be important since the rate of activation was found to be linear with p_i up to 250 μbar bar^{-1} (Mott and Woodrow, 1993). This suggests that the rate of Rubisco activation will be slower in leaves with a low stomatal conductance than in leaves with open stomata. This CO_2 dependence of the rate of activation may function to provide better coordination between the rate of Rubisco activation and stomatal conductance during transient photosynthesis.

The light-dependent increase in Rubisco activity appears to be facilitated by another enzyme, Rubisco activase (Salvucci *et al.*, 1986). The exact reaction catalyzed by this enzyme is uncertain, but evidence suggests that it may involve removal of ribulose-1,5-bisphosphate (RuBP) from decarbamylated (inactive) Rubisco, thereby allowing carbamylation to proceed (Portis, 1992). Evidence indicates that this enzyme may also be regulated by light and that this may influence the rate of increase in Rubisco activity and hence induction (Jackson *et al.*, 1991; Woodrow and Mott, 1992). Rates of induction in the Rubisco-limited portion of the response are slower if induction starts from a very low PFD as compared to a higher PFD, which may be due to the light-dependent activation of Rubisco activase.

Evidence for an RuBP regeneration limitation during the first 1–2 min of induction was first observed in fast transient gas-exchange measurements in which the kinetics of the increase in assimilation did not match the expected kinetics for an increase in Rubisco activity. The CO_2 dependence of this limitation was consistent with a limitation by RuBP supply to Rubisco (Kirschbaum and Pearcy, 1988a). Measurements of metabolite pools have established that during this time, RuBP concentrations were below 1.5 times binding site concentrations (Sassenrath-Cole and Pearcy, 1992), which is generally taken as an indication of an RuBP limitation to this enzyme (Sharkey, 1989; von Caemmerer and Edmonson, 1986). By 60 s, RuBP concentrations had increased to values above binding site concentrations, indicating removal of this limitation. After this time, the further increase in assimilation rates closely matches the increase in stomatal conductance and Rubisco activity as discussed earlier. The RuBP limitation is greatest after 5 to 10 min of shade, in agreement with the greater fast-induction limitation of gas exchange (Kobza and Seemann, 1988; Seemann *et al.*, 1988; Woodrow and Mott, 1989).

The mechanisms underlying the RuBP regeneration limitation are still not well understood. Four enzymes in the RuBP regeneration path are light activated (Edwards and Walker, 1983). In addition, this limitation could be explained by a requirement for autocatalytic buildup of metabolites (Leegood and Walker, 1980). However, the available evidence from metabolite pool studies in soybean leaves suggests that there is no requirement for autocatalytic buildup (Sassenrath-Cole and Pearcy, 1992). Although the cyclic nature of photosynthetic carbon metabolism makes identification of a specific site of limitation difficult, the light activation requirement for fructose-1,6-bisphosphatase is correlated with the fast-induction phase. Light activation/deactivation of other enzymes in the pathway may also play a role.

2. The Role of Stomata   The role of the stomata in limiting the use of lightflecks is likely to vary widely among species and with environmental conditions. The limitation imposed by stomata during induction depends strongly on the initial conductance established in the prior low-light period. Low humidity, for example, causes a much greater stomatal limitation (Kirschbaum and Pearcy, 1988a; Tinoco-Ojanguren and Pearcy, 1993a) and therefore reduces assimilation during the initial phases of induction. Estimation of the stomatal limitation is difficult because cuticular conductances may be significant and patchy stomatal behavior during induction may interfere with the estimate of p_i (Daley *et al.*, 1989). Some species such as aspen (*Populus tremuloides*) appear to maintain high conductances even in the shade and therefore exhibit little stomatal limitation during induction (Roden and Pearcy, 1993). On the other hand, any observed increases in assimilation after 10 min of induction can only be due to an increase in stomatal conductance since by this time Rubisco is essentially fully active.

Unless a lightfleck duration is long (>10 min), stomatal conductance (g_s) responds too slowly to an individual lightfleck to provide much benefit in terms of utilization of that lightfleck itself. It can, however, greatly influence use of subsequent lightflecks. Stomatal response to lightflecks is distinctly hysteretic, typically exhibiting a 2- to 3-min lag phase followed by a more rapid opening and then by a slow closing response after the lightfleck (Kirschbaum *et al.*, 1988). Maximum g_s may not be reached until 15–20 min after the lightfleck.

Mechanistically, the hysteretic response of g_s to lightflecks is probably related to the blue light response of guard cell protoplasts (Zeiger *et al.*, 1985; Assmann *et al.*, 1985; Assmann, 1988). Kirschbaum *et al.* (1988) successfully modeled the response of the g_s of *Alocasia* leaves to lightflecks using a three-component model, with each having different time constants. The first component, which was assumed to be a biochemical signal transduction mechanism, possibly related to activation of an enzyme, had a short time constant for the increase (<0.5 min) but a much slower time constant (7 min) for the decrease. This caused a buildup of osmotica and then an uptake of water, which were modeled by the second and third components, respectively. Both of these components had long (10–20 min) but symmetrical time constants for the increases and decreases.

The hysteretic response appears to be modified by other environmental factors. High water vapor pressure deficit reduces the hysteresis by causing a faster decrease in conductance, and may also cause the increase to be somewhat more rapid (Assmann and Grantz, 1990; Tinoco-Ojanguren and Pearcy, 1993a). It therefore also reduces the utilization of subsequent lightflecks. Mild water stress causes a significant lengthening of the induction response, probably because of slower stomatal responses (R. W. Pearcy, unpublished observations).

Wide variations in the rate of stomatal opening and closing have been reported among species. Shade-tolerant tree species were shown to have faster stomatal opening in response to sunflecks than shade-intolerant species (Woods and Turner, 1971). Grasses, in general, have much faster stomatal responses than dicots (Johnsson *et al.*, 1976). *Hedera helix* has been shown to have much more rapid stomatal opening in response to a blue light increase than *Vicia faba,* but still not as fast as that of *Triticum aestivum* (Karlsson and Assmann, 1990). Changes in g_s in response to lightflecks were much slower in the tropical forest gap species *Piper auritum* when it was grown in the shade as compared to when it was in the sun (Tinoco-Ojanguren and Pearcy, 1992). In contrast, the opposite occurred in an understory species, *Piper aequale.* These differences influenced carbon gain during lightflecks by creating a larger stomatal limitation to use of subsequent lightflecks in a series for shade-grown *P. auritum.* Data of Küppers and Schneider (1993) are consistent with a greater stomatal limitation to induction in partial-shade as compared to full-shade beech leaves, which then caused a greater limitation to lightfleck use.

Knapp and Smith (1987, 1989, 1990) have examined species differences in stomatal response to alternating sun/shade periods (5–12 min period length) created by moving clouds. Some species exhibited relatively rapid responses of g_s to sun and shade periods, which appeared to limit the increase in assimilation during the sun periods. Species exhibiting rapid responses were primarily herbs with relatively high assimilation rates that also underwent rather large fluctuations in water potential during the sun/shade transitions. By contrast, the g_s of woody species with low assimilation rates did not respond to sun/shade transitions. These woody plants had a greater water capacitance and hence experienced less fluctuation in water potential. g_s in these species appeared to be much less limiting to the increase in assimilation during the sun periods. More recent studies indicate that g_s of woody species with relatively high assimilation rates respond rapidly to light changes even though water potentials tend to remain rather static (Knapp, 1992). Thus, photosynthetic capacity or possibly the maximum g_s, which varies with photosynthetic capacity, may be more important than growth form itself. Water use efficiency appeared to be enhanced at the expense of some assimilation in species with high assimilation rates, whereas assimilation was maximized at the expense of water use efficiency in species with low assimilation rates. The measurements did not allow an assessment of whether biochemical induction limitations were also important, and hence these may have also differed between species.

B. Transient Responses to Short Lightflecks

The transient response of assimilation to lightfleck also depends on Rubisco kinetics and their interaction with the buildup and depletion of RuBP. The initial increase in assimilation must be caused by an increase in RuBP concentration resulting from the accelerated electron transport. This initial increase is limited under the appropriate conditions by the fast-induction component (see Section IV,A,1). The decrease in assimilation must also be related to the depletion of RuBP by Rubisco, coupled with the reduced supply because of slower electron transport. Although this simple model would appear to be sufficient to account for the essence of the increases and decreases in transient assimilation, the actual responses are more complicated because they depend not only on the rates of supply but also on the pool sizes of metabolites like RuBP. In addition, the delay in photorespiratory CO_2 release can significantly alter the dynamics of gas exchange after long lightflecks when pools of photorespiratory metabolites build up to high levels.

Postlightfleck CO_2 fixation occurs because of a capacity for rapid reduction of 3-phosphoglyceric acid (PGA) to triose-phosphate (TP) and other reduced, "high-energy" metabolites during the lightfleck. These can then be utilized for continued CO_2 fixation after the lightfleck. In effect, the metabolite pools act like a capacitor that allows electron transport to run transiently faster at the beginning of the lightfleck and then allows CO_2 fixation to continue after the lightfleck. The transiently more rapid electron transport is evident as a burst of O_2 evolution at the begining of a lightfleck (Kirschbaum and Pearcy, 1988c). During steady-state photosynthesis, the rate of electron transport is constrained by the rate of CO_2 fixation since they are inexorably linked by the use and regeneration of ATP and $NADPH_2$. Pools of PGA can build up during the shade periods between sunflecks to levels two to three times higher than the steady-state levels in high light (Sharkey *et al.*, 1986; Sassenrath and Pearcy, 1992). During the lightfleck, PGA is rapidly reduced and TP and RuBP pools build up rapidly (within 5 s). The decrease in these pools following the lightfleck is consistent with the observed postlightfleck CO_2 assimilation (Sharkey *et al.*, 1986).

V. Photosynthesis in Natural Sunfleck Regimes

Despite their potentially large contribution to the carbon gain of understory plants, only a few studies have been carried out in the field to assess their actual contribution. Most of the available data are from forest understories and consist of diurnal response curves collected for one to at most a few days. Integration of these curves for understory plants from shaded tropical forests has revealed that 30 to 60% of the daily assimilation on clear days is due to utilization of sunflecks (Björkman *et al.*, 1972; Pearcy and Calkin, 1983; Pearcy, 1987; Pfitsch and Pearcy, 1989b). On the other hand, results from deciduous forests reveal only a 10 to 20% contribution of sunfleck utilization (Schulze, 1972; Weber *et al.*, 1985). This difference may be due to the higher diffuse light level in the latter.

The most detailed studies of the contribution of sunflecks are those conducted on the redwood forest understory plant *Adenocaulon bicolor* (Pfitsch and Pearcy, 1989a,b). Essentially all the variation in the daily carbon gain in different microsites and days was found to be due to sunflecks (Figure 6). In the microsites with the most sunfleck activity, 70% of the daily carbon gain was attributable to sunfleck utilization. The measured carbon gain was, however, 20 to 30%, less than a daily carbon gain predicted from a steady-state model utilizing the light measurements (Figure 7). Since the model did not include the dynamic responses it did not account for either induction limitations or postillumination CO_2 fixation. Thus, the lower measured totals suggest that induction limitations are much larger than the contribution of postillumination CO_2 fixation. This is consistent with the much greater contribution of long but relatively infrequent sunflecks to the available light in the understory of this forest.

The role of sunflecks within canopies, such as crops or individual tree canopies, as opposed to under forest canopies, has been little studied. The extreme patchiness of sunfleck and shade in these canopies more often than not results in part of the leaf being in a sunfleck and part in the shade. Without proper spatial averaging of PFD and gas exchange, the contribution of the sunflecks cannot be readily separated. The problem is lessened but not eliminated in forest understories, where the sunflecks are larger and the penumbra cause gradual transitions so that a few sensors around the chamber can adequately average the PFD (Pfitsch and Pearcy, 1989b). In contrast, the role of sunflecks in crop canopies can so far only be inferred from the induction state of the leaves within the canopy and from the nature of the light environment. The high frequency of short but bright sunflecks suggests that there

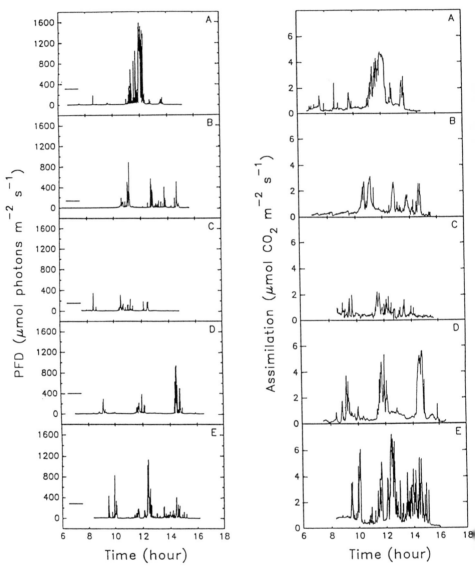

Figure 6   Diurnal course of photon flux density and assimilation by *Adenocaulon bicolor* leaves in different microsites and days in a redwood forest understory. (From Pfitsch and Pearcy, 1989b.)

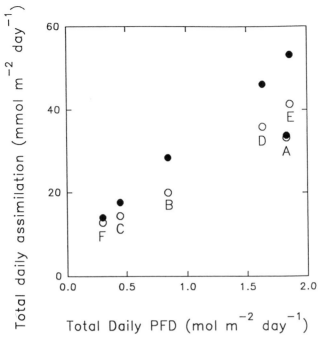

Figure 7 Relationship between total daily assimilation and total daily photon flux for the diurnal courses shown in Figure 6. The solid circles show predictions from a steady-state model. F is for a day when the leaf received only diffuse light. (From Pfitsch and Pearcy, 1989b).

could be a substantial contribution of postillumination CO_2 fixation. On the other hand, induction limitations also appear to be substantial. Soybean leaves within a canopy exhibited an induction state (IS_{60}) of only 0.52 on average, because of both reduced Rubisco activities and g_s of leaves (Pearcy and Seemann, 1990). Shading of already fully induced leaves for periods of only 5 to 10 min, a condition occurring frequently within the canopy, caused a significant induction loss.

VI. Modeling the Response to Sunflecks

Here we use a dynamic model of CO_2 assimilation to predict the response to sunflecks and to estimate the consequences of both postillumination CO_2 fixation and induction under different light regimes. As the model has been reported in detail in Gross *et al.* (1991), we will cover only the essential features and recent modifications. The basis of the structure of the model is the steady-state model of Farquhar and von Caemmerer (1982), which was modified to be explicitly dynamic by including metabolite pool sizes as well as activation and deactivation of Rubisco by light. These components vary dynamically in the model. As in the Farquhar and von Caemmerer (FvC) model, p_i is retained as a key variable. Since p_i is a function of g_s and assimilation rate, it is a dynamic variable here whereas it is static in the FvC model. We used the dynamic model of stomatal conductance of Kirschbaum *et al.* (1988) to calculate the time course of stomatal conductance in varying light.

The version of the model used here includes a modification to expressly account for the fast-induction component. Whereas the Gross *et al.* (1991) model had only a single pool (R) that includes all reduced metabolites, the new version breaks this into two pools, the precursors to RuBP (still called R) and RuBP itself. The rate of conversion of R to RuBP is a function of the activity of an enzyme corresponding to one of the light-activated steps in RuBP regeneration, which is viewed as analogous to fructose-1,6-bisphosphatase. It may, however, represent a composite of the dynamics of several light-activated enzymes in RuBP regeneration (Sassenrath-Cole and Pearcy, 1992). Light activation and deactivation were modeled in a manner analogous to that for Rubisco, except that the time constants were much smaller. Inclusion of this step gives better fits to the dynamics of assimilation during lightflecks and a better simulation of induction. Clearly, the model is still a vast simplification of the biochemistry of photosynthetic carbon metabolism. However, the objective was not to incorporate all reactions, as the dynamics of many less important reactions are even more imperfectly known. Instead, it was to account for the known dynamics of CO_2 assimilation in variable light.

Running the model involves first initializing it to a steady-state condition at the beginning PFD and ambient CO_2 pressure (assumed to be constant throughout). Then the model is stepped to the next time for input of the current PFD. The Rubisco activity is calculated from the past Rubisco activity and the activation or deactivation over the time step. The R pool size is calculated from the past size and the rate of addition or removal. Similarly, the current stomatal conductance is calculated dynamically. The current value of p_i is then calculated so as to ensure that the assimilation rate obtained from the biochemical portion of the model is equal to that obtained from the stomatal conductance model. This process is repeated for each time step. For comparison, the steady-state values of all variables were obtained at the PFD present at each time step. This allows comparison of the dynamic model to the situation occurring if a leaf were able instantaneously to adjust to a particular PFD value.

Parameters for the model have been obtained either from the literature, such as the kinetic parameters and rates of activation and deactivation of Rubisco (Farquhar and von Caemmerer, 1982; Seemann *et al.*, 1988), or from measures of gas-exchange parameters. Most parameters in the model remained fixed for all runs and all species while a subset was varied, depending on photosynthetic capacity, stomatal conductance, etc., of the particular leaf. In practice an initial guess of the parameter values was made from the steady-state light and CO_2 responses. The values of the parameters were then adjusted to give a good fit to the induction response and the measured lightfleck use efficiencies. Table I shows the values of the parameters that were varied for *Alocasia* and soybean leaves.

In general, the model gave excellent agreement between the observed dynamic responses and the computed responses (compare Figure 8 with Figures 2 and 4). The actual shapes of the responses do not agree perfectly but the assimilation rates during time courses of induction and induction loss as well as the LUEs predicted for lightflecks of different duration or at different induction states agreed within ± 10–15%.

We examined the consequences of different frequencies and durations of lightflecks for assimilation of *Alocasia* leaves by creating simulated lightfleck regimes. Each regime consisted of a series of lightflecks

Table I Parameters Used in the Dynamic Model for *Alocasia* and Soybean[a]

	Alocasia	Soybean
Stomatal submodel		
Maximum g_s (mmol m^{-2} s^{-1})	50	500
Minimum g_s in dark (mmol m^{-2} s^{-1})	2	35
Time constant for K increase (min)	0.37	0.37
Time constant for K decrease (min)	7.0	7.0
Time constant for change in osmotica (min)	13	18
Time constant for change in water content (min)	15	10
Photosynthesis submodel		
Apparent K_m for RuBP (μmol)	60	30
Electron transport capacity (V_{jmax}, μeq m^{-2} s^{-1})	30	90
Carboxylation capacity (V_{cmax}, μmol CO_2 m^{-2} s^{-1})	60	70
Maximum pool size of R (μmol m^{-2})	15	40
Maximum pool size of RuBP (μmol m^{-2})	90	150
Time constant for decrease in Rubisco activity (min)	6	5
Time constant for increase in Rubisco activity (min)	20	20
Time constant for change in RuBP regeneration capacity (min)	1	1
Ratio of maximum RuBP regeneration capacity to Rubisco activity	1.5	1
Apparent K_m of RuBP regeneration capacity (μmol)	10	20
Day respiration (μmol CO_2 m^{-2} s^{-1})	0.1	0.4

[a] Parameters that were not varied are not presented.

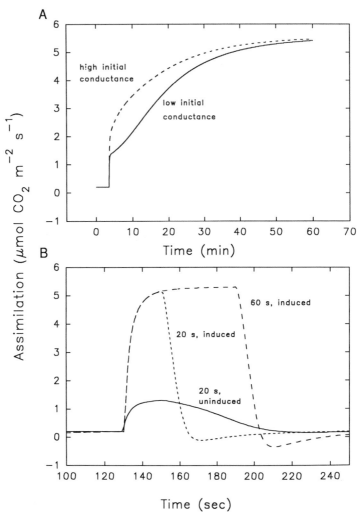

Figure 8 Predicted assimilation from the dynamic model showing the induction response (A) under either a high or a low initial conductance or (B) to lightflecks similar to those given in Figure 4. The simulation is for an *Alocasia macrorrhiza* leaf.

(500 μmol photons m^{-2} s^{-1}) superimposed on a background of 12 μmol photons m^{-2} s^{-1} over the course of an 8-h day. The lightflecks were either 5 s or 5 min in duration, and were either uniformly spaced throughout the run or clustered into groups with intervening longer low-light periods. Each regime had exactly the same total PFD as well as the same proportional contribution of sunflecks to this total. Thus,

comparisons of the output (assimilation) of the model indicate the effects of the dynamics of assimilation due to different lightfleck distributions and durations.

The model predicted a 22% greater carbon gain for the 5-s lightfleck regime as compared to the 5-min lightfleck regime (Table II). This difference was due both to the higher induction state maintained in the 5-s lightfleck regime as well as to greater contribution of postlightfleck assimilation in this regime. The assimilation in the 5-s lightfleck regime was 2% higher than the steady-state prediction and reflects the balance between enhancement of assimilation by postillumination CO_2 assimilation and induction limitations in the model at this lightfleck frequency and duration. Clustering the 5-min into a 1-h period at midday increased assimilation gain by 8% because of carryover of the induction state between lightflecks. By contrast, clustering of the 5-s lightflecks had little effect on the total assimilation because induction effects are less important for these short lightflecks.

Table II Simulation of Photosynthesis of an *Alocasia macrorrhiza* Leaf under Regimes with Different Timing and Duration of Lightflecks and with Removal of Different Dynamic Limitations[a]

	Assimilation (mol CO_2 m^{-2} × 10^{-1})				
Simulation	All	No stomatal limitation	No Rubisco limitation	No stomatal or Rubisco limitation	No stomatal, Rubisco, or RuBP reg. limitation
Lightfleck regime					
Uniform					
5-min lightflecks	0.1175	0.1407	0.1198	0.1564	0.1601
	(−22)	(−10)	(−21)	(−1)	(0)
5-s lightflecks	0.1541	0.1588	0.1513	0.1586	0.1797
	(+2)	(+1)	(0)	(+1)	(+12)
Clustered					
5-min lightflecks	0.1268	0.1445	0.1312	0.1564	0.1601
	(−16)	(−8)	(−14)	(−1)	(0)
5-s lightflecks	0.1535	0.1618	0.1493	0.1622	0.1802
	(+1)	(+3)	(−2)	(+2)	(+12)
Steady state	0.1505	0.1569	0.1511	0.1576	0.1600

[a] The numbers in parentheses below the assimilation values are the percentage differences from the steady-state model simulation.

The uniform regimes consisted of three 5-min lightflecks or 169 5-s lightflecks uniformly distributed in an 8.6-h period. The clustered regimes consisted of the same lightflecks but clustered into a 1-h period (5 min) or into three 1-h periods spaced at 2-h intervals (5 s). The shade and lightfleck PFDs were 15 and 500 μmol photons m^{-2} s^{-1}, respectively.

We examined the consequences of specific limitations in the model to utilization of sunflecks by altering the parameters in a way that removed a particular dynamic limitation. The dynamic stomatal limitation was altered by changing the light response of g_s so that it remained high (equal to the steady-state value in high PFD) at both 500 and 10 μmol photons m^{-2} s^{-1}. This caused a 19% increase for the 5-min lightfleck regime but had little effect in the 5-s lightfleck regime. Clustering of the 5-min lightflecks in this case caused only a 2.7% increase. Removing the dynamics of Rubisco in a similar manner while maintaining the dynamic stomatal limitations had a markedly different effect. In this case, there was a negligible increase in assimilation in the 5-min lightfleck regime despite full activation of Rubisco. Assimilation actually decreased slightly in the 5-s regime. The explanation for the differences in behavior is that full activation of Rubisco when g was low caused p_i to decrease substantially, resulting in greater oxygenase activity. Additionally, dynamic limitations in RuBP regeneration would still be limiting to some extent, although these are small under the conditions of the model and rather rapidly overcome. This prediction suggests that maximizing utilization of lightflecks requires a coordinated dynamic response.

Elimination of both the stomatal and Rubisco limitations increased assimilation by 33 and 11% over that predicted with all limitations, or all except the stomatal limitation included, respectively. Removal of the RuBP regeneration limitation caused only a small (2.3%) further increase in assimilation. Thus, in terms of relative importance, the dynamic responses of stomata and then Rubisco were by far the most significant under these light regimes. With removal of the RuBP regeneration limitation, the only dynamic component remaining was the postlightfleck assimilation due to the buildup and utilization of metabolite pools. Under these conditions, assimilation was 36% higher than the full-limitation prediction and 10% higher than the steady-state prediction.

A. Simulation of Photosynthesis under Natural Lightfleck Regimes

To further explore the consequences of the dynamics of photosynthetic CO_2 exchange, the response of *Alocasia* leaves to natural understory light regimes was simulated. The light regimes were obtained from measurements at 1-s intervals using Campbell Scientific 21X microloggers and gallium arsenide phosphide photosensors (GaAsP sensors; Gutschick *et al.*, 1985) in the Curtain Fig forest near Atherton, Queensland, Australia, in late February 1987 (see Pearcy, 1987). The sensors were mounted either directly onto *Alocasia* leaves or on metal stakes at about the same height as the leaves. Assimilation was simulated for a total of 32 sensor-days representing a range of light environments from open (canopy top; 3 records), to a gap where 1.5 h of potential direct sunlight was received

(5 records), to the shaded forest understory (24 records). An example of a simulated diurnal course is shown in Figure 9. Note the much lower maximum assimilation rates achieved during lightflecks with the dynamic as compared to the steady-state simulation. Also note the very different time responses for metabolite pool sizes, Rubisco activity, and g_s, and the faster increases than decreases in the latter two parameters.

Little difference (<2.5%) was found between the steady-state and the dynamic simulation for the open or gap environment. Greater than 90% of the available light and assimilation occurred during the high-light periods in these environments and a steady-state assumption appeared to be adequate to predict photosynthesis. In the understory, however, the dynamic model predicted daily assimilation totals from 1.2 to 25.5%

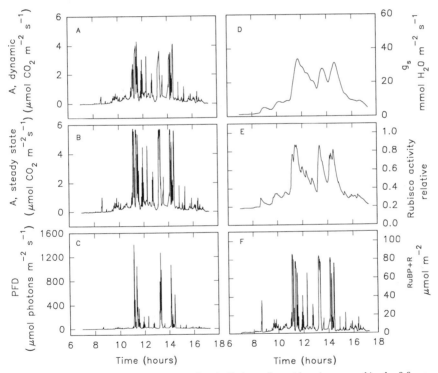

Figure 9 Simulated daily course of assimilation of an *Alocasia macrorrhiza* leaf from the (A) dynamic model and for comparison (B) the steady-state model under a natural sunfleck regime measured with a GaAsP photosensor attached to the leaf (C). Also shown are the simulated daily courses of (D) stomatal conductance, (E) Rubisco activity, and (F) the sum of the R and RuBP pool sizes. The model calculated these outputs at 1-s intervals based on the input of the PFD as recorded by the datalogger at this interval.

(mean = 12.6%) lower than those predicted by the steady-state model (Table III). Predictions from the steady-state and dynamic model were close when sunflecks contributed less than 25% of the total daily PFD. In contrast, the steady-state and dynamic models generally differed the most when sunflecks contributed more than 50% of the total daily PFD. However, even when sunflecks made a large contribution, there was a substantial variation in the effects due to inclusion of the dynamic responses. No relationship was apparent with any particular average measure of sunfleck characteristics, such as the mean sunfleck duration, the mean time between sunflecks, or the mean peak PFD in the sunflecks.

We examined the effects of specific dynamic limitations, as done earlier for the defined lightfleck regimes, for four natural understory regimes; two where the dynamic model predicted significantly lower carbon gain than the steady-state model and two where the predictions were close together. Removal of the dynamic stomatal limitation had the greatest effect in all cases, followed by removal of the Rubisco limitation (data not shown). Consistent with the results under defined lightfleck regimes,

Table III Characteristics of the Light Environment and Simulated Assimilation of *Alocasia* Leaves in Different Understory Microsites and of Soybean Leaves at Different Positions within a Canopy[a]

Total PFD (mol photons m^{-2} day^{-1})	Number of sunflecks	Mean duration(s)	Percentage of sunflecks	Predicted assimilation (μmol CO_2 m^{-2} day^{-1})		Percentage difference
				Dynamic	Steady state	
			Alocasia			
1.55	106	36	71	0.193	0.233	−17
1.15	76	42	73	0.136	0.179	−24
1.59	120	50	59	0.283	0.300	−5
1.00	42	52	43	0.173	0.197	−12
0.52	68	19	23	0.098	0.105	−7
			Soybean			
20.7	1701	9	92	2.21	2.18	−2
1.75	236	3	46	0.233	0.215	−7
3.83	984	2	45	0.671	0.676	+1
5.22	529	6	83	0.499	0.465	−7
3.13	154	21	91	0.282	0.240	−15
4.07	126	26	79	0.495	0.392	−21

[a] The light records are from photosensors mounted directly on the *Alocasia* leaves and on stationary posts at different heights within the soybean canopy.

removal of the Rubisco limitation alone had only a small effect. Removal of the stomatal, Rubisco, and RuBP regeneration limitations together resulted in virtually identical predictions from the steady-state and dynamic models. Since the only dynamic part left is the metabolite pools, these results indicate that postlightfleck CO_2 fixation makes almost no contribution to the assimilation in these understory light environments. This is in agreement with the small proportion of the available PFD contributed by short sunflecks (see Section II) for which postillumination CO_2 fixation is most significant.

We also ran the model, after parameterizing it for soybean leaves, with the light regimes from a soybean canopy (see Pearcy *et al.*, 1990) as an alternative that was vastly different from the forest understory. These light regimes measured at different depths in the canopy had from 150 to 1701 sunflecks that contributed 45 to 92% of the PFD. The sunfleck duration varied from a mean of 2–3 s on a windy day to 24 s on the next day that was still. We used steady-state and dynamic gas-exchange responses to parameterize the model. Parameter values chosen for the dynamic components were those that best fit the efficiency of lightfleck utilization and the induction gain and loss characteristics reported for soybeans (Pons and Pearcy, 1992; Pons *et al.*, 1992). The relevant parameters for the model are shown in Table I. Assimilation was simulated for eight PFD regimes measured at different positions on three different days.

Comparison of the dynamic and steady-state model outputs showed that the dynamic model predicted daily assimilation rates that ranged for 101 to 79% (mean = 92%) of the steady-state prediction (Table III). Thus soybean leaves, like *Alocasia* leaves in the understory, may be limited by induction. This agrees with the finding of a strong induction limitation within the soybean canopy (Pearcy and Seemann, 1990). Removal of the stomatal limitation, however, caused only a small increase in the dynamic prediction. Removal of the Rubisco limitation caused a slightly greater increase in assimilation. Removal of the Rubisco, stomatal, and RuBP regeneration limitation caused the largest increase in assimilation on the windy days with short sunfleck durations and resulted in predictions of 107 to 119% of the steady-state prediction. On the still day, however, the dynamic and steady-state carbon gains with these limitations removed were within 2%. These results suggest that the RuBP regeneration limitation may be more significant in soybean canopies than in *Alocasia* leaves in the understory. Moreover, the results show that under windy conditions when there is strong canopy movement resulting in many short lightflecks, postlightfleck CO_2 assimilation can make a significant contribution to the carbon gain.

VII. Long-Term Effects of Sunfleck Activity: Carbon Balance and Growth

Extending the knowledge of the role of sunflecks in the ecology of understory plants requires insight into their role in carbon balances and growth. Pearcy and Pfitsch (1991) estimated the relative annual contribution to carbon gain of *Adenocaulon bicolor* of direct and diffuse light from $\delta^{13}C$ values and found the contribution of direct light to range from 9% in microsites with the least sunfleck activity to 44% in microsites with the most sunfleck activity. These values were lower than single-day measurements discussed in the foregoing because they integrate over both cloudy and clear days. Growth studies have demonstrated a correlation between estimated direct PFD from hemispherical photographs and growth of tree seedlings and saplings (Pearcy, 1983; Oberbauer *et al.*, 1988). Pfitsch and Pearcy (1992), however, found no relationship for *A. bicolor* growth in different microsites despite the clear impact on the proportion of annual carbon gain discussed earlier. However, when sunflecks were removed with shadow bands that had minimal impact on diffuse light, a marked reduction in plant size and reproductive output occurred. The lack of a response in different microsites may be due to other stresses that occur with sunflecks, or it may be because these plants had relatively little potential to grow larger as compared to the trees discussed earlier. The stresses associated with sunflecks include reduced water potentials that can lead to wilting (Elias, 1983), elevated leaf temperatures (Young and Smith, 1979), and possibly nutrient stress because of increased competition in the brighter microsites. Photoinhibition may have also constrained photosynthetic performance in the brightest microsites, where it could be expected to interact with temperature and water stress (Mulkey and Pearcy, 1992). Most sunflecks, however, have too low PFD or are too short to cause any lasting photoinhibition. Le Goullac *et al.* (1991) found increasing photoinhibition during successive 10-min exposures of 700 μmol photons m^{-2} s^{-1} in the tropical forest understory herb *Elatostema repens*. However, the cumulative exposure was much greater than normally experienced by a leaf in the understory. Studies with *Adenocaulon bicolor* revealed no lasting photoinhibition during either natural sunflecks or lightflecks lasting from 5 to 45 min (R. W. Pearcy and W. A. Pfitsch, unpublished results). Although assimilation rate was suppressed immediately after the sunfleck, it recovered within 90 min. Moreover, there was no constraint on the ability to respond to a subsequent sunfleck or lightfleck.

Sims and Pearcy (1993) examined the growth responses of *Alocasia* plants to lightflecks of different frequencies and durations created by a shutter system in a glasshouse. Relative growth rates of plants grown under short (6 s) but frequent lightflecks were about 50% greater than those grown at long (8 min), but infrequent lightflecks. Whitewash on the glasshouse roof kept the maximum PFD (500 μmol photons m^{-2} s^{-1}) below levels that might have been photoinhibitory. Since these regimes were matched to be identical in daily PFD and the total contribution of lightfleck and diffuse light PFD, the relative growth rate differences were due solely to the different efficiencies of utilization of short and long lightflecks. Utilization of the long lightflecks was limited by induction with little contribution of postlightfleck CO_2 fixation, whereas in the short regime postlightfleck CO_2 fixation made a significant contribution. However, diffuse light alone of the same daily total PFD as in the lightfleck regimes yielded a 50% higher relative growth rate than the short lightfleck regime. Since the diffuse light PFD was relatively high but still below saturation, it was used more efficiently than the lightflecks, which exceed light saturation. These results confirm that the dynamic constraints on lightfleck utilization observable at the level of leaf gas exchange carry over into growth responses. They also suggest that penumbral effects that spread the sunflecks and in effect enhance the partial shade PFD over a larger area are an important factor.

VIII. Conclusions

Of the scales of environmental patchiness of concern to ecologists, sunflecks are at one extreme. They exist on time scales of seconds to minutes and vary spatially on length scales of centimeters. Accordingly, the specific mechanisms governing the utilization of sunflecks are unique. Chazdon (1988) has summarized the spatial and time scales over which variation in light is important to ecological process. Whereas acclimation and morphological adjustment are important for scales of heterogeneity in treefall gaps, regulatory processes at the single-leaf level are the primary mechanisms important at the scale of sunflecks. The dynamics of photosynthesis on the scale of seconds to minutes are primarily a manifestation of the regulatory processes of carbon metabolism and of stomata. These regulatory processes can either act to limit photosynthesis during sunflecks, as in the case of induction, or enhance it via postlightfleck CO_2 fixation. The extent of limitation or enhancement depends on the characteristics of the light environment as well as species and acclimation environment. In understory microenvironments where most of the PFD is contributed by a few long sunflecks, induction limitations predominate. These limitations may reduce assimilation as much as 10 to 30% as shown by both measurements and simulation modeling. Because carbon gains are already strongly limited, further reductions of this magnitude further impact growth and reproductive success. In crop canopies, the larger contribution of short sunflecks results in a more substantial contribution of postillumination CO_2 fixation. In this case it may nearly compensate for the induction limitations or actually increase assimilation over that expected on the basis of a steady-state prediction.

Despite the strong limitations imposed by induction, it is clear that utilization of sunflecks is critically important to understory plants. Up to 60% of the carbon gain may be attributed to their utilization. Moreover, spatial variations in sunfleck activity influence growth in the understory, both positively and negatively. Negative effects may come from constraints on sunfleck utilization in the brightest microsites, such as those imposed by water stress, high leaf temperatures, or increased competition for nutrients. These effects of sunflecks combined with the enhanced assimilation under other circumstances most likely are responsible for highly patchy performance of understory plants and tree seedlings on forest floors.

References

Assmann, S. M. (1988). Enhancement of the stomatal response to blue light by red light, reduced intercellular concentrations of CO_2, and low vapor pressure differences. *Plant Physiol.* **87**, 226–231.

Assmann, S. M., and Grantz, D. A. (1990). Stomatal response to humidity in sugarcane and soybean—Effect of vapour pressure difference on the kinetics of the blue light response. *Plant, Cell Environ.* **13**, 163–169.

Assmann, S. M., Simoncini, L., and Schroeder, J. I. (1985). Blue light activates electrogenic ion pumping in guard cell protoplasts of *Vicia faba. Nature (London)* **318**, 285–287.

Björkman, O., Ludlow, M., and Morrow, P. (1972). Photosynthetic performance of two rain-forest species in their habitat and analysis of their gas exchange. *Year Book—Carnegie Inst. Washington* **71**, 94–102.

Chazdon, R. L. (1988). Sunflecks and their importance to forest understory plants. *Adv. Ecol. Res.* **18**, 1–63.

Chazdon, R. L., and Fetcher, N. (1984). Photosynthetic light environments in a lowland tropical forest in Costa Rica. *J. Ecol.* **72**, 553–564.

Chazdon, R. L., and Field, C. B. (1987). Determinants of photosynthetic capacity in six rainforest *Piper* species. *Oecologia* **73**, 222–230.

Chazdon, R. L., and Pearcy, R. W. (1986a). Photosynthetic responses to light variation in rain forest species. I. Induction under constant and fluctuating light conditions. *Oecologia* **69**, 517–523.

Chazdon, R. L., and Pearcy, R. W. (1986b). Photosynthetic responses to light variation in rain forest species. II. Carbon gain and light utilization during lightflecks. *Oecologia* **69**, 524–531.

Chazdon, R. L., and Pearcy, R. W. (1991). The importance of sunflecks for forest understory plants. *BioScience* **41**, 760–766.

Daley, P. F., Raschke, K., Ball, J. T., and Berry, J. A. (1989). Topography of photosynthetic activity of leaves obtained from video images of chlorophyll fluorescence. *Plant Physiol.* **90**, 1233–1238.

Edwards, G., and Walker, D. (1983). "C$_3$, C$_4$: Mechanisms, and Cellular and Environmental Regulation of Photosynthesis." Univ. of California Press, Berkeley.

Elias, P. (1983). Water relation pattern of understory species influenced by sunflecks. *Biol. Plant.* **25,** 68–74.

Evans, G. C. (1939). Ecological studies on the rainforest of southern Nigeria. II. The atmospheric environmental conditions. *J. Ecol.* **27,** 436–482.

Farquhar, G. D., and Sharkey, T. D. (1982). Stomatal conductance and photosynthesis. *Annu. Rev. Plant Physiol.* **33,** 317–345.

Farquhar, G. D., and von Caemmerer, S. (1982). Modelling of photosynthetic response to environmental conditions. *In* "Encyclopedia of Plant Physiology" (A. P. Göttingen and M. H. Zimmerman, eds.), New Ser., Vol. 12B, pp. 549–588. Springer-Verlag, Berlin.

Gildner, B. S., and Larson, D. W. (1992). Photosynthetic response to sunflecks in the desication-tolerant fern *Polypodium virginianum. Oecologia* **89,** 390–396.

Gross, L. J., Kirschbaum, M. U. F., and Pearcy, R. W. (1991). A dynamic model of photosynthesis in varying light taking account of stomatal conductance, Cl$_3$-cycle intermediates, photorespiration and RuBisCO activation. *Plant, Cell Environ.* **14,** 881–893.

Gutschick, V. P., Barron, M. H., Waechter, D. A., and Wolf, M. A. (1985). A portable monitor for solar radiation that accumulates irradiance histograms for 32 leaf-mounted sensors. *Agric. For. Meteorol.* **33,** 281–290.

Jackson, R. B., Woodrow, I. E., and Mott, K. A. (1991). Nonsteady-state photosynthesis following an increase in photon flux density (PFD): Effects of magnitude and duration of initial PFD. *Plant Physiol.* **95,** 498–503.

Johnsson, M., Issaias, S., Brogardh, T., and Johnsson, A. (1976). Rapid, blue-light indiced transpiration response restricted to plants with grass-like stomata. *Physiol. Plant.* **36,** 229–232.

Karlsson, P. E., and Assmann, S. M. (1990). Rapid and specific modulation of stomatal conductance by blue light in ivy (*Hedera helix*)—An approach to assess the stomatal limitation of carbon assimilation. *Plant Physiol.* **94,** 440–447.

Kirschbaum, M. U. F., and Pearcy, R. W. (1988a). Gas exchange analysis of the relative importance of stomatal and biochemical factors in photosynthetic induction in *Alocasia macrorrhiza. Plant Physiol.* **86,** 782–785.

Kirschbaum, M. U. F., and Pearcy, R. W. (1988b). Gas exchange analysis of the fast phase of photosynthetic induction in *Alocasia macrorrhiza. Plant Physiol.* **87,** 818–821.

Kirschbaum, M. U. F., and Pearcy, R. W. (1988c). Concurrent measurements of O$_2$ and CO$_2$ exchange during lightflecks in *Alocasia macrorrhiza* (L.) G. Don. *Planta* **174,** 527–533.

Kirschbaum, M. U. F., Gross, L. J., and Pearcy, R. W. (1988). Observed and modelled stomatal responses to dynamic light environments in the shade plant *Alocasia macrorrhiza. Plant, Cell Environ.* **11,** 111–121.

Knapp, A. K. (1992). Leaf gas exchange in *Quercus macrocarpa* (Fagaceae): Rapid stomatal responses to variability in sunlight in a tree growth form. *Am. J. Bot.* **79,** 599–604.

Knapp, A. K., and Smith, W. K. (1987). Stomatal and photosynthetic responses during sun/shade transitions in subalpine plants: Influence on water use efficiency. *Oecologia* **74,** 62–67.

Knapp, A. K., and Smith, W. K. (1989). Influence of growth form on ecophysiological responses to variable sunlight in subalpine plants. *Ecology* **70,** 1069–1082.

Knapp, A. K., and Smith, W. K. (1990). Contrasting stomatal responses to variable sunlight in two subalpine herbs. *Am. J. Bot.* **77,** 226–231.

Kobza, J., and Seemann, J. R. (1988). Mechanisms for the light-dependent regulation of ribulose-1,5-bisphosphate carboxylase activity and photosynthesis in intact leaves. *Proc. Natl. Acad. Sci. U.S.A.* **85,** 3815–3819.

Küppers, M., and Schneider, H. (1993). Leaf gas exchange of beech (*Fagus sylvatica* L.)

seedlings in lightflecks: Effects of fleck length and leaf temperature in leaves grown in deep and partial shade. *Trees* **7**, 160–168.

Leegood, R. C., and Walker, D. A. (1980). Photosynthetic induction in wheat protoplasts and chloroplasts. Autocatalysis and light activation of enzymes. *Plant, Cell Environ.* **4**, 59–66.

Le Gouallec, J.-L., Cornic, G., and Briantais, J.-M. (1991). Chlorophyll flourescence and photoinhibition in a tropical rainforest understory plant. *Photosyn. Res.* **27**, 135–142.

Lundegarth, L. (1921). Ecological studies in the assimilation of certain forest plants. *Sven. Bot. Tidskr.* **15**, 46.

Miller, E. E., and Norman, J. M. (1971a). A sunfleck theory for plant canopies. I. Lengths of sunlit segments along a transect. *Agron. J.* **63**, 735–738.

Miller, E., and Norman, J. M. (1971b). A sunfleck theory for plant canopies. II. Penumbra effect: Intensity distributions along sunfleck segments. *Agron. J.* **63**, 739–748.

Mott, K. A., and Woodrow, I. E. (1993). Effects of O_2 and CO_2 on nonsteady-state photosynthesis. Further evidence for ribulose-1,5-bisphosphate carboxylase oxygenase limitation. *Plant Physiol.* **102**, 859–866.

Mulkey, S. S., and Pearcy, R. W. (1992). Interactions between acclimation and photoinhibition of photosynthesis of a tropical forest understory herb, *Alocasia macrorrhiza*, during simulated canopy gap formation. *Funct. Ecol.* **6**, 719–729.

Myneni, R. B., Asrar, G., Kanemasu, E. T., Lawlor, D. J., and Impens, I. (1986). Canopy architecture, irradiance distribution on leaf surfaces and consequent photosynthetic efficiencies in heterogeneous plant canopies. I. Theoretical considerations. *Agric. For. Meteorol.* **37**, 189–204.

Oberbauer, S. F., Clark, D. B., Clark, D. A., and Quesada, M. A. (1988). Crown light environments of saplings of two species of rain forest emergent trees. *Oecologia* **75**, 207–212.

Oker-Blom, P. (1984). Penumbral effects of within-plant shading on radiation distribution and leaf photosynthesis: A Monte-Carlo simulation. *Photosynthetica* **18**, 522–528.

Osterhout, W. J., and Hass, J. R. C. (1919). On the dynamics of photosynthesis. *J. Gen. Physiol.* **1**, 1–16.

Pearcy, R. W. (1983). The light environment and growth of C_3 and C_4 tree species in the understory of a Hawaiian forest. *Oecologia* **58**, 19–25.

Pearcy, R. W. (1987). Photosynthetic gas exchange responses of Australian tropical forest trees in canopy, gap and understory micro-environments. *Funct. Ecol.* **1**, 169–178.

Pearcy, R. W. (1988). Photosynthetic utilization of lightflecks by understory plants. *Aust. J. Plant Physiol.* **15**, 223–238.

Pearcy, R. W. (1990). Sunflecks and photosynthesis in plant canopies. *Annu. Rev. Plant Physiol. Plant Mol. Biol.* **41**, 421–453.

Pearcy, R. W., and Calkin, H. (1983). Carbon dioxide exchange of C_3 and C_4 tree species in the understory of a Hawaiian forest. *Oecologia* **58**, 26–32.

Pearcy, R. W., and Pfitsch, W. A. (1991). Influence of sunflecks on the delta-c-13 of *Adenocaulon bicolor* plants occurring in contrasting forest understory microsites. *Oecologia* **86**, 457–462.

Pearcy, R. W., and Seemann, J. R. (1990). Photosynthetic induction state of leaves in a soybean canopy in relation to light regulation of ribulose-1,5-bisphosphate carboxylase and stomatal conductance. *Plant Physiol.* **94**, 628–633.

Pearcy, R. W., Osteryoung, K., and Calkin, H. W. (1985). Photosynthetic responses to dynamic light environments by Hawaiian trees. The time course of CO_2 uptake and carbon gain during sunflecks. *Plant Physiol.* **79**, 896–902.

Pearcy, R. W., Roden, J., and Gamon, J. A. (1990). Sunfleck dynamics in relation to canopy structure in a soybean (*Glycine max* (L.) Merr.) canopy. *Agric. For. Meteorol.* **52**, 359–372.

Pfitsch, W. A., and Pearcy, R. W. (1989a). Daily carbon gain by *Adenocaulon bicolor*, a redwood forest understory herb, in relation to its light environment. *Oecologia* **80**, 465–470.

Pfitsch, W. A., and Pearcy, R. W. (1989b). Steady-state and dynamic photosynthetic response of *Adenocaulon bicolor* in its redwood forest habitat. *Oecologia* **80**, 471–476.

Pfitsch, W. A., and Pearcy, R. W. (1992). Growth and reproductive allocation of *Adenocaulon bicolor* following experimental removal of sunflecks. *Ecology* **73**, 2109–2117.

Pons, T. L., and Pearcy, R. W. (1992). Photosynthesis in flashing light in soybean leaves grown in different conditions. II. Lightfleck utilization efficiency. *Plant, Cell Environ.* **15**, 577–584.

Pons, T. L., Pearcy, R. W., and Seemann, J. R. (1992). Photosynthesis in flashing light in soybean leaves grown in different conditions. I. Photosynthetic induction state and regulation of ribulose-1,5-bisphosphate carboxylase activity. *Plant, Cell Environ.* **15**, 569–576.

Portis, A. R. (1992). Regulation of ribulose 1,5-bisphosphate carboxylase oxygenase activity. *Annu. Rev. Plant Physiol. Plant Mol. Biol.* **43**, 415–437.

Raschke, K. (1979). Movements of stomata. *In* "Encyclopedia of Plant Physiology" (W. Haupt and M. E. Feinleib, eds.), New Ser., Vol. 7, pp. 383–441. Springer-Verlag, Berlin.

Roden, J. S., and Pearcy, R. W. (1993). Effect of leaf flutter on the light environment of poplars. *Oecologia* **93**, 201–207.

Salvucci, M. E., Portis, A. R., and Ogren, W. L., Jr. (1986). A soluble chloroplast protein catalyzes ribulosebisphosphate carboxylase/oxygenase activation in vivo. *Photosynth. Res.* **7**, 193–201.

Sassenrath-Cole, G. F., and Pearcy, R. W. (1992). The role of ribulose-1,5-bisphosphate regeneration in the induction requirement of photosynthetic CO_2 exchange under transient light conditions. *Plant Physiol.* **99**, 227–234.

Schulze, E.-D. (1972). Die Wirking von Licht und Temperatur auf den CO_2-Gaswechel verschiedener Lebensformen aus der Kractschicht eines montanen Buchenwaldes. *Oecologia* **9**, 223–234.

Seemann, J. R., Kirschbaum, M. U. F., Sharkey, T. D., and Pearcy, R. W. (1988). Regulation of ribulose 1,5-bisphosphate carboxylase activity in *Alocasia macrorrhiza* in response to step changes in irradiance. *Plant Physiol.* **88**, 148–152.

Seemann, J. R., Kobza, J., and Moore, B. D. (1990). Metabolism of 2-carboxyarabinitol 1-phosphate and regulation of ribulose-1,5-bisphosphate carboxylase activity. *Photosynth. Res.* **23**, 119–130.

Sharkey, T. D. (1989). Evaluating the role of rubisco regulation in photosynthesis of C_3 plants. *Philos. Trans. R. Soc. London, Ser. B* **323**, 435–448.

Sharkey, T. D., Seemann, J. R., and Pearcy, R. W. (1986). Contribution of metabolites of photosynthesis to postillumination CO_2 assimilation in response to lightflecks. *Plant Physiol.* **82**, 1063–1068.

Sims, J. A., and Pearcy, R. W. (1993). Sunfleck frequency and duration affects growth rate of the understory plant, *Alocasia macrorrhiza* (L.) G. Don. *Funct. Ecol.* **10**, (in press).

Smith, W. K., Knapp, A. K., and Reiners, W. A. (1989). Penumbral effects on sunlight penetration in plant communities. *Ecology* **70**, 1603–1609.

Tang, Y. H., Wasitani, I., Tsuchiya, T., and Iwaki, H. (1988). Fluctuation of photosynthetic photon flux density within a *Miscanthus sineusis* canopy. *Ecol. Res.* **3**, 253–266.

Tinoco-Ojanguren, C., and Pearcy, R. W. (1992). Dynamic stomatal behavior and its role in carbon gain during lightflecks of a gap phase and an understory *Piper* species acclimated to high and low light. *Oecologia* **92**, 222–228.

Tinoco-Ojanguren, C., and Pearcy, R. W. (1993a). Effects of vapor pressure deficit on the stomatal response to steady-state and transient light conditions in two *Piper* species. *Oecologia*, **94**, 388–394.

Tinoco-Ojanguren, C., and Pearcy, R. W. (1993b). Stomatal versus biochemical limitations to CO_2 assimilation during transient light in sun and shade-acclimated *Piper* species. *Oecologia* **94**, 395–402.

Vines, H. M., Tu, Z.-P., Armitage, A. M., Chen, S.-S., and Black, C. C., Jr. (1983). Environmental responses of the post-lower illumination CO_2 burst as related to leaf photorespiration. *Plant Physiol.* **73**, 25–30.

von Caemmerer, S., and Edmondson, D. L. (1986). Relationship between steady-state gas exchange, *in vivo* ribulose bisphosphate carboxylase activity and some carbon reduction cycle intermediates in *Raphanus sativus. Aust. J. Plant Physiol.* **13**, 669–688.

Weber, J. A., Jurik, T. W., Tenhumen, J. D., and Gates, D. M. (1985). Analysis of gas exchange in seedlings of *Acer saccharum:* Integration of field and laboratory studies. *Oecologia* **65**, 338–347.

Woodrow, I. E., and Berry, J. A. (1988). Enzymatic regulation of photosynthetic CO_2 fixation in C_3 plants. *Annu. Rev. Plant. Physiol. Plant Mol. Biol.* **39**, 533–594.

Woodrow, I. E., and Mott, K. A. (1989). Rate limitation of non-steady-state photosynthesis by ribulose-1,5-bisphosphate carboxylase in spinach. *Aust. J. Plant Physiol.* **16**, 487–500.

Woodrow, I. E., and Mott, K. A. (1992). Biphasic activation of ribulose bisphosphate carboxylase in spinach leaves as determined from nonsteady-state CO_2 exchange. *Plant Physiol.* **99**, 298–303.

Woods, D. B., and Turner, N. C. (1971). Stomatal response to changing light by four tree species of varying shade tolerance. *New Phytol* **70**, 77–84.

Young, D. R., and Smith, W. K. (1979). Influence of sunflecks on the temperature and water relations of two subalpine understory congeners. *Oecologia* **43**, 195–205.

Zeiger, E., Iino, M., and Ogawa, T. (1985). The blue light responses of stomata: Pulse kinetics and some mechanistic implications. *Photochem. Photobiol.* **42**, 759–763.

7

Signals for Seeds to Sense and Respond to Gaps

Carlos Vázquez-Yanes and Alma Orozco-Segovia

I. Introduction

If we were capable of visualizing the landscapes of the world's surface devoid of any human influence, all the different plant communities would offer the image of a continuous or scattered cover of vegetation with sporadic heterogeneities produced by natural disturbances in the plant cover. Those disturbances are a natural part of every plant community but their frequency, magnitude, structure, and plant composition vary enormously, depending on plant community structure, the growth pattern of the constituent plants, and the nature of the disturbance agents. These disturbed spaces may or may not be characterized by drastic local changes in the environmental conditions. In most cases they are promptly occupied by new plants descended from the same components of the community or from specialized colonizing pioneer species. Colonization may start from regrowth of parts of previously existing plant individuals, established seedlings or saplings, or the germination of seed—either newly arriving by dispersal or already present in the soil.

The existing literature contains numerous examples of gap colonization by plants coming from seed. Seed physiologists have described several mechanisms that induce germination in response to sudden changes in the environmental conditions. Some ecophysiologists have attempted to integrate both kinds of information to recognize the evolutionary consequences of environmental heterogeneity produced by disturbances on the development of physiological responses.

Exploitation of Environmental Heterogeneity by Plants   209

In this review we define and describe gaps and their environmental peculiarities that represent conditions that act as cues or signals for seed germination. At the same time, we review the factors that may determine germination in response to sudden environmental changes that take place when a gap is formed.

Gaps represent areas with a marked reduction in living plant biomass in a community. As a result of the reduced competition for resources, new plants may take advantage of the newly available resources. Gaps need to be distinguished from normal phenological cycles that may produce similar profound environmental changes and that can trigger seed germination in a similar fashion as in a newly opened gap (Pons, 1984). For example, in tropical dry forests, temperature and light at the soil surface change abruptly during the leafless period (Lee, 1989; Barradas, 1991). Physiological mechanisms promoting germination during the leafless season in forests are sometimes similar to those occurring in gaps (Pons, 1983).

Newly germinated seed represents the most hazardous moment during the life of a plant and it is expected that in the course of plant evolution, many plants have acquired physiological sensors to respond to the environmental cues characteristic of a gap environment in order to take advantage of available resources.

The purpose of this review is to describe and define the physiological sensors for timing seed germination in an ecological context by using examples from the current literature. We will first describe environmental changes that characterize gaps in different communities and what appropriate sensors should be.

II. Gaps and Plant Communities

Almost every plant community experiences local disturbances due to natural local events or catastrophic events of different magnitudes. These changes or catastrophes usually occur sufficiently regularly that there are patterns of colonization and/or regeneration that characterize each community (White, 1979).

At a small scale, several kinds of disturbance can occur, for example, falling branches, death of standing plants, harvester ant activity, fecal deposits in prairies, animal burrows, or disease outbreaks. Large-scale catastrophes may include large treefalls, wind-throws, fires, elephant resting places, large-scale plague outbreaks, hurricanes, tornadoes, or landslides. Meandering rivers produce regular changes in alluvial plains in which the newly deposited alluvial soil on the beaches of the river represents a primary succession condition. Patterns of colonization in this situation may be more repeatable than in the gaps of a mature community.

In deserts, where there is not a continuous canopy of plants but rather scattered patches of vegetation interspersed with bare ground, the established plants may act as a sort of "gap" where lower soil temperatures and shade promote the germination and establishment of plants (Smith *et al.*, 1987; Valiente-Banuet *et al.*, 1991). This may also occur in other sparse plant communities like the heathlands of Scotland, where gaps are so poor in resources that plant establishment starts from the shaded periphery of established plants and progresses toward the center of the gaps (de Hullu and Gimingham, 1984).

The most frequent gap-forming disturbance of each community probably determines the biological characteristics of the gap specialists or pioneers of the community. These species might also often take advantage of less frequent large-scale disturbances. For example, in tropical forests, many gap colonizers are the same species that border roads when new roads are opened in the forests. In most communities there are usually characteristic gap-forming agents, for example, fecal deposits in grasslands with large populations of ruminants, fires in Californian chaparral, and falling trees in tall tropical humid forests. There are also regional differences; for example, rain forest gaps are much more frequent in Central America than in the plains of Borneo (Whitmore, 1975).

Of course, gaps existed in plant communities before human colonization. Man's influence has probably been to increase the magnitude and frequency of disturbances and the persistence of the disturbed environment—often for indefinite periods, and to change some of the main players in the disturbance (e.g., cattle for bison in the prairies).

A. Plant Establishment from Seed in Gaps

Gaps vary enormously in size and in the amount of residual vegetation after the gap is first formed. These two characteristics determine the manner in which the gaps will be filled with new vegetation. Some gaps can be filled primarily with new branches or ramets growing from surrounding vegetation or from vegetative growth from broken trunks, rhizomes, bulbs, etc. Other gaps or parts of gaps may be filled by growth of previously existing seedlings or with newly germinated plants. Of course, combinations of these processes may occur.

Ecologists working in forest communities have long been aware of the importance of gaps in forest dynamics. Recently there has been increased understanding of the crucial role played by the gaps in forest structure, regeneration, and species diversity and of the evolution of plant germination, establishment, and growth patterns.

Early in the study of tropical rain forest gap dynamics, gap size was initially thought to be crucial in the determination of how gaps would be refilled. Smaller gaps would be filled by vegetative growth of previously existing plants and large gaps would be occupied by new plants from seedlings or seed (Bazzaz, 1984). According to Marks (1974) there are basically two ways in which a plant community responds to a disturbance: (1) "response through the reorganization of the vegetation established prior to disturbance; and (2) response through vegetation that becomes established following disturbance." Recent research on tropical and temperate forest indicates that both of these processes may occur in gaps regardless of their size. In the temperate forest of Pennsylvania, factors like soil disturbance, uprooting of trees, and animal activity were more important in determining gap colonization from seed than gap size alone (Collins and Pickett, 1987, 1988). In tropical forests, gap size may be only one of several factors that determine the colonization. In a nutrient-poor virgin forest in Venezuela, large and small gaps were filled with previously established plants and newly germinating plants did not participate in the colonization (Uhl *et al.*, 1988). On the other hand, medium- to large-sized gaps of many Central American forests are rapidly colonized by germinating pioneer trees (Brokaw, 1985). Thus, communities differ in response to disturbance not simply because of gap size, but because of factors like the frequency of disturbance, availability of propagules, distance from the edge of undisturbed vegetation, and the degree of soil and litter removal.

Plants that are more likely to colonize newly opened gaps by seed dispersal are obviously those fruiting when the gap originates. Endozoochorous dispersal by birds or bats attracted by gaps is frequent among plants specialized in gap colonization through seed dispersal (Murray, 1986). In a study performed in Costa Rica, the tropical forest pioneer tree *Cecropia peltata* was efficiently dispersed by bats attracted to the gaps by the fruits of this species that were already present in the gaps (Fleming and Williams, 1990). But there is evidence that most gaps in tropical forest are not very attractive to animal seed dispersal vectors because they offer few resources to frugivores (Schupp *et al.*, 1989). Fleshy fruited gap plants have been found to have a variety of dispersal agents that ensure an extensive and homogeneous seed rain over the community (Estrada *et al.*, 1984).

Among anemochorous plants, one might expect ripe fruits to be borne during the time of the year when the gaps are more likely to appear, such as when windstorms are more frequent in forested areas. In the "Los Tuxtlas" rain forest in southeastern México, all the pioneer trees with anemochorous dispersal, such as *Heliocarpus appendiculatus* and *Ochroma lagopus*, disseminate their small, dry propagules during the driest days of early spring when very strong, dry winds from the south ensure a widespread and efficient distribution of the seeds over the plant community. Some gaps formed during the winter storms are still available for colonization in spring (C. Vázquez Yanes, personal observation).

A peculiar example of seed dispersed by the same disturbance agent that creates the gap is the fecal patches produced by ruminants in some prairies. The feces kills part of the plant biomass where they are deposited and at the same time provides some of the seed of the new colonizers (Brown and Archer, 1987).

B. Dormant Seed in the Soil

The relative importance of newly dispersed seed as opposed to the resident seed bank in filling of gaps by colonizing plants has been controversial, but only recently has the subject attracted much serious research. For the purpose of this review, we will consider that a seed is part of the seed bank if it is dormant in the soil before the disturbance that creates the gap takes place. It does not matter how long the seed has been lying in the soil. Most researchers define the seed bank as the dormant seed present in the soil at a given moment without any reference to the moment of arrival (Leck *et al.*, 1989). In fact, the longer that a seed remains alive in the soil under an enforced dormancy mechanism, the greater probability that the seed will be present when a gap is formed.

The structure, dynamics, density, and composition of seed banks in different plant communities have been extensively reviewed (Leck *et al.*, 1989). In many communities, the most abundant and permanent members of the seed bank are those of species specialized for gap colonization from seed (Whitmore, 1983). But there are differences among communities, and even of the same vegetation type but in different regions, in the diversity and relative importance of the gap colonizer flora. This might be the result of the structure and nature of the gaps in different geographical regions.

A comparison of viable seed quantity in seed banks of tropical forests in America, Africa, and Asia showed great variability that was mainly related to the magnitude and frequency of canopy disturbance in each locality (Putz and Appanah, 1987). In northern Australia, seed bank size and composition in tropical forests vary greatly, depending mainly on the structural differences among forest types and distance from disturbed vegetation (Graham and Hopkins, 1990).

Studies in the paleotropics and neotropics indicate that the soil seed bank is the main source of the seed-originated vegetation that gets established after a disturbance. In Panama, the composition of the seed bank of the rain forest is the same as that of the pioneer vegetation that emerges when a forest edge is produced (Williams-Linera, 1990). A study in Malaysia showed that even in the presence of a scant seed bank, the seed bank's contribution to the future vegetation in gaps is greater than that of the newly dispersed seed (Putz and Appanah, 1987).

A viable soil seed bank with an appropriate kind of environmentally enforced dormancy, which allows immediate response to the sudden change of environmental conditions, is an advantage for gap colonization. Also, since imbibition will have already occurred, any short-lasting innate dormancy would have been eliminated. On the other hand, the risk of predation would be greater for seed in the soil than for seed dispersed and germinating immediately upon arrival. In any case, the population represented by most viable seed at the beginning of gap recolonization, no matter how or when those germinating seeds arrived, will be favored.

C. Environmental Signals in Gaps for Germination

The magnitude and nature of the environmental change that takes place in a gap depends on the density of the vegetation layer that has been destroyed and the degree of soil disturbance. In a grassland, it would be expected that the destruction of the grass cover would influence factors like temperature, light, and relative humidity much less than would the opening of a large gap in the canopy of a dense forest (Goldberg and Werner, 1983).

Most gaps are essentially alterations of the plant canopy but many also have some soil disturbance due to uprooting and litter removal. There is usually an increase in the photon flux density (PFD) at the soil level. Daily fluctuations in temperature, either at the soil-surface level or in the air, increase and, as a consequence, relative humidity and soil moisture are usually reduced. The bare soil following uprooting of vegetation suffers more drastic environmental change than a litter-covered soil (Putz, 1983; Beatty and Sholes, 1988).

Examples of the environmental changes taking place in gaps have been described. Temperature changes at the soil level have been measured in tropical and temperate forests (e.g., Grubb and Whitmore, 1966; Denslow, 1980; Vázquez-Yanes and Orozco-Segovia, 1982; Collins and Pickett, 1987; Washitani and Takenaka, 1987). The basic pattern is a change from relatively stable soil temperatures beneath the canopy to considerable temperature fluctuation in the bare parts of the gap, reaching a maximum in the exposed soil where vegetation has been uprooted (Figure 1).

The time of year can also affect the magnitude of environmental change. Gaps created during the leafless part of the year cause little change in temperature or light. Winter and dry-season gaps remain uncolonized until the following growing season. Also, gaps on north-facing slopes in the Northern Hemisphere have a smaller increase in

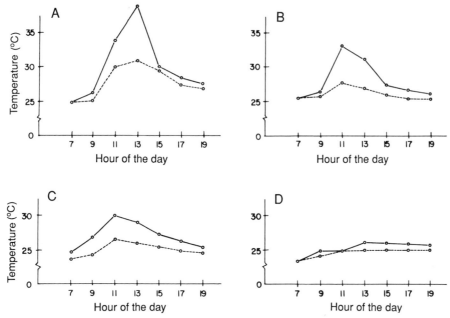

Figure 1 Soil temperatures in four sites along a light gap in a tropical forest canopy: (A) in the center of the gap, (B) near the center of the gap, (C) at the border of the gap, and (D) in the nearby forest. Solid lines indicate maximal temperatures during the period of observation and dashed lines indicate mean temperatures. (Redrawn from Vázquez-Yanes and Orozco-Segovia, 1982.)

PFD compared to those on south-facing slopes (Collins and Pickett, 1987).

Photon flux density and light quality have also been measured in gaps in communities ranging from herbaceous vegetation to forests. Canopy removal always produces an increase in PFD and a reduction in the red/far-red ratio of the light (R/FR) at the soil surface (Wilson, 1965; Goldberg and Werner, 1983; Nakashizuka, 1985; Lee, 1987; Vázquez-Yanes *et al.*, 1990) (Figure 2).

Relative air humidity and soil moisture fluctuations have not been well characterized in most communities but it is logical that in most cases the soil surface and air should be drier in gaps (Pinker, 1980; Fetcher *et al.*, 1985; Collins and Pickett, 1987). However, in herbaceous communities, Goldberg and Werner (1983) did not find differences in soil moisture among gaps of different sizes.

Other environmental changes taking place in gaps that might affect the seed are of a more chemical or biological nature and less well known. These include substances leached from the plant debris, modifications

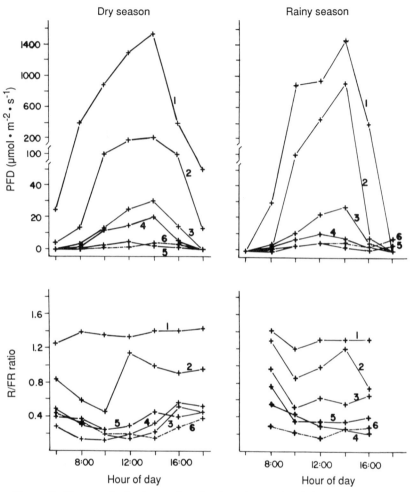

Figure 2 Photosynthetic photon flux density values and red/far red ratio of diffuse light throughout the day in six places with different degrees of canopy development in a tropical rain forest. Measurements were made during a sunny day during the dry season and a cloudy day during the rainy season (1) outside the forest in the open; (2) in a small gap in the forest without woody vegetation; (3) in a small gap with pioneer trees; (4) in secondary vegetation 10 years old; and (5, 6) in places inside mature forest. (Redrawn from Vázquez-Yanes *et al.*, 1990.)

of the soil mineral solutes or of microorganism populations, or the activity of seed pathogens and predators. The existing evidence for tropical forest indicates that seed pathogens and predators may be less active in the sunny parts of gaps (Foster, 1986). Campbell *et al.* (1989) found that in the gaps produced by falling of the Amazonian tree *Duroia hirsuta*, the debris has strong inhibitors of germination and growth of lettuce seed and seedlings. From these studies, the authors concluded that similar inhibition occurs for several gap colonizers of the area that do not grow as well in the gaps produced by the falling of this tree compared to in other gaps.

In experiments with *Plantago lanceolata* in a chalk grassland, Pons (1989) found that increased soil nitrate concentration in gap soils was sufficient to break dormancy and promote seed germination. But not all gaps in other communities show an increase in nitrate or other plant nutrients (Denslow, 1980; Collins and Pickett, 1987; Uhl *et al.,* 1988).

III. Environmentally Enforced Dormancy of Seeds

Gaps created under favorable conditions for growth are soon filled with plants that gradually change the environmental conditions described earlier. Space for new plants is reduced quickly and, therefore, one might expect that seed of plants adapted to gap colonization should have sensors that initiate germination as soon as a gap is produced.

The primary function of the several kinds of seed dormancy described in the literature (Harper, 1959) is to help segregate in time and/or space the growth of new plants of a given species. Gap colonizers must mainly rely on environmentally enforced dormancy, also known as exogenous or ecological dormancy. In some cases, innate dormancy (primary or endogenous) might be disadvantageous for gap colonizers because the seed might still be going through the physiological process of dormancy withdrawal when the gap is formed, and therefore would be unable to respond. Any innate dormancy should be restricted to the unfavorable seasons for plant growth and then disappear at the beginning of the growing season to be followed by an environmentally enforced dormancy.

Internal cycles of production and degradation of growth-regulating substances (hormones) produce innate dormancy. Such is the case during the cold stratification period of some temperate plants. Sometimes this innate dormancy is followed by enforced dormancy as in the case of *Linum catharticum* from chalk grasslands in the Netherlands (van Tooren and Pons, 1988).

A. Temperature-Regulated Dormancy

Hard-coat innate dormancy is a common characteristic of the seed of several families of plants like Fabaceae, Malvaceae, Bombacaceae, and others. Plants from these families frequently behave as gap colonizers either in tropical moist and dry forest or even in temperate areas. The mechanism that makes the coat permeable to water is purely mechanical (Barton, 1965). Coat breakage leading to seed germination takes place by abrasion, transit through guts of animals, or heat. Many seeds with an impermeable coat have a suberized layer either in the coat scleren-chyma or at the micropyle. Heat may melt this layer, allowing water to penetrate the seed (Cushwa *et al.*, 1968) (Figure 3).

Following fire, heat is the germination-triggering mechanism for these seed. The seed of *Ulex europaeus* from shrublands in northwestern Spain was found to be abundant in the soil seed bank under unburned vegeta-tion (Puentes *et al.*, 1989) and seedlings appear abundantly after fire (Pereiras *et al.*, 1985). Heat treatments around 80°C increase permeability of the seed coat and stimulate germination.

Some seed requires no more than 45°C, which can easily develop at the bare soil surface during a sunny day, to become permeable to water. Seed of *Rhus javanica*, a pioneer tree from temperate forests in Japan,

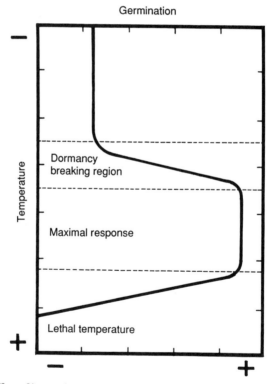

Figure 3   Effect of increasing temperatures on seed germination of hard-coated seeds. (Modified from Cushwa *et al.*, 1968.)

germinates after the removal of water-impermeable dormancy by the heat of direct insolation (Washitani and Takenaka, 1986).

The common tropical forest gap tree *Ochroma lagopus* (balsa tree: Bombacaceae) has this type of dormancy. A suberized clear layer of the palisade sclerencyma typical of the seed coat of this tree becomes permeable following heating of the soil, by superficial fires, or by marked alternations in temperature produced by the direct insolation of the soil surface (Vázquez-Yanes, 1974).

Soil temperature fluctuations characteristic of many gap environments also a cue for germination for seed lacking a hard seed coat. Thompson *et al.* (1977) and Thompson and Grime (1983) have proposed that daily temperature fluctuations, with or without the combined effect of light, provide a mechanism for sensing the presence of gaps or depth-sensing by buried seed. This can initiate establishment of plants from seed within canopy gaps in different kinds of habitats.

The internal sensor for temperature fluctuations has been related to the kinetics of certain enzymes or to the permeability of the cell membrane of the embryo (Hand *et al.*, 1982). A pioneer tree from Japanese temperate forests, *Mallotus japonicus*, requires daily temperature fluctuations of 12 to 40°C for complete germination in gaps (Washitani and Takenaka, 1987). Similar results were found for *Heliocarpus donnellsmithii*, which requires more than 10°C of daily temperature fluctuation for full germination in the Mexican tropical rain forest (Vázquez-Yanes and Orozco-Segovia, 1982). There is some evidence that a requirement for temperature fluctuations may develop with time in naturally buried seed in the field (Thompson and Whatley, 1984).

B. Phytochrome in Seed

Plants must synchronize their growth and phenological changes with the modifications of the light environment produced by seasonal changes or by structural changes in the plant community. To meet these requirements, plants have evolved various mechanisms. The most prominent of them involves phytochrome, which has been implicated in several light responses of plants such as photoperiodism and related physiological changes, photomorphogenesis, leaf movements, and light sensitivity of seed, which is called photoblastism (see Ballaré, Chapter 3, this volume).

Phytochrome operates in nature as a signal-transducing photoreceptor enabling the plant to acquire information on the light environment and so allowing modulation of physiological and developmental processes according to environmental change (Smith and Morgan, 1983). Light-controlled seed germination has been associated with phytochrome since the studies of Borthwick *et al.* (1952).

Phytochrome exists in two main forms that are interconvertible by exposure to different light qualities (wavelengths). The active (germina-tion-inducing) form, Pfr, is converted from the inactive form Pr by exposure to red light (655–665 nm). Far-red light (725–735 nm) converts Pfr to Pr, often reversing the effect of red exposure. Seed disseminated with Pr needs a light stimulus of the proper R/FR ratio, such as sunlight, after imbibition for germination. Otherwise they remain dormant, in some cases for many years (Kivilaan and Bandurski, 1973). Seed may also be disseminated with enough Pfr for dark germination, but later acquire a light requirement after burial (Wesson and Wareing, 1969) or just after exposure to low R/FR ratio light on the soil surface under a canopy (van der Veen, 1970).

Phytochrome-mediated light-quality sensitivity in seeds is quite fre-quent in nature, not only among light gap colonizers but also in many kinds of herbaceous weeds (see Ballaré, Chapter 3, this volume). Many of these weeds originally may have evolved as gap colonizers but are now abundant in the permanent "gaps" characteristic of the managed or human-transformed ecosystems (van Tooren and Pons, 1988).

Phytochrome structure and function in seed have been the object of extensive research but in some aspects remain poorly understood. Three current theories, not necessarily mutually exclusive, claim to provide an explanation of how phytochrome transduces a light stimulus result-ing in a physiological response (Colbert, 1988). Phytochrome has been proposed (1) to have an enzyme kinase activity in the Pfr form, (2) to regulate gene expression by direct interaction of Pfr with the genome, and (3) to change cell-membrane permeability. None of these hypothe-ses provides a complete explanation of all aspects of the role phyto-chrome plays in plants. Several characteristics of the mode of action of phytochrome *in vivo* manifest dimmerlike behavior (Brockmann *et al.*, 1987).

Other, poorly understood aspects of phytochrome function are the kinetics of the dark reversion and effects of blue light and light near the absorption peaks of phytochrome (Frankland, 1976; Tanno, 1983). Intermediate forms of phytochrome are currently the object of active research and the findings may provide the cue for understanding many aspects of phytochrome function (Casal and Smith, 1989).

In nature, activation of phytochrome conversion in seed takes place primarily in moist or totally imbibed conditions. If dry seed responds, it requires very high PFD for conversion of Pr to Pfr, whereas often conversion of the Pfr to Pr occurs in the dark or at low PFD with low R/FR ratios (McArthur, 1978). This ensures that the seed maturing in sunlight becomes light requiring after dehydration or dispersal onto soils covered with vegetation.

C. Light-Regulated Dormancy

Seed requiring light for germination is called positive photoblastic seed. Positive photoblastism has essentially two ecological roles: the preservation of dormancy of buried seed or of seed of heliophiles dispersed to shaded sites. Photoblastic species differ in their sensitivity to a particular Pfr/Pt (active phytochrome/total phytochrome) photoequilibrium established by a given R/FR ratio (Smith, 1982). Some seed may germinate under light of a very low R/FR ratio but not in darkness, indicating that the Pfr/Pt required to trigger germination can be attained beneath green canopies but not in the darkness produced by deep soil burial (van Rooden *et al.*, 1970; Vázquez-Yanes and Orozco-Segovia, 1990a).

Seed beneath green canopies or litter may still receive enough light to promote germination, indicating that only extremely low PFD is required for phytochrome-mediated light sensitivity. For example, *Digitalis purpurea* can be stimulated to 70% germination by exposure to light of only 0.026 μmol m^{-2} s^{-1} (Bliss and Smith, 1985). In this case, the factor inhibiting germination under canopies is light quality, because light penetrating through plant green foliage or litter is enriched in Fr relative to R. Sunlight has a mean R/FR ratio of 1.2, but beneath green canopies the ratio may be reduced to levels below 0.5 (Frankland, 1981; Smith, 1982).

There are numerous examples of the germination-inhibiting effect of light filtered by green canopies in various plant communities from tropical forests to temperate grasslands (Figure 4) (Cumming, 1963; Black, 1969; Taylorson and Borthwick, 1969; Popay and Roberts, 1970; van der Veen, 1970; King, 1975; Gorski, 1975; Vázquez-Yanes, 1976; Valio and Joly, 1979; Fenner, 1980a,b, Silvertown, 1980; Amminuddin and Ng, 1982; Vázquez-Yanes and Smith, 1982; Pons, 1983; Washitani and Saeki, 1984; Gross, 1985; Washitani, 1985; Orozco-Segovia and Vázquez-Yanes, 1989; Raich and Gong, 1990).

Light penetration through soil is greatly affected by factors such as soil moisture, particle size, and color but it appears that significant amounts of physiologically active light rarely penetrate more than 4 or 5 mm and often less. Some soils act as filters reducing the R/FR ratio (Tester and Morris, 1987).

Buried dormant seed may respond to the light stimulus created by their sudden exhumation as may take place during the plowing of agricultural land or in other forms of soil disturbance. In this case the environmental dormancy is enforced mainly by darkness and a short burst of light may trigger germination (Sauer and Struik, 1964).

During gap formation the light environment affecting light-mediated seed germination may change in three ways. First, buried seed may be

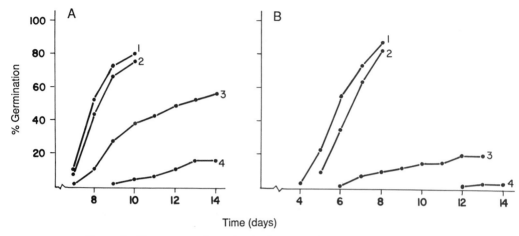

Figure 4   Time course of germination in a natural gap in a tropical rain forest. Positions 1 and 2 were in the center of the light gap and positions 3 and 4 were progressively closer to the periphery of the gap. No germination occurred after 1 month in any of the remaining positions inside the forest. (A) *Cecropia obtusifolia* and (B) *Piper auritum.* (Redrawn from Vázquez-Yanes and Smith, 1982.)

exhumed by uprooting of plants. Second, smaller soil disturbance may expose seeds to more light, and third, canopy and/or litter disturbance may drastically alter the R/FR ratio of the light reaching the soil surface (Vázquez-Yanes and Orozco-Segovia, 1992).

One of the characteristics of phytochrome-mediated germination is its photoreversibility if this occurs before the so-called escape time, which can be defined as the time after which the effect of R can no longer be reversed by FR (Duke *et al.*, 1977). Phytochrome activation by a high R/FR ratio can be inactivated by a subsequent exposure to low R/FR, provided the escape time has not elapsed. The ecological consequences of the escape time may be related to detection of true light gaps as opposed to sunflecks (Vázquez-Yanes and Orozco-Segovia, 1987; Orozco-Segovia *et al.*, 1987; Orozco-Segovia and Vázquez-Yanes, 1989). Sunfleck incidence on a given soil patch is brief and unpredictable depending on canopy structure, weather, and season (Chazdon, 1988). On the other hand, gaps, depending on their size, result in more predictable occurrences of direct sunlight lasting for longer periods of time. High-R/FR light reaches the soil for long and repeatable periods of time (Vázquez-Yanes *et al.*, 1990), bringing a larger fraction of the seed to the escape time than in sunflecks. This is the case for the tropical American pioneer tree *Cecropia obtusifolia*, in which the escape time involves hours (Vázquez-Yanes and Smith, 1982). In *Piper* species with contrasting habitat specificities, the escape time for each species was determined by manipulating light quality and quantity using white light, far red light, and darkness in the laboratory and, in the field, forest shade light, nongreen shade light outside the forest, and darkness. The results showed that each one of the four species has a different response to the treatments, but in all cases the escape time is reached after long periods of irradiation with light of the appropriate R/FR ratio for achieving the phytochrome photoequilibrium for the response threshold characteristic of each species (Orozco-Segovia and Vázquez-Yanes, 1989) (Figure 5).

The differences in light responses between herbaceous plants in open places and pioneer plants from the forest are related to the two primary roles of photoblastism: (1) the response to momentary light stimulus in the sudden exposure of buried seed to the surface light environment and (2) the detection of a repeatable change in the forest light environment, by changes in R/FR ratio resulting from opening of canopy gaps (Vázquez-Yanes and Orozco-Segovia, 1990a).

D. Light-Controlled Germination inside Natural Plant Communities

Most seed research has been performed with seed collected directly from the mother plant and used either immediately or after storage in artificial conditions. There is little information about seed normally dispersed and incorporated into the soil seed bank. We know very little about the effect of endozoochorous dispersal on subsequent seed behavior, or how quickly the dormancy mechanisms are modified when seed is in the soil and or exposed to the light environment in the plant community.

Dispersal by animals may alter seed dormancy by chemical and physical treatment during transit through animal guts. One simple change during endozoochorous dispersal is the increased moisture content that the seed acquires. Experiments performed with *C. obtusifolia* seed indicate that the short transit period in bat intestines is less deleterious than the longer passage through monkey intestines. Passage through monkey intestines alters the optical properties of the seed coat and therefore the photoblastic response (Vázquez-Yanes and Orozco-Segovia, 1986).

Seed that remains dormant for a period in the soil may develop induced (secondary) dormancy, altering its response to environmental conditions that enforced its original dormancy. Examples of this come from experiments on weedy species that are not necessarily related to light gap responses (reviewed by Karssen, 1980).

Seeds may lose resistance to germination or become light insensitive at certain periods of the year, often corresponding with times of establishment (e.g., Marks and Nwachuku, 1986; Baskin and Baskin, 1987).

Water content of seed when dispersed to the soil surface affects the R/FR ratio required to induce germination. Preimbibed seed of *Piper*

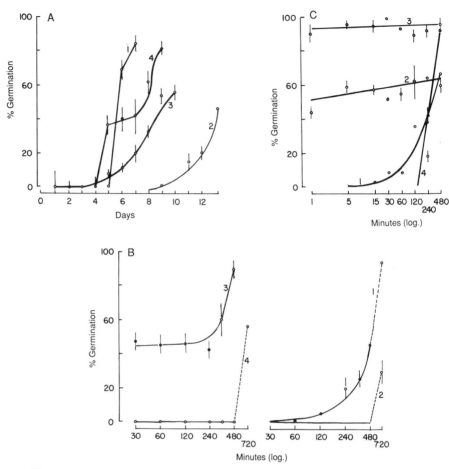

Figure 5 (A) Germination after exposure to days of white light in 12-hour photoperiods. (B) Response to different periods of white light and far red in filtered Plexiglas chambers (narrow-band transmittance). (C) Germination after different periods of full diffuse light and light filtered by a green canopy. In (B) and (C), the seed was in far red for 12-hour photoperiods during 13 days. Against this background, there were daily applications of white light during the periods indicated along the logarithmic abscissa. The discontinuous line of (B) indicates germination in the field outside the green canopy under full diffuse light conditions. (1) *Piper auritum;* (2) *P. aequale;* (3) *P. hispidum;* and (4) *P. umbellatum.* (Redrawn from Orozco-Segovia and Vázquez-Yanes, 1989.)

auritum in a Mexican rain forest germinates at lower R/FR ratios than does dry seed. It could be that phytochrome sensitivity depends on the degree of hydration when the seed is first exposed to the forest light environment (A. Orozco-Segovia, unpublished data).

Light sensitivity and photoreversion timing allow the seed of gap pioneers to remain dormant for a period after arrival in the community. Subsequently, the seed may gradually change its physiological properties and germinate, or alternatively remain dormant under the control of factors such as light quality and quantity beneath the litter or in the soil or chemical characteristics of the soil such as partial anaerobiosis.

Only a few experiments on light sensitivity have been performed with seed coming from the soil seed bank. Experiments with *Piper* species in a Mexican rain forest, in which imbibed seed was buried for up to 1 year before testing for germination, showed that the light requirements for germination changed in different ways through time for different species. Seed may become either more sensitive or indifferent to light or change in its response to the R/FR ratio (Vázquez-Yanes and Orozco-Segovia, 1986; Orozco-Segovia and Vázquez-Yanes, 1989) (Figure 6).

E. Interactions of Light and Temperature

Many laboratory experiments have demonstrated that the germination of seed with positive light photoblastism can be modulated by temperature conditions, especially diurnal temperature fluctuations (Thompson and Watley, 1984). Formation of gaps leads to greater fluctuation of soil surface temperatures because of the direct insolation. These temperature fluctuations may alter the response of seed light conditions in four ways: (1) the speed of germination may be altered by shortening the escape time (Orozco-Segovia *et al.*, 1987); (2) higher temperatures may affect membrane permeability in the same way as the active form of phytochrome; (3) some of the phytochrome intermediates are known to be temperature dependent (Kendrick and Spruit, 1977); and (4) temperature may regulate the formation of products with which light-promoted germination mechanisms interact (Erasmus and van Staden, 1986). Because of this, some seeds that are dormant in darkness or in low R/FR ratios of light at constant temperature may germinate in those conditions under a fluctuating temperature regime (Thompson *et al.*, 1977; Thompson and Grime, 1983; Williams, 1983). The tropical pioneer tree *C. obtusifolia* showed partial germination under a fluctuating temperature regime of 20–30°C even in darkness. Most seed samples of this species do not germinate at constant temperatures without light (Vázquez-Yanes and Orozco-Segovia, 1990a).

Although light sensitivity experiments in relation to formation of light gaps have been performed for only few species and in only a few locations, one might speculate that this must be the most common seed dormancy mechanism of light gap pioneers in communities with continuous green canopies during the growing season. In temperate forests, a similar dormancy mechanism of some understory weeds can induce germination when the temperature increases in the spring but the com-

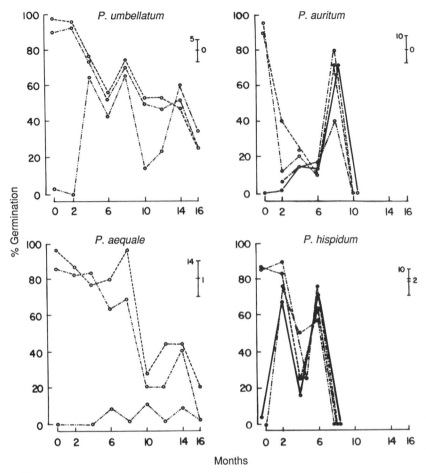

Figure 6 Effect of light quality on the germination of four *Piper* species. The seed was buried for different periods of time. Germination in white light (---), red light (-··-), far red (-···-), and darkness (—). Treatments were given at 25°C. (Redrawn from Orozco-Segovia and Vázquez-Yanes, 1989.)

munity still lacks a green canopy (Fenner, 1980b; Pons, 1983, 1984). Thus, photoblastism may have another evolutionary origin besides opportunistic colonization of gaps.

F. Dormancy and Germination beneath Litter

The litter layer may become an important element in controlling germination and establishment. Litter can be a physical barrier for the emergence of seedlings derived from small seed that lack the reserves to penetrate the litter. Litter can prevent seed from reaching the soil and provide an inappropriate germination bed where seedlings fail to establish. On the other hand, litter may act as a nursery that holds enough moisture for the germination of medium- to large-sized recalcitrant seed (Day *et al.*, 1986; Carson and Peterson, 1990). The litter inhibits germination by blocking light, by reducing temperature fluctuations (Beatty and Sholes, 1988) (Figure 7), or by chemical inhibitors in the litter (de Jong and Klinkhamer, 1985; Facelli and Pickett, 1991). Litter also preferentially transmits longer wavelengths that may be strongly inhibitory for germination of photoblastic seeds (Figure 8). (Goldberg and Werner, 1983; Bliss and Smith, 1985; Vázquez-Yanes *et al.*, 1990).

Although the physiological ecology of germination as affected by litter has received little study, there is much evidence that litter disturbance is an important factor in determining the germination of gap pioneers. In temperate forests, a greater appearance of pioneers in gaps occurs in the areas where the litter has been removed by the uprooting of trees or by human activities (Carson and Peterson, 1990). In tropical forests, pits and mounds of gaps resulting from roots and crowns of felled trees offer different opportunities for the establishment of seedlings due in part to litter disturbance (Putz, 1983). The presence of litter may be an important factor in determining the failure of members of the seed bank in colonizing some gaps in tropical forests (Putz and Appanah, 1987; Williams-Linera, 1990).

Some tropical and temperate forests are characterized by a low frequency of disturbance. Gaps in these forests normally do not exhibit much litter and soil disturbance because the dead trees are not large enough, they do not fall to the ground, or they fall without uprooting. True pioneers are rare or nonexistent in these communities and the

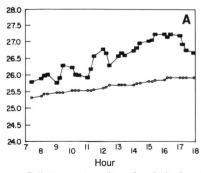

Daily temperature alternations in the forest

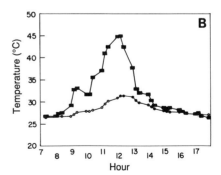

Daily temperature alternations in the open

Figure 7   Insulating effect of litter on temperature fluctuations (A) inside the forest and (B) in a gap, beneath litter (O) and above the litter (■). (C. Vázquez-Yanes, unpublished.)

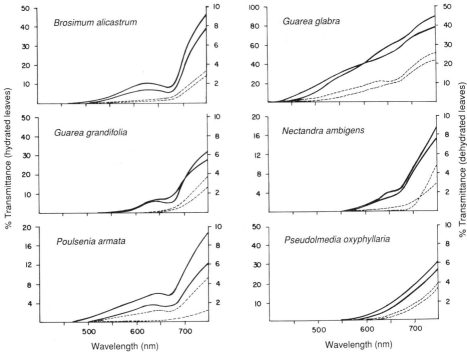

Figure 8   Percentage transmittance of two wet (—) and two dry (----) leaf litter layers of several rain forest species. There is a shift to longer wavelengths that reduces the red/far red ratio. (Redrawn from Vázquez-Yanes *et al.*, 1990.)

species in these forests lack the germination-controlling mechanisms characteristic of pioneers (Shimizu, 1984; Hara, 1985; Putz and Appanah, 1987; Armesto and Fuentes, 1988; Uhl *et al.*, 1988). This could indicate that the evolution of the seed physiology that characterizes light-gap-germinating pioneers occurred mainly under the joint effect of light-gap formation and litter and soil disturbance.

G. Seed Longevity in Soil

The longevity of seed in controlled storage has been the subject of considerable research that has led to the definition of two basic kinds of seed: the orthodox seed that can be dehydrated and the recalcitrant seed that cannot. Each has two subtypes. True orthodox seed can be dehydrated to levels close to 5% (wet basis) moisture content and stored at temperatures below 0°C for very long periods. The suborthodox seed differs in that it has lipid reserves and its survival in cold storage is relatively short because of chemical changes of the lipids. The temperate recalcitrant seed cannot be dehydrated below 20% moisture content but can be stored in temperatures close to 0°C for up to 3 years. The tropical recalcitrant seed, or true recalcitrant, must remain moist, does not tolerate low temperatures, and remains viable for only a few months (Roberts and Ellis, 1989). There is an intermediate category between the orthodox and the true recalcitrant seed produced by some tropical plants like coffee and papaya that can be stored after specific dehydration treatments (Ellis *et al.*, 1990, 1991).

It would seem that the previous classification is only relevant for seed storage purposes, but it also reflects the adaptation of seed to germinate under specific conditions as well as the differences in ecological longevities that might exist in nature. Tropical recalcitrant seed maintains a certain level of respiration and metabolic activity along with a short quiescence after dissemination and a minimal dormancy period. All of these characteristics promote germination soon after arrival to the soil and the seed has little longevity in nature. This is typical of most large-seeded tropical rain forest trees. Temperate recalcitrant seed is similar but it probably remains quiescent through the relatively mild winter before germination.

On the other hand, orthodox seed is adapted to go through a complex rearrangement of cellular components with removal of all free water in the cells during dehydration and cessation of all metabolic activity. Such changes allow cold storage at very low temperatures. These properties reflect the fact that many orthodox seeds in nature go through periods of dormancy during inappropriate environmental conditions for germination and establishment. These orthodox seeds are typical of species of arid, semiarid, and very seasonal regions, but they are also prevalent among opportunistic colonizers of many types. Probably most species that form seed banks and are adapted to exploiting gaps belong to this group. In the tropical rain forest of "Los Tuxtlas," the typical pioneer tree species produce orthodox seed whereas most other tree species produce primarily recalcitrant seed (Vázquez-Yanes and Orozco-Segovia, 1990b). Most seed may not achieve deep dehydration in continuously moist soils, but cessation of respiratory metabolism is probably essential for survival in the dormant state for any length of time. Otherwise seed reserves would be rapidly consumed, leading to seed death or prevention of establishment because the seedlings are so weak. Some orthodox, photoblastic seeds from tropical rain forest plants may remain dormant for more than 3 years in imbibed dark conditions at room temperature (Orozco-Segovia and Vázquez-Yanes, 1990).

Longevity of seed in the soil has been studied mostly in the context of weed control strategies. In most experiments, the seed is buried in containers, mesh bags, and other protected conditions that may modify soil factors and change the real ecological longevity of the seed (Priestley, 1986).

The longevity of dormant seed naturally disseminated to the soil and incorporated into the seed bank has always been difficult to determine but some studies have been reviewed (Priestley, 1986). Arrival time of seed dated by independent evidence like historical records, such as known ages of construction, etc., indicates periods of survival in the soil of several years, decades, or even centuries, primarily for photoblastic weed seed from temperate areas (Odum, 1965). Such longevity is probably the exception rather than the rule. Studies based on population dynamics of seed of the rain forest gap pioneer *Cecropia obtusifolia* in México suggest that it usually survives in the soil for much less than a year (Alvarez-Buylla and Martínez-Ramos, 1990).

IV. Conclusions

After reviewing the extensive literature on ecology and physiological ecology of seed germination, there is little doubt that one of the primary functions of light and temperature fluctuation sensors in seed is for the regulation of germination in gaps. These control systems extend the survival time of seed beneath green canopies and/or litter before a light gap or litter gap is formed.

However, most of this evidence is from controlled laboratory studies that may not adequately represent field conditions. Many field experiments performed with the goal of detecting the factors inducing dormancy and germination in natural conditions lack appropriate controls or a detailed knowledge of actual light, temperature, and humidity conditions. This is particularly apparent when small changes from the true natural condition have to be made in the field in order to conduct a tractable experiment (Raich and Gong, 1990). On the other hand, physiological laboratory experiments are generally not designed to represent the natural environment where seed is actually adapted to germinate or not germinate.

For more insightful research into the physiological ecology of light- and temperature-regulated seed germination, it is necessary to design experiments that would fulfill several requirements such as careful monitoring of environmental conditions of undisturbed and disturbed communities at canopy and ground levels. In field experiments, very often the change of one factor like light is accompanied by changes in other factors that are not measured such as temperature, air, and soil humidity, etc. The determination of the light environment must include precise light quality and R/FR ratio measurements.

Experiments should be performed simultaneously in the laboratory and in the field, but laboratory experiments should be designed with environmental factors simulating those properties that characterize the natural environment. Some experiments should also be performed with seed that has undergone natural dispersal and soil environmental conditions to detect changes in dormancy mechanisms through time.

Most of the plants that colonize light gaps are poorly known with respect to many aspects of their life history. This is often a source of confusion for those who attempt to study light-gap colonizing mechanisms. The main characteristic that separates true pioneers from other plants that grow in gaps is their seed biology. Species to be studied should be true pioneers that become established from seed in light gaps (Whitmore, 1989).

References

Alvarez-Buylla, E., and Martínez-Ramos, M. (1990). Seed banks versus seed rain in the regeneration of a tropical pioneer tree. *Oecologia* **84,** 314–325.

Amminuddin, B. M., and Ng, F. S. P. (1982). Influence of light on germination of *Pinus caribaea* and *Vitex pinnata. Malays. For.* **45,** 62–68.

Armesto, J. J., and Fuentes, E. R. (1988). Tree species regeneration in a mid-elevation, temperate rain forest in Isla de Chiloé, Chile. *Vegetatio* **74,** 151–159.

Barradas, V. L. (1991). Radiation regime in a tropical dry deciduous forest in western Mexico. *Theor. Appl. Climatol.* **44,** 57–64.

Barton, L. V. (1965). Dormancy in seeds imposed by the seed coat. *In* "Encyclopedia of Plant Physiology" (W. Rühland, ed.), Vol. 15, Part 2, pp. 727–745. Springer-Verlag, Berlin and New York.

Baskin, J. M., and Baskin, C. C. (1987). Seasonal changes in germination responses of buried seeds of *Portulaca smalii. Bull. Torrey Bot. Club* **114,** 169–172.

Bazzaz, F. A. (1984). Dynamics of wet tropical forest and their species strategies. *In* "Physiological Ecology of Plants of the Wet Tropics" (E. Medina, H. A. Mooney, and C. Vázquez-Yanes, eds.), pp. 233–243. Dr. W. Junk Publishers, The Hague, The Netherlands.

Beatty, S. W., and Sholes, O. D. V. (1988). Leaf litter effect on plant species composition of deciduous forest treefall pits. *Can. J. For. Res.* **18,** 553–559.

Black, M. (1969). Light controlled germination of seeds. *Symp. Soc. Exp. Biol.* **23,** 193–217.

Bliss, D., and Smith, H. (1985). Penetration of light into soil and its role in the control of seed germination. *Plant, Cell Environ.* **8,** 475–483.

Borthwick, H. A., Hendricks, S. B., Toole, E. H., and Toole, V. K. (1952). A reversible photoreaction controlling seed germination. *Proc. Natl. Acad. Sci. U.S.A.* **38,** 662–666.

Brockmann, J., Rieble, S., Kazarinova-Fukshansky, N., Seyfried, M., and Schäfer, E. (1987). Phytochrome behaves as a dimmer in vivo. *Plant, Cell Environ.* **10,** 105–111.

Brokaw, N. M. (1985). Gap-phase regeneration in a tropical forest. *Ecology* **66,** 682–687.

Brown, J. R., and Archer, S. (1987). Woody plant seed dispersal and gap formation in a North American subtropical savanna woodland: The role of domestic herbivores. *Vegetatio* **73,** 73–80.

Campbell, D. G., Richardson, P. M., and Rosas, A., Jr. (1989). Field screening for allelopathy in tropical forest trees, particularly *Duroia hirsuta,* in the Brazilian Amazon. *Biochem. Syst. Ecol.* **17,** 403–407.

Carson, W. P., and Peterson, C. J. (1990). The role of litter in an old-field community: Impact of litter quantity in different seasons on plant species richness and abundance. *Oecologia* **85,** 8–13.

Casal, J. J., and Smith, H. (1989). The function, action and adaptive significance of phytochrome in light-grown plants. *Plant, Cell Environ.* **12,** 855–862.

Chazdon, R. L. (1988). Sunflecks and their importance to forest understory plants. *Adv. Ecol. Res.* **18,** 1–63.

Colbert, J. T. (1988). Molecular biology of phytochrome. *Plant, Cell Environ.* **11,** 305–318.

Collins, B. S., and Pickett, S. T. A. (1987). Influence of canopy opening on the environment and herb layer in a northern hardwoods forest. *Vegetatio* **70,** 3–10.

Collins, B. S., and Pickett, S. T. A. (1988). Demographic responses of herb layer species to experimental canopy gaps in a northern hardwoods forest. *J. Ecol.* **76,** 437–450.

Cumming, B. G. (1963). The dependence of germination photoperiod light quality and temperature in *Chenopodium* spp. *Can. J. Bot.* **41,** 1211–1233.

Cushwa, C. T., Martin, R. E., and Miller, R. L. (1968). The effects of fire on seed germination. *J. Range Manage.* **21,** 250–254.

Day, A. D., Ludeke, K. L., and Thames, J. L. (1986). Revegetation of coal mine soil with forest litter. *J. Arid Environ.* **11,** 249–253.

de Hullu, E., and Gimingham, C. H. (1984). Germination and establishment of seedlings in different phases of the *Calluna* life cycle in a Scottish heathland. *Vegetatio* **58,** 115–121.

de Jong, T. J., and Klinkhamer, G. L. (1985). The negative effects of litter of parent plants of *Cirsium vulgare* on their offspring: Autotoxicity or immobilization? *Oecologia* **65,** 153–160.

Denslow, J. S. (1980). Gap partitioning among tropical rain forest trees. *Biotropica* **12,** Suppl., 47–55.

Duke, O. S., Egley, G. H., and Reger, B. J. (1977). Model for variable light sensitivity in imbibed dark-dormant seeds. *Plant Physiol.* **59,** 244–249.

Ellis, R. H., Hong, T. D., and Roberts, E. H. (1990). An intermediate category of seed storage behaviour? *J. Exp. Bot.* **41,** 1167–1174.

Ellis, R. H., Hong, T. D., and Roberts, E. H. (1991). Effect of storage temperature and moisture on the germination of papaya seeds. *Seed Sci. Res.* **1,** 69–72.

Erasmus, D. J., and van Staden, J. (1986). Germination of *Chromolaena odorata (L.) K. & R.* achenes: Effect of temperature, imbibition and light. *Weed Res.* **26,** 75–81.

Estrada, A., Coates-Estrada, R., and Vázquez-Yanes, C. (1984). Observation on fruiting and dispersers of *Cecropia obtusifolia* at Los Tuxtlas, Mexico. *Biotropica* **16,** 315–318.

Facelli, J. M., and Pickett, S. T. A. (1991). Plant litter: Its dynamics and effects on plant community structure. *Bot. Rev.* **57,** 1–32.

Fenner, M. (1980a). The inhibition of germination of *Bidens pilosa* seeds by leaf-canopy shade in some natural vegetation types. *New Phytol.* **84,** 95–101.

Fenner, M. (1980b). The induction of a light requirement in *Bidens pilosa* seeds by a leaf canopy shade. *New Phytol.* **84,** 103–106.

Fetcher, N., Oberbauer, S. F., and Strain, B. R. (1985). Vegetation effects on microclimate in lowland tropical forest in Costa Rica. *Int. J. Biometeorol.* **29,** 145–155.

Fleming, T. H., and Williams, C. F. (1990). Phenology, seed dispersal, and recruitment in *Cecropia peltata* (Moraceae) in Costa Rican tropical dry forest. *J. Trop. Ecol.* **6,** 163–178.

Foster, S. A. (1986). On the adaptive value of large seeds for tropical moist forest trees: A review and synthesis. *Bot. Rev.* **52,** 260–299.

Frankland, B. (1976). Phytochrome control of seed germination in relation to the light environment. *In* "Light and Plant Development" (H. Smith, ed.), pp. 477–491. Butterworth, London.

Frankland, B. (1981). Germination in shade. *In* "Plants and the Daylight Spectrum" (H. Smith, ed.), pp. 187–204. Academic Press, London.

Goldberg, D. E., and Werner, P. A. (1983). The effects of size of opening in vegetation and litter cover on seedling establishment of goldenrods (*Solidago* spp). *Oecologia* **60**, 149–155.

Gorski, T. (1975). Germination of seeds in the shadow of plants. *Physiol. Plant.* **34**, 342–346.

Graham, A. W., and Hopkins, M. S. (1990). Soil seed banks of adjacent unlogged rainforest types in north Queensland. *Aust. J. Bot.* **38**, 261–268.

Gross, K. L. (1985). Effects of irradiance and spectral quality on the germination of *Verbascum thapsus* L. and *Oenothera biennis* L. seeds. *New Phytol.* **101**, 531–541.

Grubb, P. J., and Whitmore, T. C. (1966). A comparison of montane and lowland rain forest in Ecuador. II. The climate and its effect on the distribution and physiognomy of the forest. *J. Ecol.* **543**, 303–333.

Hand, D. J., Craig, G., Takaki, M., and Kendrick, R. E. (1982). Interactions of light and temperature on seed germination of *Rumex obtusifolius* L. *Planta* **156**, 457–468.

Hara, M. (1985). Forest response to gap formation in a climax beech forest. *Jpn. J. Ecol.* **35**, 337–343.

Harper, J. L. (1959). The ecological significance of dormancy and its importance in weed control. *Proc. Int. Congr. Crop Prot. 4th, 1957,* Vol. 1, pp. 415–420.

Karssen, C. M. (1980). Environmental conditions and endogenous mechanisms involved in secondary dormancy of seeds. *Isr. J. Bot.* **29**, 45–64.

Kendrick, R. E., and Spruit, C. J. P. (1977). Phototransformations of phytochrome. *Photochem. Photobiol.* **26**, 201–214.

King, T. J. (1975). Inhibition of seed germination under leaf canopies in *Arenaria serpyllifolia, Veronica arvensis* and *Cerastum halostioides. New Phytol.* **75**, 87–90.

Kivilaan, A., and Bandurski, R. S. (1973). The ninety-year period for Dr. Beal's seed viability experiment. *Am. J. Bot.* **60**, 140–145.

Leck, L. A., Simson, R. L., and Parker, V. T. (1989). "Ecology of Seed Banks." Academic Press, London.

Lee, D. W. (1987). The spectral distribution of radiation in two neotropical rainforest. *Biotropica* **19**, 161–166.

Lee, D. W. (1989). Canopy dynamics and light climates in a tropical moist deciduous forest in India. *J. Trop. Ecol.* **5**, 65–79.

Marks, M. K., and Nwachuku, A. C. (1986). Seed-bank characteristics in a group of tropical weeds. *Weed Res.* **26**, 151–157.

Marks, P. L. (1974). The role of pin cherry (*Prunus pennsylvanica* L.) in the maintenance of stability in northern hardwood ecosystems. *Ecol. Monogr.* **44**, 73–88.

McArthur, A. J. (1978). Light effects upon dry lettuce seeds. *Planta* **144**, 1–5.

Murray, K. G. (1986). Consequences of seed dispersal for gap-dependent plants: Relationships between seeds shadows, germination requirements and forest dynamic processes. *In* "Frugivores and Seed Dispersal" (A. Estrada and T. H. Fleming, eds.), pp. 187–198. Dr. W. Junk Publishers, Dordrecht, The Netherlands.

Nakashizuka, T. (1985). Diffused light conditions in canopy gaps in a beech (*Fagus crenata* Blume) forest. *Oecologia* **66**, 472–474.

Odum, S. (1965). Germination of ancient seeds. *Dan. Bot. Ark.* **24**, 1–70.

Orozco-Segovia, A., and Vázquez-Yanes, C. (1989). Light effect on seed germination in *Piper* L. *Acta Oecol. Oecol. Plant.* **10**, 123–146.

Orozco-Segovia, A., and Vázquez-Yanes, C. (1990). Effect of moisture in some rain forest seed species. *Biotropica* **22**, 215–216.

Orozco-Segovia, A., Vázquez-Yanes, C., Coates-Estrada, R., and Pérez-Nasser, N. (1987). Ecophysiological characteristics of the seed of the tropical forest pioneer *Urera caracasana* (Urticaceae). *Tree Physiol.* **3**, 375–386.

Pereiras, J., Puentes, M. A., and Casal, M. (1985). Efecto de las altas temperaturas sobre la germinación de las semillas del tojo (*Ulex europeaus* L.). *Stud. Oecol.* **6**, 125–133.

Pinker, R. (1980). The microclimate of a dry tropical forest. *Agric. Meteorol.* **22,** 249–265.

Pons, T. L. (1983). Significance of inhibition of seed germination under the leaf canopy in ash coppice. *Plant, Cell Environ.* **6,** 385–392.

Pons, T. L. (1984). Possible significance of changes in the light requirements of *Cirsium palustre* seeds after dispersal in ash coppice. *Plant, Cell Environ.* **7,** 263–268.

Pons, T. L. (1989). Breaking of seed dormancy by nitrate as a gap detection mechanism. *Ann. Bot. (London)* [N.S.] **63,** 139–143.

Popay, A. I., and Roberts, E. H. (1970). Ecology of *Capsella bursa-pastoris* (L.) Medik. and *Senecio vulgaris* L., in relation to germination behaviour. *J. Ecol.* **58,** 123–139.

Priestley, D. A. (1986). "Seed Aging." Cornell Univ. Press, Ithaca, NY.

Puentes, M. A., Pereiras, J., and Casal, M. (1989). Estudio del banco de semillas de *Ulex europaeus* L. en matorrales de Galicia (NW España). I. Primeros resultados. *Rev. Écol. Biol. Sol* **26,** 1–10.

Putz, F. E. (1983). Treefall pits and mounds, buried seeds, and the importance of soil disturbance to pioneer tree species in the tropics. *Ecology* **64,** 1069–1074.

Putz, F. E., and Appanah, S. (1987). Buried seeds, newly dispersed seeds, and the dynamics of a lowland forest in Malaysia. *Biotropica* **19,** 326–333.

Raich, J. W., and Gong, W. K. (1990). Effects of canopy openings on tree seed germination in a Malaysian dipterocarp forest. *J. Trop. Ecol.* **22,** 146–154.

Roberts, E. H., and Ellis, R. H. (1989). Water and seed survival. *Ann. Bot. (London)* [N.S.] **63,** 39–52.

Sauer, J., and Struick, G. (1964). A possible ecological relation between soil disturbance, light flash and seed germination. *Ecology* **45,** 884–886.

Schupp, E. W., Howe, H. F., Augspurger, C. K., and Levey, D. J. (1989). Arrival and survival in tropical treefall gaps. *Ecology* **70,** 562–569.

Shimizu, Y. (1984). Regeneration of the subtropical evergreen broad-leaved forest at Chichijima in the Bonin (Ogasawara) Islands with reference to an environmental gradient and canopy gaps. *Jpn. J. Ecol.* **34,** 87–100.

Silvertown, J. (1980). Leaf-canopy-induced seed dormancy in a grassland flora. *New Phytol.* **85,** 109–118.

Smith, H. (1982). Light quality, photoperception, and plant strategy. *Annu. Rev. Plant Physiol.* **33,** 481–518.

Smith, H., and Morgan, D. C. (1983). The function of phytochrome in nature. *In* "Encyclopedia of Plant Physiology" (O. L. Lange, P. S. Nobel, C. B. Osmond, and H. Ziegler, eds.), New Ser., Vol. 16b, pp. 491–512. Springer-Verlag, Berlin.

Smith, S. D., Patten, D. T., and Monson, R. K. (1987). Effects of artificially imposed shade on a Sonoran Desert ecosystem: Microclimate and vegetation. *J. Arid Environ.* **13,** 65–82.

Tanno, N. (1983). Blue light induced inhibition of seed germination: The necessity of the fruit coats for the blue light response. *Physiol. Plant.* **58,** 18–20.

Taylorson, R. B., and Borthwick, H. A. (1969). Light filtration by foliar canopies, significance for light controlled weed seed germination. *Weed Sci.* **17,** 359–361.

Tester, M., and Morris, C. (1987). The penetration of light through soil. *Plant, Cell Environ.* **10,** 281–286.

Thompson, K., and Grime, J. P. (1983). A comparative study of germination responses to diurnally-fluctuating temperatures. *J. Appl. Ecol.* **20,** 141–156.

Thompson, K., and Whatley, J. C. (1984). A thermogradient bar apparatus for the study of the germination requirements of buried seeds *in situ*. *New Phytol.* **96,** 459–471.

Thompson, K., Grime, J. P., and Mason, G. (1977). Seed germination in response to diurnal fluctuations of temperature. *Nature (London)* **267,** 147–149.

Uhl, C., Clarck, K., Dezzeo, N., and Maquirino, P. (1988). Vegetation dynamics in Amazonian treefall gaps. *Ecology* **69,** 751–763.

Valiente-Banuet, A., Bolongaro-Crevenna, A., Briones, O., Ezcurra, E., Rosas, M., Nuñez, H., Barnard, G., and Vázquez, E. (1991). Spatial relationships between cacti and nurse shrubs in a semi-arid environment in central Mexico. *J. Veg. Sci.* **2,** 15–20.

Valio, I. F. M., and Joly, C. A. (1979). Light sensitivity of the seeds on the distribution of *Cecropia glaziovi* Snethlange (Moraceae). *Z. Pflanzenphysiol.* **91,** 371–376.

van der Veen, R. (1970). The importance of the red far-red antagonism in photoblastic seeds. *Acta Bot. Neerl.* **19,** 809–812.

van Rooden, J., Akkermans, L. M. A., and van Der Veen, R. (1970). A study on photoblastism in seeds of some tropical weeds. *Acta Bot. Neerl.* **19,** 257–264.

van Tooren, B. F., and Pons, T. L. (1988). Effects of temperature and light on the germination in chalk grassland species. *Funct. Ecol.* **2,** 303–310.

Vázquez-Yanes, C. (1974). Studies on the germination of seeds of *Ochroma lagopus* Swartz. *Turrialba* **24,** 176–179.

Vázquez-Yanes, C. (1976). Estudios sobre ecología de la germinación en una zona cálido-húmeda de México. *In* "Regeneración de selvas" (A. Gómez-Pompa, C. Vázquez-Yanes, S. del Amo, and A. Butanda, eds.), pp. 279–387. Ed. Continental, México.

Vázquez-Yanes, C., and Orozco-Segovia, A. (1982). Seed germination of a tropical rain forest tree *Heliocarpus donnell-smithii* in response to diurnal fluctuations of temperature. *Physiol. Plant.* **56,** 295–298.

Vázquez-Yanes, C., and Orozco-Segovia, A. (1986). Dispersal of seeds by animals: Effect on light controlled dormancy in *Cecropia obtusifolia. In* "Frugivores and Seed Dispersal" (A. Estrada and T. H. Fleming, eds.), pp. 71–77. Dr. W. Junk Publishers, Dordrecht, The Netherlands.

Vázquez-Yanes, C., and Orozco-Segovia, A. (1987). Light gap detection by the photoblastic seeds of *Cecropia obtusifolia* and *Piper auritum,* two tropical rain forest trees. *Biol. Plant.* **29,** 234–236.

Vázquez-Yanes, C., and Orozco-Segovia, A. (1990a). Ecological significance of light controlled seed germination in two contrasting tropical habitats. *Oecologia* **83,** 171–175.

Vázquez-Yanes, C., and Orozco-Segovia, A. (1990b). Seed dormancy in a tropical rain forest. *In* "Reproductive Biology of Tropical Plants: Man and the Biosphere Series" (M. Hadley, ed.), pp. 247–259. UNESCO, Paris/Parthenon Publishing, Carnforth, Lancaster.

Vázquez-Yanes, C., and Orozco-Segovia, A. (1992). Effect of litter from a tropical rain forest on tree seed germination and establishment under controlled conditions. *Tree Physiol.* **11,** 391–400.

Vázquez-Yanes, C., and Smith, H. (1982). Phytochrome control of seed germination in the tropical rain forest pioneer trees *Cecropia obtusifolia* and *Piper auritum* and its ecological significance. *New Phytol.* **92,** 447–485.

Vázquez-Yanes, C., Orozco-Segovia, A., Rincón, E., Sánchez-Coronado, M. E., Huante, P., Toledo, J. R., and Barradas, V. L. (1990). Light beneath the litter in a tropical forest: Effect on seed germination. *Ecology* **71,** 1952–1958.

Washitani, I. (1985). Field fate of *Amaranthus patulus* seeds subjected to leaf canopy inhibition of germination. *Oecologia* **66,** 338–342.

Washitani, I., and Saeki, T. (1984). Leaf canopy inhibition of germination as a mechanism for the disappearance of *Amaranthus patulus* Bertol. in the second year of secondary succession. *Jpn. J. Ecol.* **34,** 55–61.

Washitani, I., and Takenaka, A. (1986). "Safe sites" for the seed germination of *Rhus javanica:* A characterization by responses to temperature and light. *Ecol. Res.* **1,** 71–82.

Washitani, I., and Takenaka, A. (1987). Gap-detecting mechanism in the seed germination of *Mallotus japonicus* (Thunb.) Muell. Arg., a common pioneer tree of secondary succession in temperate Japan. *Ecol. Res.* **2,** 191–201.

Wesson, G., and Wareing, P. F. (1969). The induction of light sensitivity in weed seeds by burial. *J. Exp. Bot.* **20,** 414–425.

White, P. S. (1979). Pattern, process, and natural disturbance in vegetation. *Bot. Rev.* **45,** 230–299.

Whitmore, T. C. (1975). "Tropical Rain Forests of the Far East." Oxford Univ. Press (Clarendon), London.

Whitmore, T. C. (1983). Secondary succession from seed in tropical rain forest. *For. Abstr.* **44,** 767–779.

Whitmore, T. C. (1989). Canopy gaps and the two major groups of forest trees. *Ecology* **70,** 536–538.

Williams, E. D. (1983). Effects of temperature fluctuation, red and far-red light and nitrate on seed germination of five grasses. *J. Appl. Ecol.* **20,** 923–935.

Williams-Linera, G. (1990). Origin and early development vegetation in Panama. *Biotropica* **22,** 235–241.

Wilson, J. W. (1965). Stand structure and light penetration. *J. Appl. Ecol.* **2,** 383–390.

8

Assessing the Heterogeneity of Belowground Resources: Quantifying Pattern and Scale

G. Philip Robertson and Katherine L. Gross

I. Introduction

That belowground resources in terrestrial plant communities are heterogeneously distributed is very nearly an ecological truism. Whether imposed by geomorphological features of the landscape such as the glacial redistribution of soil parent material (e.g., Walker and Ruhe, 1968), by patchy plant distributions across an otherwise uniform geomorphology such as early sand dune succession (e.g., Olson, 1958), or by microtopographic variation within old fields (e.g., Reader and Best, 1989), water and nutrients are rarely homogeneously distributed in soils. Nevertheless while spatial heterogeneity is well recognized, the scale or extent to which it occurs, and how this might differ among communities, is very poorly understood. Although spatial variation of belowground resource distributions has been the explicit subject of a number of studies in recent decades (e.g., Downes and Beckwith, 1957; Snaydon, 1962; Pigott and Taylor, 1964; Zedler and Zedler, 1969; Allen and MacMahon, 1985; Folorunso and Rolston, 1985; Robertson *et al.*, 1988), our understanding of the scale of variability of belowground resources is in most cases at a qualitative level that does not lend itself well to generalizations about patterns nor—more importantly—to generalizations about controls on or consequences of these patterns.

To at least some degree this lack of understanding is due to a lack of appropriate quantitative tools for detecting and accurately describing spatial patterning. Although variability in resource levels across a site can be assessed by replicated sampling, whether the variability observed can be interpreted as spatially structured heterogeneity depends on the distribution of the variability. In general, assessments of spatial heterogeneity in resources implicitly address three questions related to variability: (1) How variable is the resource in question; that is, is the range of variability likely to be biologically significant? (2) What proportion of the total variation observed can be accounted for by spatial factors; that is, how predictable is the pattern in space? (3) At what scale or scales is the patterning, if present, expressed?

Answers to the first question—the general degree of sample variability— can be addressed well by standard parametric statistics. Answers to the latter questions, however, presuppose that we can detect spatial patterning of a resource and identify the scales at which it is expressed. From at least a quantitative standpoint this is rarely a straightforward exercise, because parametric statistics are poorly suited for analyses of autocorrelated data and because the sampling effort required to quantify the spatial scale of the variation can be daunting.

At a qualitative level the detection of pattern and scale in soil resources can be relatively easy, especially at large, highly aggregated scales such as the regional or continental (Figure 1). Variation in soil characteristics and fertility at large scales is in general coincident with patterns of climate and vegetation. At smaller scales, however—where variation is likely to affect the local distribution and abundance of plant species and the performance of individual organisms—pattern detection and its implications can be more difficult to address. First, measurements of soil resources at these scales (Figure 2) tend to be more analytically demanding than at larger scales where one can rely on surrogate measures such as slope and soil texture (e.g., Burke, 1989; Schimel *et al.*, 1985). Second, soil resource patterning within communities may be subtle and also temporally variable, with small differences at the scale of individuals having important ecophysiological effects that can be amplified to the community and ecosystem levels (e.g., Snaydon, 1962; Woldendorp, 1983; Jackson and Caldwell, 1989; Tilman and Wedin, 1990).

To rigorously address questions linking variation in plant population and community patterns to soil resource heterogeneity, we need statistically robust, spatially explicit quantitative tools that will allow us (1) to detect and test hypotheses about environmental controls on the patterning and scale of soil resource distributions and (2) to design field sampling programs that capture appropriate levels of variation with a reasonable degree of sampling effort. The development of such methods are important for addressing a broad range of questions in ecology, both from the standpoint of experimental design and rigor—for example, how do we avoid bias due to spatial autocorrelation in field experiments—and from the standpoint of developing a mechanistic under-

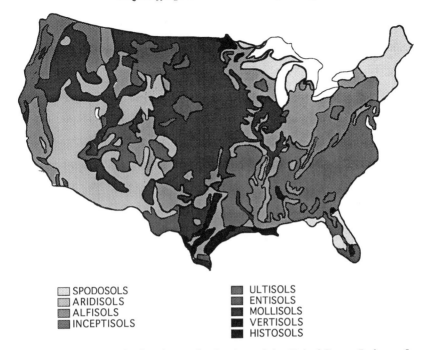

□ SPODOSOLS	▓ ULTISOLS
▓ ARIDISOLS	▓ ENTISOLS
▓ ALFISOLS	▓ MOLLISOLS
▓ INCEPTISOLS	▓ VERTISOLS
	▓ HISTOSOLS

Figure 1 Patterns of soil orders and suborders of the United States. Redrawn from a map provided by the Soil Survey Division, U.S. Soil Conservation Service.

standing of the causes and consequences of these patterns across different levels of ecological organization. In particular, attempts to understand the consequences of global environmental change require that we examine the appropriate levels of aggregation for describing ecological phenomena at regional and global scales.

In this chapter we describe the potential applications of geostatistics for detecting and quantifying spatial heterogeneity in ecological communities, and illustrate how geostatistics might be used to provide ecologically useful information about potential controls on soil processes and their concomitant effects on plant species distributions. Geostatistics is a powerful and readily interpretable technique for quantifying autocorrelation within a spatial domain and for then using this information to identify scales and patterns of spatial heterogeneity. Only recently have these methods been used to quantify soil resources (e.g., Webster, 1985; Trangmar *et al.*, 1985; Webster and Oliver, 1990) or biological processes in the context of plant community patterns (Robertson, 1987; Robertson *et al.*, 1988; Jackson and Caldwell, 1993). Rossi *et al.* (1992) reviewed the

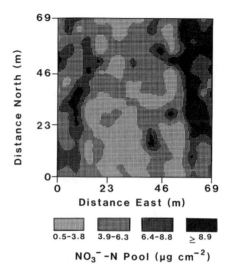

Figure 2   Nitrate nitrogen concentrations in the top 20 cm of soil from a southeast Michigan old field. Values were interpolated by kriging at 1-m intervals across the site after first establishing spatial dependence (see discussion of kriging in text); $n = 300$ sample locations. (From Robertson *et al.,* 1988.)

development of geostatistics and potential applications and pitfalls for ecology in general; our goal here is to evaluate the advantages and constraints of using geostatistical approaches to quantify spatial heterogeneity in soil resources and to relate this to ecological patterns at various scales.

II. Detecting Spatial Dependence: Autocorrelation Analysis

Environmental gradients are a basic feature of ecological communities: few biologically significant resources are either homogeneously or randomly distributed at scales that affect individual organisms. Gases such as CO_2 and O_2, for example, may be more or less uniformly distributed in the earth's boundary layer, but gradients across a leaf surface or through a soil aggregate can profoundly affect rates of biological activity for individual organisms. Likewise, precipitation over a given time interval may be randomly distributed across a semiarid landscape, but localized soil properties and differential uptake rates within communities will quickly result in moisture gradients at scales relevant to individual plants. Detecting such gradients, and in particular detecting over what scales they occur, is a question best addressed by autocorrelation analysis.

Spatial autocorrelation is based on the simple premise that near neighbors are more likely to experience similar environments than are far neighbors; for example, that soil nitrogen levels at a given location are better predicted by nitrogen levels at locations 1 m distant than by levels at locations 100 m distant. Whether this is in fact the case—or whether nitrogen levels are independent of sample location at even 1-m distances—can be determined only by examining changes in the degree of autocorrelation for a property over a range of distance intervals.

There are a number of statistical methods available for detecting autocorrelation in environmental data; these include conventional autocorrelogram approaches (e.g., Webster and Cuanalo, 1975), Moran's I autocorrelation analysis (e.g., Sokal and Oden, 1978; Slatkin and Arter, 1991; Heywood, 1991), Mantell's test (e.g., Heywood, 1991), fractal analysis (e.g., Tyler and Wheatcraft, 1990a,b), spectral analysis (e.g., McBratney and Webster, 1981), and semivariance analysis (e.g., Burgess and Webster, 1980). Of these techniques, semivariance analysis has seen the most application in the study of belowground resources, probably in part because of its robust simplicity, but also because of its now widespread use and rapid development in the geological sciences.

A. Semivariance Analysis

Semivariance analysis provides a versatile and unbiased means for examining autocorrelation in environmental data. Semivariance is evaluated by the statistic

$$\gamma(h) = \frac{1}{2N(h)} \sum_{i=1}^{n} [z(x_i) - z(x_{i+h})]^2,$$

where $\gamma(h)$ is the semivariance for all locations in a spatial domain separated by the distance interval h, $z(x_i)$ represents the value of the property at location x_i, and $z(x_{i+h})$ the value at distance interval h from x_i; N is the total number of pairs in the domain that are separated by distance h. In practice, h is defined as a distance class interval, representing all sample locations within a domain that are separated by a specific class of distances such as 0–5 m or 25–30 m.

The primary advantage of the semivariance statistic over other measures of autocorrelation (e.g., Moran's I) is that semivariance does not require second-order stationarity. Regionalized variable theory (Matheron, 1971), from which the semivariance statistic is derived, allows the assumption that variance and covariance are homogeneously distributed throughout the sample region—an assumption central to parametric statistics—to be relaxed; semivariance analysis requires only that covariance among samples be finite and independent of position within the region (Webster, 1985). This allows the unbiased evaluation of auto-correlation where variance and covariance are heterogeneously distributed throughout a domain—the usual case for environmental variates such as soil properties. Semivariance is, however, sensitive to skewed distributions and clustering (see Krige, 1981; Rossi *et al.*, 1992); often environmental data must be transformed to approach normality before geostatistical analysis.

A secondary advantage of semivariance is its direct use in kriging, a means for optimally interpolating values for locations not sampled across a site. As discussed in the following, semivariance (as a measure of spatial dependence) is used to weight sample points when deriving kriged interpolation estimates.

1. The Semivariogram   Calculating semivariance for all possible distance intervals or classes within a domain yields the semivariogram or variogram (Figure 3), a graphical representation of the degree of spatial dependence within a region. Semivariograms are derived by fitting models to the semivariances calculated for the various distance classes (h) in the domain. By modeling semivariance one may estimate autocorrelation

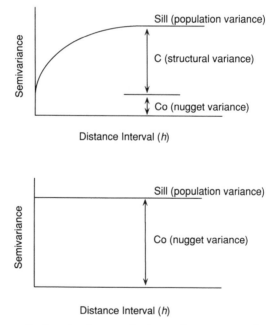

Figure 3   Generalized semivariograms. In the top figure structural or spatially dependent variance makes up a large proportion of total population variance. In the bottom figure there is no spatial dependence at the scales examined.

for all possible distance intervals in a region in spite of the fact that relatively few distance intervals may be available for calculating semivariance.

Where spatial dependence is present (Figure 3, top), γ typically rises to some asymptote (often termed "sill"), which should be roughly equivalent to the population variance. At least theoretically, semivariance should approach zero at a distance interval close to zero—any given sample location should be perfectly autocorrelated with itself—but in practice this is rare. The difference, estimated as the y intercept of the semivariogram (termed "nugget" variance by early geostatisticians because occasional gold nuggets were encountered away from seams), represents either random sampling/analytical error or spatial dependence at distance intervals less than the minimum interval sampled. The proportion of the total variance accounted for by structural or spatially dependent variance ($C/[C + C_0]$; see Figure 3) is a useful index of the spatially dependent predictability of the resource.

From an ecological perspective, then, a semivariogram documents whether there is a spatial component to the variability (is there patchiness) and the robustness of the pattern (how distinct the patches are). In addition, the semivariogram also reveals the spatial scale over which autocorrelation occurs (patch size). These properties can be quantified and compared among sites or experimental treatments by statistically parameterizing semivariogram models. Though variability may be characteristic of all environments, knowing the degree of spatial dependence and its extent is critical for evaluating its ecological significance. For some soil resources, for example, for soil respiration within a 0.5-ha Michigan old field (Figure 4, bottom) (Robertson *et al.*, 1988), there may be no detectable spatial component to the variance at the scales examined—in this case 1 m to 50 m. This suggests that at the time this field was sampled, controls on soil CO_2 fluxes were either randomly distributed (spatially independent) or influenced by ecological processes operating at smaller scales than those examined (in this case 1 m), for instance, at scales of individual plants or even rhizospheres. In most cases reported to date for soil resources, however, structural or spatially dependent variance represents a major proportion of the total population variance. For example, in this same old field over 70% of the variation in nitrogen availability (as indexed by nitrogen mineralization potentials, Figure 4, top) was spatially autocorrelated (Robertson *et al.*, 1988).

Where autocorrelation is present, the semivariogram (and semivariance statistics) will define the range or distance over which spatial dependence is expressed. More specifically, the distance at which the semivariogram reaches an asymptote—approximately equivalent to population variance—is the distance over which geographic points in the community

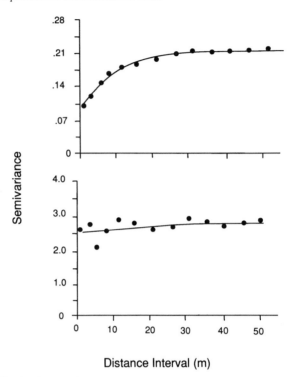

Figure 4   Semivariogram for soil N mineralization potentials (top) and soil respiration (bottom) in the top 20 cm of soil from a Michigan old field community; $n = 300$ sample locations. (From Robertson *et al.*, 1988.)

are (on average) spatially related. Beyond this range, points are spatially independent of one another. Knowledge of the range for a particular soil resource, and how (or if) it differs among communities, may be equally or more important than the mean level of the resource for understanding the effects on plant population and community patterns. For example, in the old field studied in Michigan, nitrogen mineralization is spatially correlated at points in the field less than ca. 25 m distant of one another (see Figure 4). Moreover, because the slope of the semivariogram is greater at smaller distances, points within 10 m of one another are far more similar to one another than are points within a 10 to 25-m radius.

That semivariance analysis can both quantify the degree of spatial dependence for a resource in a community and explicitly define the scale over which dependence is expressed, makes the analysis especially valuable for inferences about controls on community resources and possible effects on associated plants. The analysis described earlier, for example, suggested that controls on soil respiration in early successional old fields may be expressed at spatial scales of <1 m. In a nearby cultivated maize (*Zea mays*) community, spatially dependent respiration was restricted to scales of <0.15 m (Figure 5) (Merrill *et al.*, 1993). This implies that CO_2 fluxes in these communities—and the biological activity associated with these fluxes—may be controlled by factors that operate at the rhizosphere or soil aggregate scale. In other, perhaps even most communities, one might expect spatial patterning for CO_2 fluxes and other soil properties to exist—and be discretely expressed—at rhizosphere, individual plant, microtopographic, and larger scales simultaneously.

2. Nested Structure   Multiple-scaling of variability is likely to be a common feature of ecological phenomena because processes that affect patterns operate at different scales (e.g., Burrough, 1981; Wiens, 1989; Milne, 1991; Palmer, 1992). Such multiple scaling can be most clearly inferred from a nested semivariogram model. For example, soil pH across a 43-ha soybean (*Glycine max*) community (Figure 6) (Robertson *et al.*, 1992) appears to be spatially dependent at two scales, suggesting that multiple factors may be influencing the level and distribution of pH at this site. The steep rise in semivariance between 1 and 5 m suggests that local, perhaps microtopographic or plant-based factors influence pH at scales of 5 m or less. A second sharp increase in semivariance at

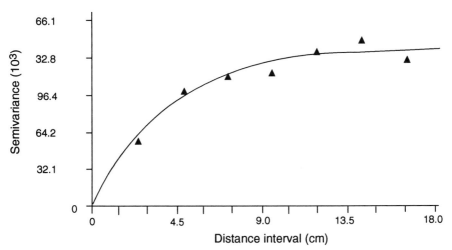

Figure 5 Semivariogram for soil respiration along a 1-m transect in an early-season corn (*Zea mays*) field; $n = 40$ 2.5-cm-diameter soil cores taken immediately adjacent to one another. (From Merrill *et al.*, 1992.)

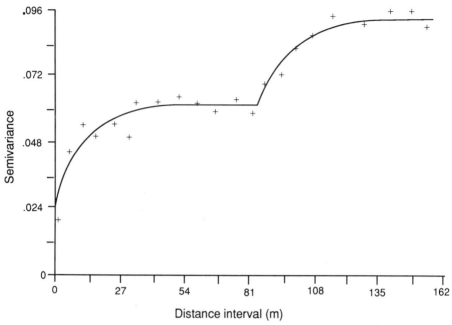

Figure 6   Semivariogram for soil pH from a 42-ha Michigan agricultural field; $n = 620$ soil cores. Note the nested semivariogram structure with spatial dependence between 0–5 m and again at 80–120 m. (From Robertson *et al.*, 1993.)

ca. 90 m suggests that a different, probably cultural, control is exerted at the larger field scale. One might expect additional jumps in semivariance at still larger scales as geomorphologic and climatic influences come to bear.

Although sampling limitations will in most cases restrict our direct knowledge of nested variation, at least in concept such an approach should be capable of sorting variation into numerous components ranging from the rhizoplane ($<10^{-4}$ m) to the landscape (10^4 m). The nested nature of such a semivariogram (Figure 7) would suggest different controls at different scales for the resource at question. There is some evidence for nested patterns of spatial dependence in terrestrial vegetation, particularly in highly dissected and/or human-altered landscapes (e.g., Sugihara and May, 1990; Milne, 1991; Palmer, 1992). Being able to detect such patterns could be especially valuable for defining appropriate levels of aggregation for estimates of regional- and global-scale soil processes such as CO_2 or other trace gas fluxes (Robertson, 1993).

3. Anisotropy   For many soil properties spatial dependence may be strongly directional, that is, spatial dependence in one compass direction

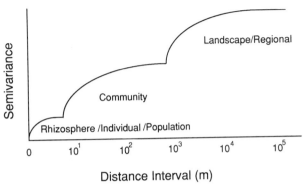

Figure 7 Generalized semivariogram denoting spatial-dependent variation at different (nested) scales ranging from the rhizoplane to the global.

may differ from dependence in another. This can be readily imagined by considering a mountainside plant community in which soil resources (e.g., soil nitrogen pools) might change much more rapidly in an up-slope—downslope direction than in a cross-slope direction. One might then expect the semivariogram for the upslope–downslope direction to have a much steeper rise to an asymptote—a much smaller range—than that for the cross-slope direction. Such geometric anisotropy (cf. zonal anisotropy, David, 1977) might also be expressed for other ecological phenomena under less obvious conditions. For example, seed dispersal patterns and consequently seed bank densities may be affected by pre-dominant wind direction, historical cultivation patterns, or topography. Similarly, organic matter distributions and subsurface/lateral water flow that might affect soil moisture can be strongly affected by topographic and subsurface geomorphologic patterns.

Anisotropy can be evaluated by comparing semivariograms calculated for sample points aligned in different individual directions—often $0°$, $45°$, $90°$, and $135°$ from a central axis. Where present, anisotropy can provide important clues for understanding patterns and underlying con-trols on belowground resources.

4. Semivariance and Fractal Analysis Fractal analysis provides another means for examining scale-dependent variation, and several recent stud-ies have used fractals to analyze patch characteristics in vegetation at the landscape level (e.g., Burrough, 1981; Naven and Lieberman, 1984; Milne, 1991; Palmer, 1992). In its simplest form, a fractal is an index of self-similarity; that is, an analysis of the degree to which complexity at one scale is repeated at others. In effect, then, the Hausdorff–Besicovich or fractal dimension D is a measure of spatial dependence and in this sense very similar to the semivariance statistic. In fact, Burrough (1981) points out that semivariance can effectively estimate D such that

$$D = 2 - (m/2),$$

where m = the slope of a log-log semivariogram.

The fractal dimension thus can provide a single-value summary of spatial complexity across a region, and in this respect it can be quite useful. But such a summary may be of limited use for evaluating the spatial complexity of community-level resources—information such as the range of spatial dependence and information about nested variation (Figure 6) can be lost or obscured by expressing complexity as a single value. Analyses of how the fractal dimension shifts at larger scales can indicate whether pattern is being influenced by factors that operate at different scales (Sugihara and May, 1990; Palmer, 1992), but often the spatial scales at which a landscape is homogeneous can be better determined by semivariance analysis. Burrough (1981) suggests avoiding fractal models altogether where controls on environmental properties are complex and may differ at different scales. This is likely to be the case for small-scale variation in soil resources, influenced by a mixture of biological, chemical, and physical features of the environment.

III. Characterizing Pattern: Kriging

Of methods currently available for characterizing pattern, that is, for generating isopleths (maps) of resource levels across spatial domains such as plant communities, none matches kriging for providing optimal, unbiased estimates of locations not sampled. Traditional approaches to such interpolation have included simple linear techniques as well as approaches that weight nearby neighbors using moving averages, localized regression surfaces, cubic splines, and inverse square distributed weights (Webster and Oliver, 1990). Kriging is similar to these in that it is a local averaging technique, but different in that it provides optimal unbiased estimates for interpolated points by calculating and minimizing the estimation error associated with these points. Estimation error is calculated using knowledge about spatial relatedness in the region to weight the neighbors used for the estimate; the semivariogram provides this knowledge. That kriging produces an estimation error term (e.g., standard error) for every interpolated point is in itself a strong advantage of the technique for attempts to understand *in situ* patterning—no other interpolation techniques provide explicit measures of statistical confidence.

The kriging interpolation estimate is defined as

$$z(B) = \sum_{i=1}^{n} \lambda_i \, z(x_i),$$

where $z(B)$ is the estimated value for the resource over a local area or block B, $z(x_i)$ is the measured sample value at point x_i, n is the number of samples within the defined estimation neighborhood, and λ_i is the weight associated with sample value $z(x_i)$. This weight is based on (1) the distance of B from x_i and (2) the degree to which samples separated by distance $B - x_i$ are spatially dependent (as estimated by the semivariogram). To avoid bias, all weights associated with an estimate sum to 1.0.

A highly simplified, hypothetical example illustrates this weighting in Figure 8. Here four sample points $(x_1 \ldots x_4)$ fall within the 4-m-radius estimation neighborhood for interpolation point B. The kriging weights associated with each point are largely determined by the shape of the semivariogram. In scenario 1, points beyond the semivariogram's range of 2.0 m are given weights close to 0 while the closer, more related points x_1 and x_2 are assigned weights of .8 and .2, respectively; x_1's weight is higher because of the relative steepness of the semivariogram curve at 0.5 m distance.

In the second scenario the semivariogram describes a linear model. All sampled points in the neighborhood are used in this estimation, with weights assigned based primarily on proximity.

In the third scenario, no spatial dependence was established over the range of distances sampled (the semivariogram exhibits a "pure nugget" effect) and all neighbors are assigned equal weights. Note that in this scenario the estimation error term for interpolated values will be substan-

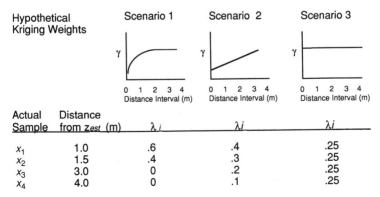

Hypothetical Kriging Weights		Scenario 1	Scenario 2	Scenario 3
Actual Sample	Distance from z_{est} (m)	λ_i	λ_i	λ_i
x_1	1.0	.6	.4	.25
x_2	1.5	.4	.3	.25
x_3	3.0	0	.2	.25
x_4	4.0	0	.1	.25

Figure 8 Hypothetical kriging weights for a single interpolate location with four sample points in its 5-m interpolation neighborhood. Scenarios illustrate different weights as a function of different semivariograms (different levels of spatial dependence).

tially higher than it would be had the semivariogram exhibited spatial dependence.

The procedure for assigning weights to sampled neighbors is not, of course, as simple as these scenarios suggest. In actual practice weights are chosen to minimize estimation variance.

$$\sigma^2(B) = 2 \sum_{i=1}^{n} \lambda_i \, \gamma(x_i,B) - \sum_{i=1}^{n} \sum_{j=1}^{n} \lambda_i \lambda_j \, \gamma(x_i,x_j) - \gamma(B,B),$$

where $\gamma(x_i,B)$ is the average semivariance between the estimation block B and sample point x_i, $\gamma(x_i,x_j)$ is the semivariance between sample points i and j, and $\gamma(B,B)$ is the average semivariance within the estimation block (for point or punctual kriging this term becomes 0 since $\gamma(X_0,X_0) = 0$). A series of linear equations is used to minimize variance using all sampled points in the interpolation neighborhood; the method is thus computationally intensive because a new set of linear equations must be solved for every interpolated point.

For mapping soil resources, block kriging will usually be more appropriate than punctual kriging. Block kriging provides an average estimate for an area around the interpolation point; punctual kriging provides an estimate for the individual point. As one might expect, the estimation error for punctual-kriged interpolates is usually much higher than that for block kriged, reflecting a greater uncertainty for estimating a value for a single point location as opposed to an average value for a small area around a point. Because of this averaging, block kriging tends to be less sensitive to analytical error than punctual kriging. Punctual kriging is a perfect interpolator such that estimated values for locations that coincide with measured value locations will evaluate to the measured values. This means that outliers will map as outliers, whereas with block kriging outliers will be somewhat smoothed. Nevertheless, even for block kriging it is important to minimize analytical error as much as possible, especially if that error is somehow distance-dependent (e.g., sample carryover during soil sampling). One must thus be careful to replicate appropriately at both the sampling and analytical levels, even though replicate values will not themselves be used directly for kriging.

It is also worth noting that kriging estimation variance is independent of measured values, depending instead only on (1) the distance between interpolation point B and sample point x_i and (2) the shape of the semivariogram. This means that once one has sampled sufficiently to define the semivariogram for a particular community property or resource, this information can be used to design an optimal sampling strategy, that is, a sampling strategy that will provide interpolated values with an acceptable and already-known degree of statistical confidence (e.g., Burgess and Webster, 1981).

The further application of kriging to ecological data should allow us to better and more quantitatively evaluate patterning in ecological communities. The technique is especially well suited to those properties of communities that exhibit high levels of spatial dependence at community-level and smaller scales. This probably includes most if not all belowground resources important to plants.

IV. Conclusions

Belowground resources are distributed heterogeneously but not, in general, at random: recent analyses of the spatial structure of the variability of these resources have shown that many soil properties are significantly autocorrelated at scales that are likely to be significant to individual plants. Moreover, available evidence, though scanty, suggests that autocorrelation in natural communities is scale-dependent, that is, not continuous from the rhizosphere (less than cm) through the landscape (km) levels.

Geostatistics offer a powerful and promising technique for analyzing spatial variability. Semivariance analysis provides a robust means for quantifying autocorrelation or dependence across a spatial (or temporal) domain, and kriging provides an optimal, unbiased means for interpolating with statistical confidence. Interpolations are useful for qualitative visual descriptions of spatial variation, for relating resource levels to the performance of individuals or distribution of species within a community, or for comparing patterns and scales of resource heterogeneity in different communities. Fractals provide an alternate means for summarizing autocorrelation that may be useful for understanding patterns of scaling across landscapes in some circumstances.

Further applications of geostatistical techniques should provide substantial power for determining scales at which controls over specific resources are expressed. Applications should also allow us to identify appropriate levels of aggregation for regional and global models of ecological phenomena.

Acknowledgments

We thank M. Cavigelli and L. Huberty for helpful comments and discussion during the preparation of this chapter. Support for this work was provided by NSF grants BSR87-02332 and BSR91-07481 to Michigan State University and by the Michigan Agricultural Experiment Station.

References

Allen, M. F., and MacMahon, J. A. (1985). Impact of disturbance on cold desert fungi: Comparative microscale dispersion patterns. *Pedobiologia* **28**, 215–224.

Burgess, T. M., and Webster, R. (1980). Optimal interpolation and isarithmic mapping of soil properties. I. The semi-variogram and punctual kriging. *J. Soil Sci.* **31**, 315–331.

Burgess, T. M., and Webster, R. (1981). Optimal interpolation and isarithmic mapping of soil properties. IV. Sampling strategy. *J. Soil Sci.* **32**, 643–659.

Burke, I. C. (1989). Control of nitrogen mineralization in a sagebrush steppe landscape. *Ecology* **70**, 1115–1126.

Burrough, P. A. (1981). Fractal dimensions of landscapes and other environmental data. *Nature (London)* **294**, 240–242.

David, M. (1977). "Geostatistical Ore Reserve Estimation." Elsevier, Amsterdam.

Downes, R. G., and Beckwith, R. S. (1951). Studies in the variation of soil reaction. I. Field variations at Barooga, N.S.W. *Aust. J. Agric. Res.* **2**, 60–72.

Folorunso, O. A., and Rolston, D. E. (1985). Spatial variability of field-measured denitrification gas fluxes. *Soil Sci. Soc. Am. J.* **48**, 1214–1219.

Heywood, J. S. (1991). Spatial analysis of genetic variation in plant populations. *Annu. Rev. Ecol. Syst.* **22**, 335–355.

Jackson, R. B., and Caldwell, M. M. (1989). The timing and degree of root proliferation in fertile-soil microsites for three cold-desert perennials. *Oecologia* **81**, 149–153.

Jackson, R. B., and Caldwell, M. M. (1993). The scale of nutrient heterogeneity around individual plants and its quantification with geostatistics. *Ecology* **74**, 612–614.

Krige, D. G. (1981). Lognormal-*d* Wijsian geostatistics for ore evaluation. *S. Afr. Inst. Min. Metall., Monog. Ser. Geostat.* **1**, 1–51.

Matheron, G. (1971). "The Theory of Regionalized Variables and its Applications," Cah. Cent. Morphol. Math., Fontainebleau, No. 5. Ecole Nationale Supérieure des Mines de Paris, France.

McBratney, A. B., and Webster, R. (1981). Detection of ridge and furrow pattern by spectral analysis of crop yield. *Int. Stat. Rev.* **49**, 45–52.

Merrill, L., Cavigelli, M., and Robertson, G. P. (1993). Spatial variability of soil respiration and microbial populations at different scales in an agricultural ecosystem. To be published.

Milne, B. T. (1991). Lessons from applying fractal models to landscape patterns. *In* "Quantitative Methods in Landscape Ecology" (M. G. Turner and R. H. Gardner, eds.), pp. 199–235. Springer-Verlag, New York.

Naven, Z., and Lieberman, A. S. (1984). "Landscape Ecology." Springer-Verlag, New York.

Olson, J. S. (1958). Rates of succession and soil changes on southern Lake Michigan sand dunes. *Bot. Gaz. (Chicago)* **119**, 125–170.

Palmer, M. W. (1992). The coexistence of species in fractal landscapes. *Am. Nat.* **139**, 375–397.

Pigott, C. D., and Taylor, K. (1964). The distribution of some woodland herbs in relation to the supply of nitrogen and phosphorus in the soil. *J. Ecol.* **52**, Suppl., 175–185.

Reader, R. J., and Best, B. J. (1989). Variation in competition along an environmental gradient: *Hieracium floribundum* in an abandoned pasture. *J. Ecol.* **77**, 673–684.

Robertson, G. P. (1987). Geostatistics in ecology: Interpolating with known variance. *Ecology* **68**, 744–748.

Robertson, G. P. (1993). Fluxes of nitrous oxide and other nitrogen trace gases from intensively managed landscapes: A global perspective. *In* "Agricultural Ecosystem Effects on Trace Gases and Global Climate Change" (L. A. Harper, A. R. Mosier, J. M. Duxbury, and D. E. Rolston, eds.), pp. 95–108. Am. Soc. Agron., Madison, WI.

Robertson, G. P., Huston, M. A., Evans, F. C., and Tiedje, J. M. (1988). Spatial variability in a successional plant community: Patterns of nitrogen availability. *Ecology* **69**, 1517–1524.

Robertson, G. P., Klingensmith, K. M., Klug, M. J., and Paul, E. A. (1993). Spatial variability of soil properties and microbial activity as related to primary productivity in an agricultural landscape. To be published.

Rossi, R. E., Mulla, D. J., Journal, A. G., and Franz, E. H. (1992). Geostatistical tools for modeling and interpreting ecological spatial dependence. *Ecol. Monogr.* **62**, 277–314.

Schimel, D. S., Stillwell, M. A., and Woodmansee, R. G. (1985). Biogeochemistry of C, N, and P on a catena of the short grass steppe. *Ecology* **66**, 276–282.

Slatkin, M., and Arter, H. E. (1991). Spatial autocorrelation methods in population genetics. *Am. Nat.* **138**, 499–517.

Snaydon, R. W. (1962). Micro-distribution of *Trifolium repens* L. and its relation to soil factors. *J. Ecol.* **50**, 133–143.

Sokal, R. S., and Oden, N. L. (1978). Spatial autocorrelation in biology. 1. Methodology. *Biol. J. Linn. Soc.* **10**, 199–228.

Sugihara, G., and May, R. M. (1990). Applications of fractals in ecology. *Trends Ecol. Evol.* **5**, 79–86.

Tilman, D., and Wedin, D. A. (1990). Species effects on nitrogen cycling: A test with perennial grasses. *Oecologia* **84**, 433–441.

Trangmar, B. B., Yost, R. S., and Uehara, G. (1985). Application of geostatistics to spatial studies of soil properties. *Adv. Agron.* **38**, 45–94.

Tyler, S. W., and Wheatcraft, S. W. (1990a). The consequences of fractal scaling in heterogeneous soils and porous media. *SSSA Spec. Publ.* **25**, 109–120.

Tyler, S. W., and Wheatcraft, S. W. (1990b). Fractal processes in soil water retention. *Water Resour. Res.* **26**, 1047–1054.

Walker, P. H., and Ruhe, R. V. (1968). Hill slope models in soil formation. II. Closed systems. *Trans., Int. Congr. Soil Sci., 9th, 1968*, Vol. 4, pp. 561–569.

Webster, R. (1985). Quantitative spatial analysis of soil in the field. *Adv. Soil Sci.* **3**, 1–70.

Webster, R., and Cuanalo, H. E. (1975). Soil transect correlograms of north Oxfordshire and their interpretations. *J. Soil Sci.* **26**, 176–194.

Webster, R., and Oliver, M. A. (1990). "Statistical Methods in Soil and Land Resource Survey." Oxford Univ. Press, New York.

Wiens, J. A. (1989). Spatial scaling in ecology. *Funct. Ecol.* **3**, 385–397.

Woldendorp, J. W. (1983). The relation between the nitrogen metabolism of *Plantago* species and characteristics of the environment. *In* "Nitrogen as an Ecological Factor" (J. A. Lee, S. McNeill, and I. H. Rorison, eds.), pp. 137–166. Blackwell, Oxford.

Zedler, J., and Zedler, P. (1969). Association of species and their relationship to microtopography within old fields. *Ecology* **50**, 432–442.

9

Causes of Soil Nutrient Heterogeneity at Different Scales

John M. Stark

I. Introduction

Soil heterogeneity has been a topic of considerable interest for decades. A large amount of research has addressed causes of variability in soil properties; however, the primary focus has been variability in soil across landscapes. Although an understanding of variability at the landscape scale is useful to community or ecosystem ecologists interested in the distribution of plant and animal populations, community composition, and ecosystem structure, it is less useful to physiological ecologists interested in how individual plants respond to environmental heterogeneity.

Since this volume is devoted to the response of individual plants to heterogeneity in the environment, the discussion in this chapter will be limited to soil heterogeneity occurring at a scale that is important to the growth and survival of an individual plant. Of the research that has addressed fine-scale heterogeneity in soil, almost all has been descriptive and little has elucidated cause-and-effect relationships. Therefore much of the discussion will be based on extrapolation of results from studies dealing with heterogeneity at larger scales.

In this chapter, I first discuss the appropriate spatial and temporal scales that should be considered and how plants may "perceive" a different degree of heterogeneity than what is measured by analytical techniques. Second, I discuss the role of environmental variables in promoting patchiness in nutrient availability. Finally, I discuss how production and consumption processes regulate nutrient pool size, and what controls segregation of nutrient pools.

II. Which Soil Heterogeneity Is Important?

Before we can discuss causes of nutrient heterogeneity we must define what constitutes heterogeneity to a plant. Just because we are able to measure variability in soil nutrient concentrations does not mean that the plant will be affected by this heterogeneity. Obviously variation that occurs beyond the reach of a plant's roots will not be detected by the plant. Therefore, spatial variability occurring at scales greater than a few meters will be unimportant. Likewise, temporal variability will only be important if it occurs within the lifetime of a plant. For example, the soil mineralogical composition is very important in determining the physical nature and nutrient-supplying capacity of the soil; however, temporal changes in mineralogy resulting from weathering generally occur too slowly to constitute heterogeneity to the plant. Emphasis will be placed on changes in soil characteristics that occur in less than a year or two.

The appropriate minimum spatial and temporal scales are not immediately obvious. For example, plants may respond to patches of nutrient enrichment by proliferation of roots (Grime, Chapter 1; Fitter, Chapter 11; and Caldwell, Chapter 12, this volume). Therefore, one could argue that patch sizes smaller than the minimum root internode distance would be unimportant because increased branching would not result in additional resource exploitation. However, if a plant modifies the number of active enzyme sites within the membranes of individual root cells (Jackson *et al.*, 1990), it could respond to patch sizes much smaller than the minimum internode distance. In addition, exploitation of extremely small patches of nutrient enrichment may occur by proliferation of mycorrhizal hyphae. Thus, patch sizes as small as a few microns may be exploitable.

Similarly, the minimum temporal scale is dependent on how rapidly a plant can respond to a localized nutrient enrichment. This may be on the order of hours in the case of enzyme induction or derepression, or days in the case of root proliferation (Jackson and Caldwell, 1989). Pulses of nutrient availability that occur more rapidly than these time scales would produce little or no response in the plant and therefore will not represent an important source of heterogeneity.

Another critical consideration in defining what constitutes nutrient heterogeneity to a plant is the relationship between measured concentrations of nutrients and plant response. Plant uptake of nutrients can often be described by mixed-order kinetics involving a saturation phase (i.e., Michaelis–Menten kinetics) (Barber, 1984). At low concentrations of nutrients, plant uptake responds approximately linearly to increases in nutrient concentration (i.e., kinetics are first order); however, at higher concentrations, enzyme systems become saturated and further increases in nutrient concentration produce no further increase in uptake rate (i.e., kinetics become zero order). If nutrient concentrations are in this range, no matter how much the concentrations vary, the plant will perceive them as uniformly "high." A better representation of the heterogeneity that a plant perceives may be obtained by filtering data on soil nutrient concentrations through a Michaelis–Menten function. Figure 1 shows that plants will be most sensitive to variability at low concentrations of nutrients and least sensitive to variability at high concentrations.

III. Factors Affecting General Soil Characteristics

In the first half of this century, Hans Jenny (1941) proposed a conceptual model that has been widely used to explain differences in soils across the landscape. With a few modifications, this same model can be used to understand why soil properties differ from one microsite to another. Jenny proposed that soil characteristics were a function of at least five

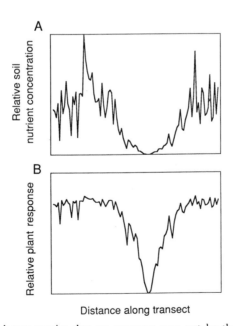

Figure 1 The heterogeneity that we measure may not be the heterogeneity that a plant perceives. (A) Changes in nutrient concentration along a soil transect. (B) How a plant might perceive this same heterogeneity (assuming plant response follows a Michaelis–Menten relationship).

variables: (i) parent material, (ii) organisms, (iii) climate, (iv) topography, and (v) the time that organisms, climate, and topography have had to exert their influence on the parent material. To apply this model to microsites, we can substitute microclimate for climate and microtopography for topography, and also consider the depth of a microsite within the soil profile. Each of these variables may act alone to produce differences in soil characteristics; however, the characteristics we see are almost always a result of fairly complex interactions among the variables. For example, vegetative cover may redistribute precipitation or modify soil temperature to create variations in microclimate. Or the microtopography may result in nonuniform leaching of salts and different distributions of salts with depth in the profile. Recognizing the complexity of the interactions, I will briefly discuss how each of the factors can affect nutrient concentrations and microsite heterogeneity.

A. Parent Material

Parent material is the original geologic material from which the soil formed. It may consist of hard rock or loose deposits of gravel, sand, silt, clay, or organic material. Both the chemical and physical nature of the parent material exert a strong influence on nutrient supply and the scale and intensity of soil heterogeneity, especially during early stages of soil development.

The chemical and physical nature of the parent material determines how rapidly weathering processes release nutrients into the soil solution. For example, parent materials high in ferromagnesian minerals, such as olivine, biotite, and hornblende, weather more rapidly and result in higher nutrient concentrations than parent materials high in quartz or certain feldspars (Birkeland, 1974). Coarse-grained rocks such as granite tend to break up more rapidly into their individual mineral grains; however, fine mineral grains weather faster because of the larger surface area exposed.

The influence of the parent material on the physical properties of the soil also affects nutrient availability. Physical properties that affect nutrient availability include soil texture and the volume of soil available for exploitation. Parent materials such as granite or sandstone tend to form coarse-textured soils with low water-holding capacities. Since the water content of a soil is an important factor controlling diffusion of nutrients to plant roots (see Section IV,C), nutrient availability may be reduced in coarse-textured soils. Parent materials that lead to shallow or rocky soils result in a smaller volume of soil (<2 mm material) capable of supplying nutrients. High rock content increases the patchiness in nutrient supply by segregating the pockets of soil. This type of heterogeneity, for example, has been shown to favor plants with extensive root systems (e.g., shrubs) rather than intensive root systems (e.g., grasses) (Walter, 1979; Stark and Redente, 1985).

Soils typically inherit all of their nutrients except nitrogen from the parent material. Therefore, the parent material may have a major influence on distribution of P, K, S, Ca, Mg, and Fe, as well as the micronutrients. However, the greater the mobility of the nutrient, the more rapidly the heterogeneity imposed by the parent material will disappear.

Although significant nitrogen is not supplied by the parent material (except in the case of certain shales or fresh sediments), heterogeneous distribution of nutrients such as phosphorus may induce a similar pattern in nitrogen distribution. Localized soil volumes with high nutrient concentrations may promote biological activity, especially biological N fixation, which would promote nitrogen accumulation. For example, Jenny (1980) found that the nitrogen content of soils formed from basic igneous rocks was significantly higher than those formed from acid igneous rocks (0.121 versus 0.074%, respectively), even though neither rock type originally contained nitrogen.

Other interactions among nutrients may occur as well. For example, high potassium concentrations may reduce magnesium and calcium uptake, but increase nitrate uptake by plants; high calcium concentrations may reduce magnesium uptake, and high phosphate concentrations may reduce zinc uptake (Barber, 1984). Therefore, heterogeneity in distribution of one nutrient may affect the ability of plants to take up other nutrients, and thus create an "illusion" of heterogeneity in the other nutrients.

All parent materials display a certain degree of heterogeneity in mineralogy and particle size. The heterogeneity may be very sharply defined, as in the case where a vein of one rock type has been extruded into another, or the heterogeneity may be diffuse, as in the case where one rock type intergrades with another as a result of gradual increase in the relative abundance of one mineral. The degree to which this heterogeneity is expressed in the resulting soil will depend on the age of the soil. Young soils have undergone little modification and thus the heterogeneity is largely a function of the heterogeneity present in the parent material. Older soils have been much more extensively modified by physical breakdown of mineral particles and redistribution of nutrients, and the heterogeneity inherited from the parent material is likely to be obscured.

B. Depth within the Soil Profile

Depth is probably the most quickly recognized source of heterogeneity in soil. Much of the depth effect can be attributed to the activity of organisms, which add organic matter to the surface layer; however, chemical and physical processes interact with biological processes to produce changes in soil characteristics with depth. With a few exceptions, surface horizons tend to be higher in organic matter, lower in clay, lower in bulk density, and more neutral in pH than subsoils. Nutrient availability tends to be higher in surface layers because of the favorable pH, the accumulation of organic matter, and the greater ease of root penetration.

Subsurface horizons may have either lower or higher pH than the surface horizons depending on the amount of water that moves through the soil. In humid regions, precipitation is high enough to leach salts completely out of the soil profile. This process, coupled with the extraction of nutrients from the subsoil by plants and deposition on the surface in litter, results in subsoil horizons that have lower pH than surface horizons. Conversely, in arid regions precipitation may be insufficient to move salts more than a few centimeters, resulting in subsoil horizons that have higher pH than surface horizons. The pH of surface soils is often closer to the optimum (pH 6.5) for nutrient availability.

Surface layers experience greater fluctuations in moisture and temperature. Frequent freezing–thawing and wetting–drying cycles promote release of nutrients by increasing turnover of microbial biomass and physically protected organic matter (Birch, 1960; Witkamp, 1969; Reid, 1974). In addition, variation in temperature and moisture conditions may result in higher rates of microbial activity than constant conditions. This effect can be shown graphically (Figure 2). Microbial response to temperature or moisture typically increases curvilinearly up to an optimum and then declines at high temperatures or soil moisture contents. However, the temperatures or moisture conditions that predominate in soils tend to be below the optimum for most microbial activity (Stark, 1991). Because the response curve is concave upward for this range, the mean response from variable conditions would be expected to be higher than the response from constant conditions (Figure 2).

Where water tables are close to the soil surface, variation in redox potential with depth may cause variation in the form and solubility of nutrients. With increased water saturation, redox potentials decline and the solubility of elements such as iron and manganese increase. Low redox potentials may make certain nutrients more available, but in some cases solubilities may increase to the point that toxicities develop. These toxicities, coupled with the lack of oxygen for root growth, may prevent root penetration very far into the saturated layer.

Fluctuating water tables can lead to substantial temporal and spatial heterogeneity in nutrients, especially when solubilities are strongly affected by redox potential. The orange and gray mottling present in subsoils that experience intermittent saturation provides visual evidence

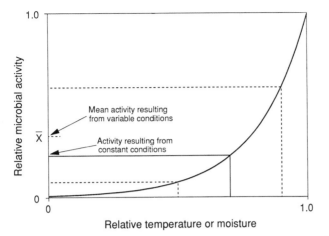

Figure 2 Effect of variation in temperature or moisture on mean rates of microbial activity. Curved line (concave upward) represents the relationship between microbial activity and temperature or moisture for the range of conditions predominantly encountered in soils. The solid straight line shows the microbial activity that would be expected at constant temperature or moisture. Dashed lines show expected activities at two different temperatures or moistures whose mean is equal to the value assumed for the constant condition.

of the scale of heterogeneity in iron availability. The patches, which typically range from 1 to 10 cm in diameter, result from localized differences in redox potential. The differences may be due to either differences in oxygen supply (resulting from air penetration in cracks between soil aggregates, higher rates of diffusion of gases in soil of lower bulk density, etc.) or differences in rates of oxygen consumption (higher carbon availability producing greater demand for oxygen by microbes, etc.). As the soils dry out and oxygen becomes more available, the color changes to a uniform orange. Though the change in color is a direct result of changes in iron oxidation state, it also implies a change in oxidation state of other nutrients as well. At neutral pH, redox potentials low enough to convert Fe III (orange color) to Fe II (gray color) are often low enough to cause NO_3^-, SO_4^{2-}, and certain micronutrients to be reduced to other forms (Bohn *et al.*, 1979).

The scale of heterogeneity that occurs with depth may be on the order of millimeters to decimeters depending on the abruptness of horizon boundaries and the thickness of horizons. All other things being equal, younger soils will have less horizon differentiation, more diffuse horizon boundaries, and less heterogeneity with depth than older soils.

C. Microtopography

The direct effects of microtopography on nutrient heterogeneity result primarily from a redistribution of moisture. Precipitation that runs off mounds into intermound areas may carry nutrients into intermound areas. Increased water input to intermound areas may also result in greater leaching of nutrients, depending on the balance between water inputs and evapotranspiration. Where mound–intermound microtopography is associated with high water tables, capillary rise and evaporation may result in accumulation of salts (usually Na, Ca, Mg, Cl, and SO_4) at the surface of intermound soils.

In many ecosystems, the effects of microtopography are confounded with the effects of variables such as organisms. For example, mound–intermound microtopography may result from the trapping of wind-blown soil by shrubs and the erosion of soil from between shrubs. The increased litterfall directly beneath the shrubs results in the mound soils having higher organic matter contents and greater nutrient availability. For this reason, the scale of microtopographical variation is often determined by the nature of the vegetation.

D. Organisms

Although initially most plant nutrients are derived from the parent material, the soil organic pool rapidly becomes the dominant sink and source of nutrients. In well-developed soils, almost all of the N, and more than half of the P and S, may be associated with the soil organic fraction (Mengel and Kirkby, 1982). Mineralization of these nutrients from organic compounds may provide almost all of the N, P, and S taken up by plants (Witkamp and Ausmus, 1976), especially in older soils. Since organisms determine the quantity, location, and decomposability of the organic matter added to soil, they have major effects in controlling soil nutrient heterogeneity.

Organisms both increase and decrease soil nutrient heterogeneity through a variety of processes. Organisms change nutrient distribution and availability either by transporting nutrients to new locations or by physically or chemically altering the material containing the nutrients. The distance over which nutrients are transported and the degree of physical or chemical alteration depend on the nature of the organism.

1. Animal Activity Animals ranging from large mammals to earthworms, nematodes, and microarthropods change nutrient distributions and the form of nutrients. Animals modify nutrient distribution by transporting soil, fresh plant material, or organic debris to new locations. The grazing behavior of many herbivores results in plant material being consumed at one location but deposited, often in the form of feces, in another. For example, domestic livestock and wildlife often graze in open areas, but bed-down beneath the cover of vegetation. Ants and termites collect organic matter from various locations and concentrate it at or near the royal chamber (Anderson *et al.*, 1981). Nematodes and microarthropods redistribute nutrients during their movement through soil. Animal burrowing activity often results in subsoil being placed on top of surface layers, and as the burrows fill in, fingers of topsoil extending down into the subsoil. This heterogeneity is readily apparent as weedy plant species colonize the subsoil mounds, and the roots of established plants concentrate in the channels of nutrient-rich topsoil extending into the subsoil. Though in the short term, soil movement by animals tends to increase heterogeneity by resulting in unlike soil materials being placed together, in the long term, it can result in greater homogeneity because of continued mixing. The large thickness of the surface horizon of many prairie soils may be attributed in part to the activity of burrowing animals.

Physical movement of soil by animals has indirect effects on heterogeneity in nutrient availablility by creating variations in microtopography and physical characteristics such as bulk density, which influence water infiltration rates, water-holding characteristics, aeration, and root penetration—all of which influence the ability of the plant to absorb nutrients from soils.

Consumption of plant material and organic debris by herbivores and detritivores speeds up release of nutrients by physically breaking down the material and, to a lesser extent, by chemically altering the compounds containing nutrients (Witkamp and Ausmus, 1976). Turnover of the organic material increases because greater fragmentation allows more rapid colonization and breakdown by microorganisms. Earthworm casts, for example, may contain nutrient concentrations that are 1.4 to 11 times the concentrations in the surrounding soil (Table I).

2. Plant Activity   Plants create heterogeneity in nutrient distribution, initially by accumulating nutrients in tissues and creating zones of depletion around plant roots. The zones of depletion may extend for a few

Table I   Nutrient Concentrations in Earthworm Casts and Adjacent Topsoil[a]

	Extractable nutrients (mg kg^{-1} soil)				
	Ca	Mg	K	P	NO$_3$-N
Topsoil	1990	162	32	9	4.7
Worm casts	2790	492	358	67	21.9

[a] Data from Russell (1973).

millimeters or more depending on the nutrient, soil physical and chemical characteristics, and soil moisture conditions (Barber, 1984). A wide variety of other processes may accompany nutrient uptake by plant roots, including release of protons, CO_2, enzymes, chelates, and other reduced carbon compounds (Rovira and Davey 1974; Lynch 1990), as well as water uptake and release (during hydraulic lift; Caldwell and Richards, 1989). These processes make the rhizosphere different both chemically and physically from the rest of the soil. Increased carbon availability in the rhizosphere, due to root sloughing and exudation, results in higher bacterial and fungal populations and higher microbial activity (Norton, 1991). Increased predation on rhizosphere populations may also increase turnover rates and produce greater nutrient availability (Robinson *et al.,* 1989). For example, Norton (1991) showed that gross rates of ammonification and microbial assimilation of NH_4^+ were 40 to 100% higher in soil adjacent to roots of pine seedlings than in soil a few millimeters away.

Rates of nutrient cycling processes, as well as the size and activity of fungal and bacterial populations, vary considerably with the age of roots (Norton, 1991). Elliott (1978) proposed a successional sequence, involving changes in root exudation, nutrient uptake, and species composition within the microbial and microfaunal communities, to explain temporal shifts in nutrient mineralization–immobilization relationships along growing plant roots (Figure 3).

Plants promote further redistribution of nutrients by translocating them above ground, incorporating them into foliage, and depositing them on the soil surface as litter. Numerous examples are available in the literature of how plant uptake and subsequent litterfall results in accumulation of nutrients at the soil surface beneath plant canopies and depletion of nutrients in interspaces (e.g., Roberts, 1950; Sharma and Tongway, 1973; Charley and West, 1975; Tiedemann and Klemmedson, 1983; and others). The canopy–interspace patterns shown in Figure 4 are typical in arid and semiarid systems where canopy cover is not contiguous. In more mesic environments, litterfall tends to be more evenly distributed horizontally, and the primary zones of depletion and accumulation occur in the subsoil and topsoil, respectively.

The vertical distribution of litter inputs within an ecosystem varies considerably depending on the vegetation. For example, the fraction of net primary production occurring below ground was estimated to be 0.4 in an arctic tundra, 0.56 in a deciduous forest, 0.69 in a cool desert shrubland, 0.83 in a shortgrass prairie (Coleman, 1976), and 0.7 to 0.8 in coniferous forests (Vogt *et al.,* 1982). These differences result in distinctly different organic matter distributions in soil profiles. Heterogeneity in organic matter distribution caused by differences in vegetation types is

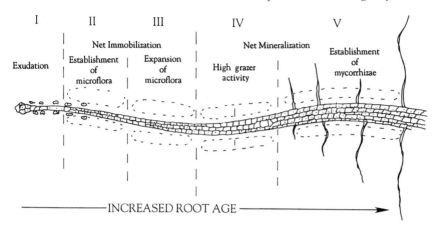

Figure 3 Hypothetical spatial and temporal changes in microbial activity and nutrient cycling along a growing plant root. Zone I is the root tip, where the greatest exudation occurs. In Zone II, microbial populations begin expanding in response to the increased carbon availability created by the exudates, and a phase of nutrient immobilization begins. In Zone III, microbial populations reach their highest levels, microbial growth slows because of depletion of carbon exudates, and grazer (predator) populations begin to increase. In Zone IV, grazing causes turnover of the microbial biomass, which initiates a phase of mineralization and release of nutrients. Finally, in Zone V, mycorrhizal fungi become established in the root tissue, which allows transport of nutrients to the root from greater distances outside the rhizosphere. This general scheme varies depending on rates of root growth and the microbial and microfaunal populations present. (Adapted from Elliott, 1978.)

fairly large scale, however, and individual plants will experience this variation only at sharp vegetation boundaries.

   The chemical and physical characteristics of litter inputs vary considerably depending on the type of vegetation and the environment. Litter may occur as dead roots, foliage, and stems, leachates from live foliage, or exudates and sloughed material from live roots. The relative proportion of litterfall occurring in each of these categories and the chemical composition vary with the type of vegetation. For example, litter from woody vegetation is thicker, has larger quantities of lignin and cellulose, and has lower nitrogen contents than litter from herbaceous vegetation. The composition of litter from a single type of vegetation may also vary with the environment. Carbon : nutrient balances or damage by insects and pollution may change the relative proportion of root and stem tissue or the quantity of secondary compounds (Jones and Coleman, 1991). There also may be significant feedback effects between the soil fertility and the chemical composition of the litterfall (Vitousek, 1982). Resorption of nutrients from senescing plant tissues may be greater in plants

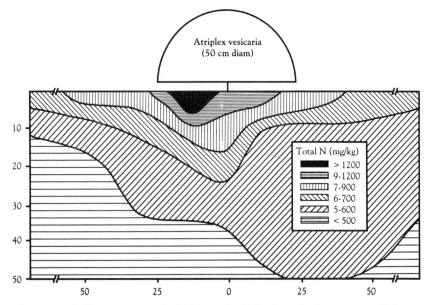

Figure 4   Influence of a single *Atriplex vesicaria* shrub on the distribution of nitrogen in soil. (From Charley, 1977, with permission.)

growing in nutrient-limited environments (Shaver and Melillo, 1984), and thus the nutrient content of litter will be lower. Generally, nitrogen and phosphorus are reabsorbed more than calcium or potassium, and thus litter concentrations of nitrogen and phosphorus would be expected to show the greatest variation with differences in soil fertility. Other environmental factors may also play an important role in the amount of resorption that occurs (Chapin, 1991).

The chemical composition, size, and location of the litter determine how rapidly it is degraded to release the nutrients. Organic matter inputs may be diffuse (e.g., small organic compounds leached from foliage by precipitation), but the majority occurs as discrete units, such as roots, leaves, branches, etc. (Coleman, 1976). Because of this, inputs of organic material will often produce hot spots of microbial activity that result in intense zones of nutrient immobilization or release (e.g., Parkin, 1987).

3. Microbial Activity   Microbial activity can result in nutrient transport from one location to another; however, the distances are usually on the order of centimeters or less. Nutrients may be translocated in fungal hyphae from areas of high nutrient availability to areas of low nutrient availability. For example, Hart (1990) found that significant amounts of

^{15}N appeared in the surface organic layers of an old-growth mixed coni-fer forest following injection of ^{15}NH$_4^+$ at a 2-cm depth into the underly-ing mineral soil. He assumed that the transport mechanism was fungal translocation and calculated the total N flux to be approximately 9 kg N ha^{-1} yr^{-1}. Others have also documented movement of elements from the mineral soil upward into surface organic layers or into decomposing woody debris and attributed it to translocation through fungal hyphae (Witkamp and Barzansky, 1968; Schimel, 1987).

Undoubtedly the greatest influence microorganisms have on nutrient heterogeneity is through their role in nutrient release and immobilization during litter decomposition. The chemical composition of the litter and environmental factors, such as temperature and moisture, control the rate of microbial activity and the fate of the nutrients contained in the litter. Of the chemical characteristics, the carbon-to-nutrient ratios in the soluble material are probably the most important in determining whether nutrients will be immobilized in microbial tissue or released to the soil solution and become available for plant uptake (Park, 1976; Paul and Clark, 1989). A common view is that microorganisms immobilize nutri-ents in sufficient quantities to maintain a particular balance in their tissues. For example, soil bacteria typically maintain a C : N : P ratio of approximately 6 : 1 : 0.2 (Paul and Clark, 1989), whereas soil fungi have slightly higher ratios. If the available substrate contains nutrients in excess of that needed to maintain the balance, these will be released into the soil solution. Conversely, if the substrate contains insufficient nutrients to maintain the balance, either additional quantities of the nutrient will be taken up from the soil solution or the microbial biomass will stop growing. Since soil microorganisms use much of the C substrate as an energy source, more than two-thirds of the C in a substrate may be released as CO$_2$ during respiration. The remaining C can be incorporated into microbial tissues (Paul and Clark, 1989). Substrates with C : N ratios of approximately 25 to 30 normally have sufficient N to balance C assimi-lation, and no net release or immobilization of N will occur. Substrates with C : N ratios greater than this generally result in net immobilization of N, and substrates with C : N ratios less than this generally result in net mineralization of N (Berg and Staaf, 1981).

Although the concept of critical carbon : nutrient ratios has proven useful when dealing with decomposition of many plant materials, it is of less use when dealing with plant materials containing large amounts of unavailable or chemically protected carbon such as lignin. Since C from lignin is fairly inert, it is not readily incorporated into microbial tissues and thus does not immediately promote immobilization of N. In addition, lignin coatings may physically protect more degradable compounds from microbial attack. Because of this, more complex equations have been developed that account for differences in the lignin content of the plant residues (e.g., Melillo *et al.*, 1982).

In environments where biological activity (i.e., decomposition and plant uptake) is limited to narrow windows of time, such as in many semiarid and arid systems, short-term changes in immobilization and mineralization by soil microorganisms may be critical to the nutrition of the plant. Over short time periods the C : N of a substrate may have little relationship to its degradability. For example, organic acids and simple sugars contain no N but are more rapidly metabolized than other components (Minderman, 1968). Thus as microorganisms deplete supplies of the most easily degradable substrates and shift to other substrates, C : N ratios fluctuate, and so does the relative dominance of mineralization and immobilization (Hart *et al.,* 1994). Unfortunately, the minimum time scale considered in most decomposition studies is months or greater, and few data are available on short-term changes in mineralization or immobilization following natural additions of organic material. The actual time course of shifts between mineralization and immobilization depends on rates of colonization of the new substrate by microbial populations and the successional processes that follow colonization. Colonization of new substrates takes different lengths of time depending on the location of the substrate. Fine roots and root hairs already in direct contact with soil are colonized rapidly. Colonization may even occur while some of the root tissue is still living (Bowen and Rovira, 1968). On the other hand, weeks may pass before litter falling on the soil surface is reached by fungal hyphae.

Once the litter is colonized, a series of population changes occurs, driven by successive depletion of substrates, interference interactions (e.g., antibiotic production), and predation. Shifts in population structure of the decomposer community results in different mineralization–immobilization relationships because different populations attack different substrates and have different nutritional requirements (Swift, 1976). For example, fungi are generally more efficient at converting organic C from substrates into biomass, and they assimilate less N per unit C, relative to bacteria (Paul and Clark, 1989). Following colonization of litter by bacteria and fungi, populations of predators such as protozoans, nematodes, collembola, etc., may increase. Even though expansion of the microbial biomass generally produces a phase of nutrient immobilization, predation on bacterial and fungal populations by protozoans, nematodes, collembola, etc., speeds turnover of the microbial biomass and initiates another phase of decomposition characterized primarily by mineralization (Fenchel and Harrison, 1976) (see also Figure 3).

The successional sequence described here may take days, weeks, or months to occur, depending on how favorable environmental conditions are. Under nearly optimum conditions in the laboratory with addition of readily decomposable substrates, minimum generation times of some common soil microorganisms were 6.5 h for *Arthrobacter* spp. and 8.8 h for *Micrococcus* spp. (Lowe and Gray, 1973), and 13 h for *Streptomyces* spp. (Williams and Mayfield, 1971); however, in the field, generation times are considerably longer. During a period of favorable moisture conditions, 2 weeks were required before populations of bacteria colonizing *Fraxinus* leaves reached maximum sizes (Gray *et al.*, 1974). Hissett and Gray (1976) provide evidence that generation times in excess of 10 days are probably more typical for natural soil conditions. In addition, the lag between increases in bacterial populations and increases in predators may be several days; thus phases of immobilization or mineralization occurring during even the initial stages of microbial succession and decomposition can be sufficiently long to be of critical importance in controlling availability of organically bound nutrients (i.e., N, P, and S) to plants.

The heterogeneity resulting from the activity of organisms may be on the scale of a few millimeters or less, in the case of earthworm casts, nutrient depletion zones around individual roots, or zones of mineralization or immobilization associated with decomposition of fine roots, root hairs, and root exudates. It may be several centimeters when large organic debris such as thick branches, roots, or crowns are decomposing, or when burrowing mammals are involved. Or it may be several meters or more in the case of litter accumulations beneath plant canopies. Likewise, temporal scales may range from a few hours when leachates from foliage are degraded, to several decades as the more resistant compounds in woody material are degraded.

E. Microclimate

Small-scale variation in microclimate is primarily due to variation in the nature and extent of the vegetative cover. Vegetation moderates fluctuations in temperatures and modifies the temporal and spatial distribution of precipitation. Thick vegetative canopies tend to delay soil wet-up because precipitation from small storms is intercepted, evaporates, and never reaches the soil surface. Dry-down may also be delayed if shading reduces evapotranspiration rates in the understory. Thus, the presence of a vegetative cover can change the coincidence of temperature and moisture conditions. For example, in the spring in the oak woodland–annual grasslands of California (Mediterranean climate), soil water potentials beneath canopies of oaks may drop below -1.5 MPa more than a month later than soil moisture in open grassy interspaces (Stark, 1991). This allows biological activity to continue under oak canopies during periods when mean soil temperatures are high. This may result in faster rates of nutrient cycling, an extended period over which nutrients are released, and greater overall nutrient availability relative to the open grassy interspaces.

Interception of precipitation by plant canopies results in a spatial redistribution of moisture. Stem-flow results in higher soil moisture next to the base of plants and less farther away beneath the canopy. The relative proportion of precipitation channeled to either through-fall or stem-flow, and thus the amount of soil moisture heterogeneity created depends on the intensity and duration of the precipitation event and the canopy architecture. Interception and redistribution of precipitation by plant canopies also directly affect heterogeneity in soil nutrients, since nutrients are dissolved in the water. In an English oak woodland, K and Mg inputs occurring as stem-flow and through-fall exceeded that occurring in litterfall. For Ca, inputs by stem-flow and through-fall were about equal to that by litterfall, and for P, inputs were 60% of litterfall (Carlisle *et al.*, 1966, 1967). Thus redistribution of precipitation by plant canopies can have substantial effects on the spatial distribution of nutrient additions.

The previous discussion describes how general variables can produce soil nutrient heterogeneity. In the following discussion I address the question of how specific nutrient production and consumption processes interact to determine how large a nutrient pool will be.

IV. Factors Regulating Nutrient Pool Sizes

A. What Determines How Large a Pool Will Be?

Although we have been measuring nutrient pool size for decades, we are still unable to accurately predict how large nutrient pools will be under a given set of conditions. The reason for this is that pool size is typically the net result of several competing processes that differ in their kinetic characteristics, and thus their response to environmental variables. We know that when the sum of nutrient production processes is higher than the sum of consumption processes, a nutrient pool will increase; however, we can rarely predict how much it will increase. The final pool size at equilibrium depends on the kinetic characteristics of the individual processes and the environmental parameters that influence them. Because of the complexity of interactions among processes, computer modeling is generally required to determine the final outcome. For example, ammonium concentrations in a soil microsite may be controlled by any or all of the production and consumption processes shown in Figure 5. Each of the fourteen processes is affected differently by

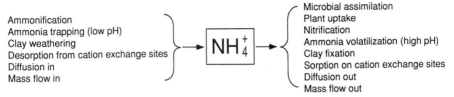

Figure 5 Potential sources and sinks of ammonium in soil.

ammonium concentration, pH, temperature, and moisture (water content, matric potential, and solute potential), as well as the availability of oxygen, reduced carbon compounds, or other nutrients. Variation in any of these factors can change the kinetics of one or more of the processes and, thus, change pool sizes. The effect of increasing temperature on ammonium pool size provides an example of the complexity of interactions. Even in a very simple system, a change in temperature can change pool size in ways that are not immediately intuitive. In the simplistic model shown in Figure 6, the ammonium pool is controlled only by ammonification and nitrification. Ammonification shows zero-order kinetics with respect to NH_4^+, whereas nitrification shows mixed-order (Michaelis–Menten) kinetics. The temperature optimum of nitrification (31°C) is in the midrange of values reported in the literature (25 to 37°C) (Stark, 1991), and the temperature optimum used for ammonification is the same as that typically reported for decomposition processes (40°C) (Hunt, 1977; Moore, 1986). Even in this simple model, NH_4^+ pools at equilibrium do not change monotonically with increasing temperature, but decline at first to a minimum at 20°C, and then increase almost exponentially above this temperature (Figure 6D). Inclusion of other processes in the model would undoubtedly complicate the response even further. If we are to understand what controls nutrient pool size in soils we must understand the kinetics of a wide variety of biotic and abiotic processes.

B. Kinetic Characteristics of Nutrient Production and Consumption Processes

Ultimately, heterogeneity in nutrient pool sizes can be traced to heterogeneity in rates of nutrient production and consumption processes. If a process is controlled by an organism, heterogeneity may be caused by differences in the quantity of enzyme present, the kinetic characteristics of the enzyme systems, the availability of reactants (e.g., O_2, ATP, etc.), or environmental variables that affect enzyme function (e.g., temperature, pH, solute potential, etc.).

When nutrient concentrations vary only slightly, processes catalyzed

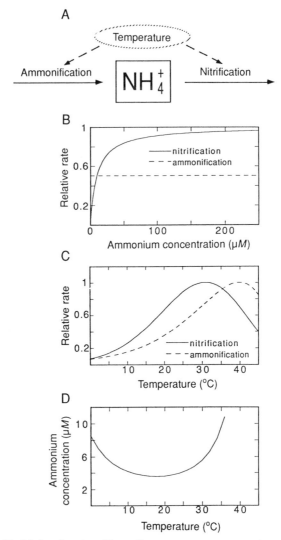

Figure 6   Model showing the effects of temperature on ammonium pool sizes, when ammonification and nitrification are the only production and consumption processes involved. (A) Flow diagram of the model. (B) Michaelis–Menten and zero-order functions used to simulate the effect of ammonium concentration on nitrification and ammonification rates. (C) Response curves used to simulate the effect of temperature on nitrification and ammonification rates. (D) Result of simulations, showing how temperature affects ammonium pool size.

by enzymes often display either first-order or zero-order kinetics; however, over a broader range of substrate concentrations these processes typically display Michaelis–Menten kinetics. For Michaelis–Menten kinetics, reaction rates are a function of K_m, V_{max}, and substrate concentration. The constant K_m is related to the affinity of an enzyme for its substrate and is determined by the structure of the enzyme molecule. K_m is thus controlled genetically; however, environmental factors such as temperature and water potential, that affect protein conformation may also influence K_m (Quinlan, 1981). Spatial or temporal variability in K_m may result from variability in plant or microbial populations or variability in environmental conditions. The constant V_{max} is largely determined by the number of active enzyme sites present. For any particular enzyme system, V_{max} may be controlled by either the number of enzymes per unit biomass or the total amount of biomass present (i.e., the population size). Because of rapid growth rates in many microbial populations, it is believed that population size rarely limits processes such as decomposition over the long term (Paul and Clark, 1989); however, population size may limit rates if the time period is short or if factors other then the substrate of interest control growth (e.g., toxicities). For example, the size of the microbial population assimilating NH_4^+ may be limited by the carbon supply. Spatial variability in carbon supply would thus result in spatial variation in the V_{max} for NH_4^+ assimilation.

Soil chemical processes fit a wide range of kinetic models (Sparks, 1989). These include zero-order, first-order, and power function models for describing reactions of potassium in soil (Havlin and Westfall, 1985), second-order for describing phosphate (Griffin and Jurinak, 1974) and aluminum (Jardine and Zelazny, 1986) reactions, fractional orders and parabolic models for describing dissolution of calcite, feldspars, oxides, and ferromagnesian minerals (Wollast, 1967; Stumm *et al.*, 1985; Bloom and Erich, 1987), mixed-order (i.e., Langmuir, Freundlich, and Brunauer–Emmett–Teller) models for describing adsorption of a variety of solutes on soil surfaces (Bohn *et al.*, 1979), and equilibrium (i.e., Kerr and Gapon) models for describing exchange reactions on soil surfaces (Bohn *et al.*, 1979).

In spite of the diversity in kinetic characteristics of abiotic processes, one can make a few generalizations about the factors that influence reaction rates in soil, and thus heterogeneity in nutrient distribution. Reaction rates of abiotic processes in soil are strongly affected by variation in the chemical constituents present, redox potential, pH, ionic strength, and temperature of the soil solution. Of the constituents present, soil colloids exert a high degree of control over nutrient availability because they act as both a source of nutrients (i.e., by decomposition, weathering, or desorption processes) and a sink for nutrients (i.e., by precipitation, condensation, and sorption processes). Rates of ion exchange also differ among colloids. For example, ion exchange may be quite rapid on kaolinite, smectite, and illite but slow on vermiculite, possibly due to a large amount of interlayer surface area with restricted access. Sawhney (1966) found that adsorption of cesium by illite and montmorillonite (a smectite) occurred within a couple of hours, whereas on vermiculite, sorption had not reached equilibrium even after 500 h.

As discussed earlier, variability in the type of colloid present occurs because of variability in the mineralogy of the parent material, the microclimate (temperature and leaching intensity), the organisms (type and quantity of organic compounds produced and released), and the soil depth. In most ecosystems, variability in colloids on a scale important to individual plants is primarily due to changes in the relative abundance of organic versus inorganic colloids; thus, the characteristics of the dominant colloids may vary more with depth than with horizontal distance.

Other major factors that influence rates of abiotic processes include temperature and ionic strength. Temperature influences reaction rates, primarily by reducing the energy barrier for activation processes, but also by affecting activity coefficients of reactants and products. Ionic strength also influences reaction rates by affecting activity coefficients. Thus changes in soil moisture by evapotranspiration will produce changes in ionic strength (Hillel, 1971), which will result in shifts in equilibria among the chemical constituents of the soil solution. Interactions among the large number of chemical constituents present in soils are so complex that realistic predictions of soil solution composition even under equilibrium conditions cannot be made without computer models.

C. What Controls Patch Segregation?

Differences in spatial distribution of nutrient production and consumption processes create soil patches with high and low nutrient concentrations; however, heterogeneity is also dependent on rates of flow between patches. In fact it is the resistance to this flow that allows heterogeneity to exist and determines the scale of the heterogeneity. For example, if there were no resistance to flow, equilibrium among microsites would be achieved instantaneously and heterogeneity would never occur. Increasing the resistance to flow slows the trend toward equilibrium and increases the isolation of microsites. Thus, transport processes are critical determinants of variability in soils.

1. Transport Processes in Soils   The three transport processes important over the scale we are interested in are (i) diffusion, (ii) mass flow, and (iii) transport by organisms. Transport of nutrients by organisms has been discussed earlier. Mass flow refers to the transport of solutes by moving water, which may occur as a result of water percolation and capillary redistribution following precipitation events, plant uptake of water, evaporation, or even changes in temperature (Hillel, 1971). Though transport by organisms and mass flow may result in either increased homogeneity or increased heterogeneity, diffusion always results in increased homogeneity. Transport by organisms and mass flow may only occur periodically or be limited to specific locations, whereas diffusion occurs continually as long as concentration gradients exist. Thus diffusion represents a potent force opposing processes leading to segregation of nutrients.

2. Factors Affecting Diffusion Rates in Soil   Diffusion of ions through a plane or through a volume where concentration changes are linear is described by Fick's first law:

$$J_z = -D\frac{dC}{dz}, \tag{1}$$

where J_z is the diffusive flux in the z direction, D is the diffusion coefficient, and dC/dz is the concentration gradient in the z direction. Diffusion coefficients have been determined for a wide variety of ions diffusing through a water matrix (Table II). In applying Fick's law to soils, however, it is necessary to use a modified diffusion coefficient to take into account various soil properties. First, soil particles are obstacles to the movement of ions and increase the path length that an ion must travel to move from one location to another. As a soil dries, soil pores drain, and water occurs only as thin coatings on the surfaces of soil particles. Since diffusion of most nutrient ions is limited to the liquid phase, the path length that an ion must follow dramatically increases as a soil dries.

Table II Diffusion Coefficients of Ions in Water at 25°C

Ion	D_0 (cm^2 s^{-1})	Reference
H^+	8.5×10^{-5}	Robinson and Stokes (1959)
Cl^-	2.0×10^{-5}	Robinson and Stokes (1959)
K^+	2.0×10^{-5}	Parsons (1959)
NH_4^+	1.9×10^{-5}	Robinson and Stokes (1959)
NO_3^-	1.9×10^{-5}	Parsons (1959)
$H_2PO_4^-$	0.9×10^{-5}	Edwards and Huffman (1959)
Ca^{2+}	0.8×10^{-5}	Parsons (1959)
Mg^{2+}	0.7×10^{-5}	Parsons (1959)
Succinate	0.1×10^{-5}	Crank (1975)

In addition, water held in layers immediately adjacent to soil colloids has a higher viscosity, which further impedes diffusion in dry soils. Nutrient cations are attracted to charged soil particles and thus diffusion is impeded, while anions are repelled and thus diffusional characteristics are more similar to those in pure water. Olsen and Kemper (1968), Clarke and Barley (1968), and others have incorporated these effects into equations that describe effective diffusion coefficients for various soil systems. One of the simplest equations is that used by Papendick and Campbell (1978) to describe the effective diffusion coefficient for NH_4^+ in a moderately textured soil:

$$D = D_0 k\theta^3,\tag{2}$$

where D_0 is the diffusion coefficient in water, θ is the soil volumetric water content, and k is a constant determined empirically (2.4 for their soil).

The important point of this discussion is that diffusion rates are strongly affected by the volumetric water content of the soil, and only moderately affected by the type of colloid and charge of the ion. This has two major implications for nutrient heterogeneity in soil. First, the rate of equilibration between adjacent microsites is strongly dependent on the soil moisture content. Equilibration occurs much more rapidly when moisture contents are high; isolation is more effective when moisture contents are low. On the other hand, biological processes are most rapid when moisture contents are moderately high (near field capacity), and thus heterogeneity due to differences between rates of production and consumption should be greatest at moderately high moisture contents. Temporal variation in moisture content also influences heterogeneity because soil drying does not occur evenly throughout the soil. Evaporation occurs more rapidly from surface layers, and plant extraction of soil moisture occurs next to roots. Therefore, the greatest heterogeneity in moisture distribution should occur sometime shortly after dry-down begins and the soil is just below field capacity. Combining the effects of these processes, one would predict that nutrient heterogeneity would be low at moisture contents near soil saturation, because rates of diffusion would be high and biological activity would be partially inhibited by low O_2 partial pressures (Figure 7). As soil moisture declines to field capacity and below, drainage and water extraction would produce nonuniform soil moisture distributions that would result both in microsites of high biological activity and in reduced rates of equilibration due to diffusional limitations. Nutrient heterogeneity should be the greatest under these conditions. At lower moisture contents, heterogeneity in distribution of moisture would be reduced by more complete extraction of moisture by plant roots and vapor equilibration. Biological activity

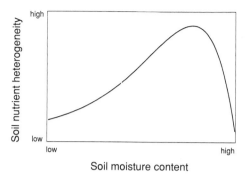

Figure 7 Proposed relationship between soil water content and nutrient heterogeneity.

would decline as a result of low water potential and depletion of substrates in diffusionally isolated microsites. In addition, plants would "perceive" less heterogeneity at low water contents because of an inability to take up nutrients at low moisture contents.

Soil moisture–diffusion relationships explain how heterogeneity in soil physical characteristics affects nutrient heterogeneity. Water movement in soil is driven by water potential gradients rather than by water content gradients. Water moves from areas of high potential to areas of low potential, and stops moving when potentials are equal. At equal water potentials, however, fine-textured soils have greater volumetric water contents than coarse-textured soils. Therefore, variability in texture within a soil can result in variability in moisture content, even under equilibrium conditions. This variability affects rates of nutrient diffusion, biological activity, and thus heterogeneity in nutrient distribution.

V. Diffusion-Reaction Equations for Describing Nutrient Patch Dynamics

Diffusion controls process rates at an even finer scale than that described earlier. Substrates must diffuse through the soil solution to the microbial cell before membrane-bound enzymes can act on them. Reactants must diffuse to mineral surfaces before weathering reactions can occur. Likewise, the nutrients released by enzyme action and weathering must diffuse away from the microbial cell or mineral crystal before plants or other microbes can take them up. Because of diffusional limitations, nutrients occur throughout the soil in gradients, with the highest concentrations occurring at sites of production and the lowest concentrations occurring at sites of consumption.

Mathematical models are available from the engineering literature that deal with the concept of nutrients existing as gradients between sites of production and consumption. Diffusion-reaction models combine equations describing supply processes (i.e., diffusion) with equations describing reaction processes, and allow one to predict overall reaction rates and to identify which process is more limiting. By linking several diffusion-reaction equations together, one could potentially describe changes in nutrient heterogeneity over time under fluctuating environmental conditions.

The simplest diffusion-reaction model combines Fick's first law (Eq. (1)) with a kinetic equation for a biological or chemical reaction. Valid use of Fick's first law requires that the concentration change is linear; thus the system must be at steady state. Although the steady-state requirement may severely limit its application, there are characteristics of the equation that make it especially useful in describing soil microsites. For example, for systems at steady state, Fick's first law can be rewritten in a form analogous to an electrical resistance equation (Campbell, 1977):

$$J = - \frac{C_b - C_m}{R}, \tag{3}$$

where J is the rate of diffusion through a volume such as a long tube, C_b is the concentration at one end of the tube, C_m is the concentration at the other end, and R is the resistance to flow. In this equation, R represents the combined effects of the inherent diffusivity of the medium and the path length that the ion must travel. The advantage to writing the equation in this form is that R can be calculated for a variety of geometries. For example, Campbell (1977) shows that

$$R = z/D \tag{4}$$

can be used to describe diffusion through a tube of uniform diameter, where z is the length of the tube and D is the effective diffusion coefficient;

$$R = \frac{s}{D} \ln \frac{z}{s} \tag{5}$$

can be used to describe diffusion between concentric cylinders (e.g., a uniform segment of root and the outer limit of the rhizosphere), where s is the diameter of the inner cylinder (the root) and z is the diameter of the outer cylinder; and

$$R = \frac{s}{D} \left(1 - \frac{s}{z} \right) \tag{6}$$

can be used to describe diffusion between two concentric spheres (e.g., between the outer edge of a microsite and a mineral crystal, a microbial cell, or colony situated in the center of the microsite) where s is the diameter of the inner sphere (the microbe) and z is the distance to the outer sphere, or the edge of the microsite. In this last equation, if s is small relative to z, s/z becomes insignificant, and the equation reduces to $R = s/D$. In terms of modeling soil microsites, this means that if the size of the microbe or mineral crystal is small relative to the size of the microsite, then the exact size of the microsite does not need to be known. This feature can be extremely useful in modeling soil systems where it is not possible to determine the size of individual microsites.

A diffusion-reaction model for microsites with any of these geometries is easily derived using Eq. (3). For this equation, C_b would be the concentration at the outer edge of the microsite (i.e., in the bulk soil solution) and C_m would be the concentration at a root or microbial surface. By solving for C_m, we can calculate the concentration of substrate that would be supplied by diffusion to the root (or microbe) for enzymatic action:

$$C_m = C_b - JR. \tag{7}$$

If we wish to model enzyme kinetics occurring at the microbe or root surface by the Michaelis–Menten equation, then C_m in Eq. (7) would be equivalent to S in the Michaelis–Menten equation. If the system is at steady state (a prerequisite for use of Fick's first law), then the rate of product consumption (described by V in the Michaelis–Menten equation) equals the rate of flux to the cell (J). Substituting the term J for V and the right side of Eq. (7) for S in the Michaelis–Menten equation, we have

$$J = \frac{V_{max}(C_b - JR)}{(K_m + C_b - JR)}, \tag{8}$$

where can V_{max} is the maximum enzymatic rate per unit of cell surface. This equation can be rearranged and solved for J, producing the equation (Papendick and Campbell, 1978)

$$J = \frac{C_b + K_m + RV_{max} - [(C_b + K_m + RV_{max})^2 - 4C_bRV_{max}]^{0.5}}{2R}. \tag{9}$$

By substituting the appropriate equation for R and using an appropriate equation for the effective diffusion coefficient (D) in soil (e.g., Eq. (2)), this last equation can be used to evaluate a variety of questions related to soil nutrient heterogeneity, such as the effect of variability in substrate supply and moisture content on nutrient release by microorganisms, the effect of variability in nutrient concentration and moisture on diffusion of nutrients to plants roots, or the effect of variability in population characteristics (i.e., V_{max} and K_m) on nutrient production or depletion.

The primary limitation of diffusion-reaction equations using Fick's first law is that they are limited to conditions where concentration gradients are linear. Non-steady-state conditions produce nonlinear gradients. Nutrient production or consumption along the diffusion pathway also results in nonlinear gradients. These applications require use of Fick's second law. For a system with spherical geometry, the diffusion-reaction equation linking Fick's second law and Michaelis–Menten kinetics is

$$\frac{\partial C}{\partial t} = D\left(\frac{\partial_2 C}{\partial z} + \frac{2\partial C}{-\partial z}\right) - \frac{V_{max}C}{(K_m + C)} \tag{10}$$

(Myrold and Tiedje, 1985) and for cylindrical geometry

$$\frac{\partial C}{\partial t} = \frac{1}{z}\frac{\partial}{\partial z}\left(zD\frac{\partial C}{\partial z}\right) - \frac{V_{max}\,C}{(K_m + C)}, \tag{11}$$

where z is the distance outward from the center of the sphere or cylinder (Barber, 1984). Unfortunately, these partial differential equations cannot be solved analytically, as can diffusion-reaction equations using Fick's first law. Instead they must be solved using methods such as numerical approximation. Because additional information on the size of the microsite and the concentration gradient is required, their use is limited to circumstances where this information can be obtained (e.g., Myrold and Tiedje, 1985).

VI. Concluding Remarks

Because very few studies have documented variation in soil nutrient concentrations at scales less than a few meters, and even fewer have determined cause-and-effect relationships at these scales, we must extrapolate from studies dealing with coarse-scale soil variability to predict what causes nutrient heterogeneity at scales important to individual plants.

In young soils, parent material has a dominant influence on soil nutrient heterogeneity; however, in most well-developed soils, heterogeneity primarily results from the activity of organisms. Organisms have an overriding influence on heterogeneity through (i) the effect of plants on the microclimate and microtopography, the development of nutrient depletion zones around roots, and the amount, location, timing, and chemical composition of organic inputs; (ii) the effect of macro- and microfauna on mixing soil and partially digesting plant and organic material; and (iii) the effect of soil microorganisms on decomposition, immobilization, and mineralization of soil nutrients from soil organic components. Since plants are the greatest source of organic matter, to a large extent the spatial scale of heterogeneity is dependent on the size of the plant part that is being decomposed, and the temporal scale depends on the quality of the material as a substrate for decomposers. In most ecosystems, the greatest input of organic material appears to be through roots, primarily as root exudates and fine root turnover. These inputs tend to promote highly localized and temporally transient zones of intense immobilization or mineralization of nutrients. Much of this heterogeneity probably occurs on a temporal scale of weeks and a spatial scale of millimeters to centimeters.

The notion that plants are primary contributors to soil nutrient heterogeneity as well as subject to that heterogeneity raises an interesting question: How well are plants adapted to exploiting the heterogeneity that they create? One would expect that if a plant species is not particularly well adapted for exploiting the type of heterogeneity that it creates, it will eventually be replaced by a species that is better adapted to that type of heterogeneity. Species replacement may be more related to the ability of a plant to exploit certain types of fine-scale nutrient heterogeneity than to its ability to tolerate temporal changes in average attributes of the environment. Thus a strong association between creation of heterogeneity and exploitation of heterogeneity may be an attribute possessed more by late-successional species than by early-successional species.

Much additional research is needed to document both the temporal and spatial scales of nutrient heterogeneity that occur in a wide variety of soils. Because plants have the ability to exploit relatively small and temporally transient patches of nutrients, emphasis should be placed on fine-scale variability in nutrient availability. Since nutrient availability is determined not only by nutrient pool sizes but also by the ability of the soil to replenish those pools, we must know more about heterogeneity in nutrient production and consumption processes as well.

Acknowledgments

I would like to thank Mary Firestone for helpful discussions that led to the development of many of the ideas presented here. I would also like to thank Martyn Caldwell and Chuck Grier for providing critical reviews and Anne Stark for providing editorial comments that greatly improved the quality of the manuscript. This work was partially supported by the Cooperative State Research Service, U.S. Department of Agriculture, under agreement No. 92-37101-7976, and by National Science Foundation Grant DEB9208212.

References

Anderson, R. V., Coleman, D. C., and Cole, C. V. (1981). Effects of saprotrophic grazing on net mineralization. *Ecol. Bull.* **33**, 201–215.

Barber, S. A. (1984). "Soil Nutrient Bioavailability." Wiley, New York.

Berg, B., and Staaf, H. (1981). Leaching, accumulation, and release of nitrogen in decomposing forest litter. *Ecol. Bull.* **33**, 163–178.

Birch, H. F. (1960). Nitrification in soils after different periods of dryness. *Plant Soil* **12**, 81–96.

Birkeland, P. W. (1974). "Pedology, Weathering, and Geomorphological Research." Oxford Univ. Press, New York.

Bloom, P. R., and Erich, M. S. (1987). Effect of solution composition on the rate and mechanism of gibbsite dissolution in acid solutions. *Soil Sci. Soc. Am. J.* **51**, 1131–1136.

Bohn, H. L., McNeal, B. L., and O'Conner, G. A. (1979). "Soil Chemistry." Wiley, New York.

Bowen, G. D., and Rovira, A. D. (1968). The influence of micro-organisms on root growth and metabolism. *In* "Root Growth" (W. J. Whittington, ed.), pp. 170–201. Butterworth, London.

Caldwell, M. M., and Richards, J. H. (1989). Hydraulic lift: Water efflux from upper roots improves effectiveness of water uptake by deep roots. *Oecologia* **79**, 1–5.

Campbell, G. S. (1977). "An Introduction to Environmental Biophysics." Springer-Verlag, New York.

Carlisle, A., Brown, A. H. F., and White, E. J. (1966). The organic matter and nutrient elements in the precipitation beneath a sessil oak (*Quercus petraea*) canopy. *J. Ecol.* **54**, 87–98.

Carlisle, A., Brown, A. H. F., and White, E. J. (1967). The nutrient content of tree stem flow and ground flora litter and leachates in a sessile oak (*Quercus petraea*) woodland. *J. Ecol.* **55**, 615–627.

Chapin, F. S. (1991). Effects of multiple environmental stresses on nutrient availability and use. *In* "Response of Plants to Multiple Stresses" (H. A. Mooney, W. E. Winner, and E. J. Pell, eds.), pp. 67–127. Academic Press, New York.

Charley, J. L. (1977). Mineral cycling in rangeland ecosystems. *In* "Rangeland Plant Physiology" (R. E. Sosebee, ed.), pp. 215–256. Soc. Range Manage., Denver, CO.

Charley, J. L., and West, N. E. (1975). Plant-induced soil chemical patterns in some desert shrub-dominated ecosystems of utah. *Utah J. Ecol.* **63**, 945–964.

Clarke, A. L., and Barley, K. P. (1968). The uptake of nitrogen from soils in relation to solute diffusion. *Aust. J. Soil Res.* **6**, 75–92.

Coleman, D. C. (1976). A review of root production processes and their influence on soil biota in terrestrial ecosystems. *Symp. Br. Ecol. Soc.* **17**, 417–434.

Crank, J. (1975). "The Mathematics of Diffusion." Oxford Univ. Press (Clarendon), London.

Edwards, O. W., and Huffman, E. O. (1959). Diffusion of aqueous solutions of phosphoric acid at 25 degrees. *J. Phys. Chem.* **63**, 1830–1833.

Elliott, E. T. (1978). Carbon, nitrogen, and phosphorus transformations in gnotobiotic soil microcosms. M. S. Thesis, Colorado State University, Fort Collins.

Fenchel, T., and Harrison, P. (1976). The significance of bacterial grazing and mineral cycling for the decomposition of particulate detritus. *Symp. Br. Ecol. Soc.* **17**, 285–299.

Gray, T. R. G., Hissett, R., and Duxbury, T. (1974). Bacterial populations of litter and soil in a deciduous woodland. II. Numbers, biomass and growth rates. *Rev. Ecol. Biol. Sol* **10**, 15–26.

Griffin, R. A., and Jurinak, J. J. (1974). Kinetics of the phosphate interaction with calcite. *Soil Sci. Soc. Am. J.* **38**, 75–79.

Hart, S. C. (1990). Control of decomposition processes and nutrient flow in a California forest and grassland. Ph.D. Dissertation, University of California, Berkeley.

Havlin, J. L., and Westfall, D. G. (1985). Potassium release kinetics and plant response in calcareous soils. *Soil Sci. Am. J.* **49**, 366–370.

Hillel, D. (1971). "Soil and Water: Physical Principles and Processes." Academic Press, New York.

Hart, S. C., Nason, G. E., Myrold, D. D., and Perry, D. A. (1994) Dynamics of gross nitrogen transformations during long-term incubation of an old-growth forest soil. *Ecology* (in press).

Hissett, R., and Gray, T. R. G. (1976). Microsites and time changes in soil microbe ecology. *Symp. Br. Ecol. Soc.* **17**, 23–39.

Hunt, H. W. (1977). A simulation model for decomposition in grasslands. *Ecology* **58**, 469–484.

Jackson, R. B., and Caldwell, M. M. (1989). The timing and degree of root proliferation in fertile-soil microsites for three cold-desert perennials. *Oecologia* **81**, 149–153.

Jackson, R. B., Manwaring, J. H., and Caldwell, M. M. (1990). Rapid physiological adjustment of roots to localized soil enrichment. *Nature (London)* **344**, 58–60.

Jardine, P. M., and Zelazny, L. W. (1986). Mononuclear and polynuclear aluminum speciation through differential kinetic reactions with ferron. *Soil Sci. Soc. Am. J.* **50**, 895–900.

Jenny, H. (1941). "Factors of Soil Formation." McGraw-Hill, New York.

Jenny, H. (1980). "The Soil Resource." Springer-Verlag, New York.

Jones, C. G., and Coleman, J. S. (1991). Plant stress and insect herbivory: Toward an integrated perspective. *In* "Response of Plants to Multiple Stresses" (H. A. Mooney, W. E. Winner, and E. J. Pell, eds.), pp. 249–280. Academic Press, San Diego.

Lowe, W. E., and Gray, T. R. G. (1973). Ecological studies on coccoid bacteria in a pine forest soil. II. Growth of bacteria introduced into soil. *Soil Biol. Biochem.* **5**, 449–462.

Lynch, J. M. (1990). "The Rhizosphere." Wiley, New York.

Mellilo, J. M., Aber, J. D., and Muratore, J. F. (1982). Nitrogen and lignin control of hardwood leaf litter decomposition dynamics. *Ecology* **63**, 621–626.

Mengel, K., and Kirkby, E. A. (1982). "Principles of Plant Nutrition." Potash Inst., Bern, Switzerland.

Minderman, G. (1968). Addition, decomposition and accumulation of organic matter in forests. *J. Ecol.* **56**, 355–362.

Moore, A. M. (1986). Temperature and moisture dependence of decomposition rates of hardwood and coniferous leaf litter. *Soil Biol. Biochem.* **18**, 427–435.

Myrold, D. D., and Tiedje, J. M. (1985). Diffusional constraints on denitrification in soil. *Soil Sci. Soc. Am. J.* **49**, 651–657.

Norton, J. M. (1991). Carbon and nitrogen dynamics in the rhizosphere of *Pinus ponderosa* seedlings. Ph.D. Dissertation, University of California, Berkeley.

Olsen, S. R., and Kemper, W. D. (1968). Movement of nutrients to plant roots. *Adv. Agron.* **20**, 91–151.

Papendick, R. I., and Campbell, G. S. (1978). Theory and measurement of water potential. *SSSA Spec. Publ.* **9**, 1–22.

Parkin, T. B. (1987). Soil microsites as a source of denitrification variability. *Soil Sci. Soc. Am. J.* **51**, 1194–1199.

Park, D. (1976). Carbon and nitrogen levels as factors influencing fungal decomposers. *Symp. Br. Ecol. Soc.* **17**, 41–60.

Parsons, R. (1959). "Handbook of Electrochemical Constants." Academic Press, New York.

Paul, E. A., and Clark, F. E. (1989). "Soil Microbiology and Biochemistry." Academic Press, New York.

Quinlan, A. V. (1981). The thermal sensitivity of generic Michaelis–Menten processes without catalyst denaturation or inhibition. *J. Therm. Biol.* **6**, 103–114.

Reid, C. P. P. (1974). Assimilation, distribution, and root exudation of ^{14}C by ponderosa pine seedlings under induced water stress. *Plant Physiol.* **54**, 44–49.

Roberts, R. C. (1950). Chemical effects of salt-tolerant shrubs on soils. *Trans. Int. Congr. Soil Sci., 4th, 1950*, Vol. 1, pp. 404–406.

Robinson, D., Griffiths, B., Ritz, K., and Wheatley, R. (1989). Root-induced nitrogen mineralisation: A theoretical analysis. *Plant Soil* **117**, 185–193.

Robinson, R. A., and Stokes, R. H. (1959). "Electrolyte Solutions." Butterworth, London.

Rovira, A. D., and Davey, C. B. (1974). Biology of the rhizosphere. *In* "The Plant Root and Its Environment" (E. W. Carson, ed.), pp. 153–204. Univ. Press of Virginia, Charlottesville.

Russell, E. W. (1973). "Soil Conditions and Plant Growth." Longman, London.

Sawhney, B. L. (1966). Kinetics of cesium sorption by clay minerals. *Soil Sci. Soc. Am. Proc.* **30**, 565–569.

Schimel, J. P. (1987). Plant/microbial competition for nitrogen in a California forest and grassland. Ph.D. Dissertation, University of California, Berkeley.

Sharma, M. L., and Tongway, D. J. (1973). Plant-induced soil salinity patterns in two saltbush (*Atriplex* sp.) communitities. *J. Range Manage.* **26**, 121–125.

Shaver, G. R., and Melillo, J. M. (1984). Nutrient budgets of marsh plants: Efficiency concepts and relation to availability. *Ecology* **65**, 1491–1510.

Sparks, D. L. (1989). "Kinetics of Soil Chemical Processes." Academic Press, San Diego.

Stark, J. M. (1991). Environmental factors versus ammonia-oxidizer population characteristics as dominant controllers of nitrification in an oak woodland-annual grassland soil. Ph.D. Dissertation, University of California, Berkeley.

Stark, J. M., and Redente, E. F. (1985). Soil–plant diversity relationships on a disturbed site in northwestern Colorado. *Soil Sci. Soc. Am. J.* **49**, 1028–1034.

Stumm, W., Furrer, G., Wieland, E., and Zinder, B. (1985). The effects of complex-forming ligands on the dissolution of oxides and aluminosilicates. *In* "The Chemistry of Weathering" (J. I. Drever, ed.), Reidel Publ., Dordrecht, The Netherlands.

Swift, M. J. (1976). Species diversity and the structure of microbial communities in terrestrial habitats. *Symp. Br. Ecol. Soc.* **17**, 185–222.

Tiedemann, A. R., and Klemmedson, J. O. (1983). Effects of mesquite on physical and chemical properties of the soil. *J. Range Manage.* **26**, 27–29.

Vitousek, P. (1982). Nutrient cycling and nutrient use efficiency. *Am. Nat.* **119**, 553–572.

Vogt, K. A., Grier, C. C., Meier, C. E., and Edmonds, R. L. (1982). Mycorrhizal role in net primary production and nutrient cycling in *Abies amabilis* ecosystems in western Washington. *Ecology* **63**, 370–380.

Walter, H. (1979). "Vegetation of the Earth and Ecological Systems of the Geo-biosphere." Springer-Verlag, New York.

Williams, S. T., and Mayfield, C. I. (1971). Studies on the ecology of actinomycetes in soil. III. The behaviour of neutrophilic streptomycetes in acid soil. *Soil Biol. Biochem.* **3**, 197–208.

Witkamp, M. (1969). Environmental effects on microbial turnover of some mineral elements. *Soil Biol. Biochem.* **1**, 167–184.

Witkamp, M., and Ausmus, B. S. (1976). Processes in decomposition and nutrient transfer in forest systems. *Symp. Br. Ecol. Soc.* **17**, 375–396.

Witkamp, M., and Barzansky, B. (1968). Microbial immobilization of ^{137}Cs in forest litter. *Oikos* **19**, 392–395.

Wollast, R. (1967). Kinetics of the alteration of K-feldspar in buffered solutions at low temperature. *Geochim. Cosmochim. Acta* **31**, 635–648.

10

Root–Soil Responses to Water Pulses in Dry Environments

Park S. Nobel

I. Introduction

The environmental variable having the greatest influence on vegetation in deserts and other water-limited environments is rainfall. The pulses of growth of annuals and of water uptake by perennials are responses to the increases in soil water potential caused by the often unpredictable rainfalls in such regions. An equally important consideration is the responses of desert vegetation to episodes or "pulses" of drought, which calls attention to the problem of preventing lethal water loss to the soil. For instance, the living tissues in the roots of a succulent agave or cactus can have a water potential (Ψ) of -0.5 MPa when Ψ_{soil} in the root zone of these shallow-rooted perennials (mean root depth generally about 0.10 m) is -10 MPa (Nobel, 1988). The driving force on water loss from roots to the soil (9.5 MPa) is then far greater than the maximal force (0.5 MPa) for water uptake that occurs when Ψ_{soil} is close to 0.0 MPa after a rainfall. What prevents the loss of substantial amounts of water from roots of living plants to a dry soil? How fast is water taken up once Ψ_{soil} in the root zone has been raised by rainfall? The ready uptake of water from a wet soil but the prevention of large fluxes of water to a dry soil have led to the suggestion that roots act as rectifiers (Nobel and Sanderson, 1984; Dirksen and Raats, 1985; Katou and Taura, 1989). Actually, the entire root–soil system is involved in rectification, without which plants would become lethally desiccated during prolonged drought.

The objective of this review is to synthesize and to expand upon two recent reports dealing with water movement between the soil and roots (Nobel and Cui, 1992a,b). The water-movement pathway will be divided into three components: (1) the soil; (2) the root–soil air gap that can develop as a root shrinks during soil drying; and (3) the root, specifically the region from the root surface to the root xylem. Emphasis will be on the root–soil air gap, which has received relatively little attention. The water conductance of the overall pathway will be determined for changes of roots and soil accompanying decreasing Ψ_{soil} during drought and the changes in root diameter and water-uptake ability that accompany soil rewetting. Transient changes in root water fluxes will be examined for succulent perennials whose shallow roots are subjected to great temporal variation in Ψ_{soil} during drying and rewetting pulses.

II. Root Hydraulic Conductivity Coefficient

A root hydraulic conductivity coefficient (L_P; m s^{-1} MPa^{-1}) is generally used to relate the volumetric flux density of water (J_V; m^3 m^{-2} s^{-1} = m s^{-1}) into roots to the water potential differences causing the flow:

$$L_P = J_V/(\Psi_{surf} - \Psi_{xylem}), \tag{1}$$

where Ψ_{surf} is the water potential at the root surface and Ψ_{xylem} is the water potential in the root xylem (Passioura, 1988; Nobel, 1991; Oertli, 1991). For very young roots with immature xylem, L_P is low. Also, L_P can be low for older roots with substantial suberization (Newman, 1974; Caldwell, 1976; Sanderson, 1983; North and Nobel, 1991). For roots of intermediate ages of up to a few months, L_P can be 1×10^{-7} to 5×10^{-7} m s^{-1} MPa^{-1} (Fiscus, 1977; Salim and Pitman, 1984; Steudle *et al.*, 1987; Nobel *et al.*, 1990). For instance, for 6-week-old roots of *Agave deserti*, *Ferocactus acanthodes*, and *Opuntia ficus-indica*, L_P averages 2.4×10^{-7} m s^{-1} MPa^{-1} under wet conditions (Figure 1).

What happens to L_P during drought? If a large decrease in L_P were to occur, then the water loss from a root to a drying soil would be greatly reduced. Specifically, when Ψ_{surf} becomes less than Ψ_{xylem}, water moves out of a root ($J_V < 0$ in Eq. (1)), and a low L_P would mean a small J_V. However, for the three desert succulents just considered, L_P of 6-week-old roots decreases on average only about threefold as the ambient Ψ is lowered from -0.01 to -10 MPa (Figure 1). Such a decrease in L_P can help reduce water loss from a plant to the soil, but it is not sufficient by itself to account for responses to drought in the field. Indeed, the overall water conductance of the root–soil system must decrease 10^5-fold to account for the observed small loss of water from *A. deserti* to a desert

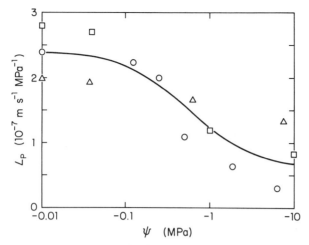

Figure 1 Changes in the root hydraulic conductivity coefficient (L_P) as the ambient water potential (ψ) is lowered. Data are for approximately 6-week-old roots of *Agave deserti* (O), *Ferocactus acanthodes* ($\triangle$), and *Opuntia ficus-indica* ($\square$) and are adapted from Nobel and Sanderson (1984), Nobel *et al.* (1990), Lopez and Nobel (1991), and North and Nobel (1991, and unpublished observations). Individual measurements are made with microcapillaries attached to the proximal end of excised roots in aqueous solutions, thus eliminating the influence on J_V of the soil and the root–soil air gap.

soil experiencing drought for 5 months (Schulte and Nobel, 1989). Thus, the major reduction in conductance during drought must occur for the soil and/or the root–soil air gap, not the root itself.

III. Soil Hydraulic Conductivity Coefficient

The volumetric flux density of water in the soil is described by Darcy's law, which in Cartesian coordinates can be represented as follows for the one-dimensional case:

$$J_V = - L_{\text{soil}} \, \partial \Psi / \partial x, \tag{2}$$

where L_{soil} (m^2 s^{-1} MPa^{-1}) is the soil hydraulic conductivity coefficient and $\delta \Psi / \delta x$ (MPa m^{-1}) represents the gradient in Ψ leading to water movement (Tinker, 1976; Campbell, 1985; Nobel, 1991). Based on the pioneering studies of Gardner (1960) and Cowan (1965), water movement into an individual root is generally considered in cylindrical coordinates with the root axis longitudinally along the cylinder axis (Newman, 1974). Also, steady-state conditions and cylindrical symmetry are generally assumed, so the volumetric flux density at a particular radial distance

(r; m) from the root axis times the accompanying circumference ($2\,\pi r$) is constant, Customarily, J_V is expressed at the root surface, which leads to the following form for Darcy's law in cylindrical coordinates (Nobel, 1991):

$$J_V = \frac{1}{r_{\text{root}}} L_{\text{soil}} \frac{(\Psi_{\text{distant}} - \Psi_{\text{gap}})}{\ln\,(r_{\text{distant}}/r_{\text{gap}})}, \tag{3}$$

where r_{root} is the root radius, r_{distant} represents the radial distance from which water moves toward a particular root, Ψ_{distant} represents the soil water potential at r_{distant}, r_{gap} is the radial distance to the outer side of the root–soil air gap, and Ψ_{gap} is the soil water potential at r_{gap} (if there is no root–soil air gap, r_{gap} is the same as r_{root}).

The value of L_{soil} depends on the soil porosity and water content. For example, L_{soil} under wet conditions can be 10^3-fold higher for a sand than for a clay (Marshall and Holmes, 1988). For the intermediate soil texture known as a sandy loam, which is characteristic of many desert soils, L_{soil} decreases 10^6-fold as Ψ_{soil} decreases from -0.01 to -10 MPa (Figure 2). Such a large decrease could explain the 10^5-fold decrease in water conductance of the root–soil system predicted to occur under extreme drought (Schulte and Nobel, 1989), but the relative magnitudes

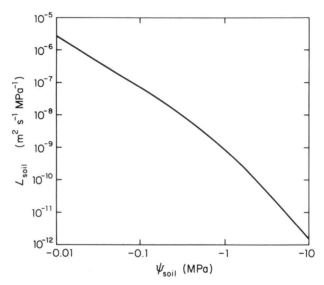

Figure 2   Decrease in the soil hydraulic conductivity coefficient (L_{soil}) as the soil water potential (ψ_{soil}) progressively decreases. Data are for a sandy loam at Agave Hill in the northwestern Sonoran Desert and are adapted from Young and Nobel (1986) and P. S. Nobel (unpublished observations).

of L_P and L_{soil} as well as the possibility of a root–soil air gap must also be considered.

IV. Root–Soil Air Gap

When a root is in wet soil (e.g., $\Psi_{soil} = -0.01$ MPa), soil water can contact the root surface (Figure 3A). The pathway for water movement is energetically downhill to r_{root} ($\Psi_{soil} > \Psi_{gap} = \Psi_{surf}$) and then energetically downhill into the root xylem ($\Psi_{surf} > \Psi_{xylem}$); additional complexities introduced by reflection coefficients less than unity, indicating that osmotic effects are not fully expressed, are ignored here (Dalton *et al.*, 1975; Fiscus, 1975, 1986; Nobel, 1991; Oertli, 1991). As the soil dries, Ψ_{soil} becomes less than Ψ_{xylem}, water loss from the root is favored, the root can shrink (Figure 3B), and an annular root–soil air gap of thickness Δx_{gap} (m) can form around a root (Figure 3B).

As soil dries or the water potential of a hydroponic solution is lowered, roots of various species shrink, reflecting water loss primarily from living cortical cells occurring outside the endodermis (Oertli, 1991). For instance, young roots of *Helianthus annuus* ($r_{root} = 0.2$ mm) shrink 20% as

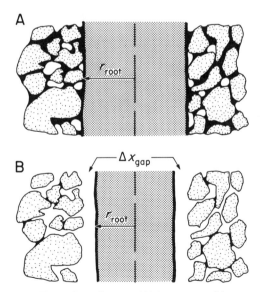

Figure 3 Schematic illustration of a root (A) in contact with a wet soil and (B) after shrinking away from a dry soil. Water is indicated by the dark shading, soil particles by the light stippling, and the root by the heavy stippling. Also indicated are the root radius (r_{root}) and the air gap that develops (Δx_{gap}) as the root shrinks away from the soil particles.

Ψ_{soil} decreases from -0.2 to -1.2 MPa (Faiz and Weatherley, 1978, 1982), 6-day-old roots of *Glycine max* shrink 30% as $\Psi_{solution}$ is lowered from 0.0 to -1.3 MPa (Taylor and Willatt, 1983), 6-day-old roots (r_{root} = 0.3 mm) of *Triticum vulgare* shrink 60% from 0.0 to -1.2 MPa (Cole and Alston, 1974), and young roots of *Vicia faba* shrink up to 60% at -1.5 MPa (Rowse and Goodman, 1981). For a mature root of *Gossypium hirsutum* (r_{root} = 0.2 mm at night), daytime shrinkage of 6 to 40% is reported in response to variations in transpiration rate (Huck *et al.*, 1970). However, changes are proposed to be less if good root–soil contact is maintained (Taylor and Willatt, 1983; Passioura, 1988) and such large daily shrinkages are not observed for other species (Newman, 1974).

For 6-week-old roots of *A. deserti, F. acanthodes,* and *O. ficus-indica,* shrinkage becomes progressively greater as Ψ is decreased from -0.01 to -10 MPa (Figure 4). For all three species, shrinkage is only about 3% at -0.3 MPa, a Ψ that is similar to the mean root water potential of certain desert succulents under wet conditions (Nobel and Lee, 1991; Nobel and Cui, 1992a). At a Ψ of -10 MPa, shrinkage is nearly 20% for all three species (Figure 4). For *O. ficus-indica,* shrinkage becomes less with root age, from 39% at -10 MPa for 1-month-old roots to 6% for 12-month-old roots (Table I). Shrinkage is similar and reversible for both attached and excised roots. Temperatures from 10 to 40°C have only minor influences on the rate or the extent of the shrinkage (Table 10.1) (Nobel and Cui, 1992b).

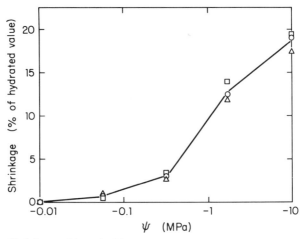

Figure 4 Shrinkage of 6-week-old roots of *Agave deserti* (○), *Ferocactus acanthodes* (△), and *Opuntia ficus-indica* (□) in response to decreases in the water potential (ψ) of the atmosphere to which the excised roots are exposed. Data are means for six roots with initial hydrated diameters of 2.07 mm for *A. deserti,* 1.79 mm for *F. acanthodes,* and 2.17 mm for *O. ficus-indica.* (Adapted from Nobel and Cui, 1992a, and unpublished observations.)

Table I Influences of Root Age, Temperature, and Water
Potential on Shrinkage of Attached Roots of *Opuntia
ficus-indica*[a]

Age (months)	Temperature (°C)	Ψ (MPa)	Shrinkage (%)
1	25	−0.3	4
1	25	−10	39
2	10	−10	18
2	25	−10	19
2	40	−10	20
4	25	−10	13
12	25	−0.3	1
12	25	−10	6

[a] The control had a Ψ of −0.01 MPa at 25°C; root diameter averaged
1.8 mm. Data are means for four roots and are adapted from Nobel and
Cui (1992b).

A. Conductance

The importance of a root–soil air gap across which water moves as a
vapor has long been debated (Bernstein *et al.,* 1959; Bonner, 1959;
Cowan and Milthorpe, 1968; Newman, 1974; Tinker, 1976; Passioura,
1988; Oertli, 1991). For instance, Philip (1958) proposed that during
periods of rapid transpiration, the soil near the root surface becomes
depleted of water so that the final water movement into a root is in the
vapor phase, which can be augmented as root shrinkage disrupts the
root–soil continuity of liquid water in a drying soil. Furthermore, reduction of the putative gap by soil vibration or mechanical deformation of
soil toward roots can increase plant water uptake (Faiz and Weatherley,
1982). An empirical model describing a root–soil contact conductance
has been proposed to help interpret water-uptake rates by roots of *Triticum aestivum* in a drying soil that are up to eightfold lower than the rates
predicted based on L_{soil} and L_P (Herkelrath *et al.,* 1977a,b). The empirical
model, which decreases the conductances of the soil and the root by the
ratio of actual soil water content to saturated soil water content, closely
predicts depletion of soil water by roots of *Glycine max* in the field (Jung
and Taylor, 1984). Also, a gap with a low conductance has been proposed
around roots of *Helianthus annuus* subjected to water stress (Faiz and
Weatherley, 1978). A highly simplified but analytical model has been
proposed to calculate the conductance of the root–soil air gap (Nobel
and Cui, 1992a).

The analytical model treats the movement of water vapor as a one-dimensional diffusion process across Δx_{gap}. The steady-state relationship describing such movement is Fick's first law (Nobel, 1991):

$$J_{wv} = \frac{D_{wv}\Delta P_{wv}}{RT\Delta x_{gap}}, \tag{4}$$

where J_{wv} is the molar flux density of water vapor (mol m^{-2} s^{-1}), D_{wv} is the diffusion coefficient of water in the vapor phase (m^2 s^{-1}), R is the gas constant (m^3 MPa mol^{-1} K^{-1}), T is the absolute temperature (K), and ΔP_{wv} is the drop in water vapor partial pressure (MPa) across Δx_{gap}. J_{wv} times the volume of liquid water per mole ($\overline{V}_w$; 18×10^{-6} m^3 mol^{-1}) equals J_V. In the air phase of a soil, P_{wv} is only slightly less than P^*_{wv}, the saturation vapor pressure (Hillel, 1982; Marshall and Holmes, 1988; Nobel, 1991). Hence, the drop in water potential of the vapor phase ($\Delta\Psi_{wv}$) across the gap can be represented under isothermal conditions as (Nobel, 1991)

$$\Delta\Psi_{wv} = \frac{RT}{\overline{V}_w}\Delta\ln\frac{P_{wv}}{P^*_{wv}} \cong \frac{RT}{\overline{V}_w}\Delta\left(\frac{P_{wv}}{P^*_{wv}}-1\right) = \frac{RT}{\overline{V}_w}\frac{\Delta P_{wv}}{P^*_{wv}}. \tag{5}$$

Based on Eqs. (4) and (5), the volumetric flux density of water across the root–soil air gap is

$$J_V = \overline{V}_w J_{wv} = \overline{V}_w\frac{D_{wv}}{RT}\frac{\Delta P_{wv}}{\Delta x_{gap}}$$

$$= \overline{V}_w^2\frac{D_{wv}P^*_{wv}}{(RT)^2}\frac{\Delta\Psi_{wv}}{\Delta x_{gap}} = \frac{\overline{V}_w^2 D_{wv}P^*_{wv}}{(RT)^2\Delta x_{gap}}(\Psi_{gap} - \Psi_{surf}). \tag{6}$$

By analogy with the definition of L_P (Eq. (1)) and using Eq. (6), the conductance of the root–soil air gap (L_{gap}; m s^{-1} MPa^{-1}) can be represented as

$$L_{gap} = J_V/(\Psi_{gap} - \Psi_{surf}) = \frac{\overline{V}_w^2 D_{wv}P^*_{wv}}{(RT)^2\Delta x_{gap}}. \tag{7}$$

Equation (7) indicates that L_{gap} depends inversely on Δx_{gap}. At 25°C, $\overline{V}_w^2 D_{wv}P^*_{wv}/(RT)^2$ is 4.18×10^{-12} m^2 s^{-1} MPa^{-1}.

B. Additional Complexities

The preceding derivation applies for isothermal conditions, whereas evaporation and condensation of water can lead to local temperature differences. When Ψ_{soil} is less than Ψ_{surf}, water tends to evaporate at the root surface and then to diffuse toward the soil. Evaporation of water can lower the temperature at the root surface and condensation of water vapor after diffusion across the root–soil air gap can increase the local soil temperature. But P^*_{wv} increases nearly exponentially with temperature (Nobel, 1991), tending to raise P_{wv} at the soil surface, where Ψ_{wv} equals Ψ_{gap} ($\Psi_{wv} = (RT/\overline{V}_w) \ln (P_{wv}/P^*_{wv})$, Eq. (5)). This can be imagined to set up a partially compensating movement of water vapor in the opposite direction back toward the root. Thus, the tendency for water vapor to move toward regions of lower temperature at a given Ψ (Campbell, 1985; Nobel, 1991) opposes the diffusion away from the root in response to $\Delta\Psi$, resulting in a lowering of the apparent L_{gap}, although the effects are complicated at the soil surface (Cass *et al.*, 1984). In one treatment, allowing for heat flow accompanying the distillation of water across a root–soil air gap decreases L_{gap} approximately three-fold (Cowan and Milthorpe, 1968).

The derivation of Eq. (7) is based on a one-dimensional approach (Eq. (4)), which is appropriate if Δx_{gap} is small relative to r_{root}. Switching to cylindrical coordinates for the more general case requires that Δx_{gap} in Eq. (7) is replaced by $r_{root} \ln (1 + \Delta x_{gap}/r_{root})$, which increases L_{gap}. For a 20% shrinkage, L_{gap} is 10% higher using cylindrical coordinates instead of the one-dimensional form (Eq. (7)). A more complicated geometrical problem is that the root may not be concentrically located within the annular air gap (Figure 5A) so that the root–soil separation (Δx_{gap}) may not be uniform all around the periphery of the root. Instead, the root may touch the soil at one side of the annulus (Figure 5B), especially if bends in the root occur, greatly increasing the water flux at the location of contact (Tinker, 1976). Such an increase in local J_V translates into a higher L_{gap}. For instance, numerical integration of L_{gap} at 45° intervals around a root indicates that L_{gap} can be increased fourfold for a root that touches the soil at one side of the annulus (Nobel and Cui, 1992b). On the other hand, the conductances of both the soil and the root are reduced by the eccentric location of the root in the air gap, which is schematically illustrated in Figure 5B by the longer pathways for water movement when the root touches the side of the gap compared with the radial pathways for water movement when the root is at the center of the gap (Figure 5A).

Roots may grow into previously existing soil passageways, such as those caused by earthworms or by soil cracking (Herkelrath *et al.*, 1977b; Faiz and Weatherley, 1982), which can create additional root–soil air gaps. Degeneration of the cortex for roots of monocotyledons during aging can also create root–soil gaps (Russell, 1977). On the other hand, soil compressed by root growth may partially recoil elastically as a root shrinks during drought, which would cause Δx_{gap} to be smaller than expected. During drying, soil may shrink toward a root, decreasing Δx_{gap} and hence raising L_{gap}. However, soil shrinkage away from roots as Ψ_{soil} becomes less is more probable (Drew, 1979), which would decrease L_{gap}.

A   B

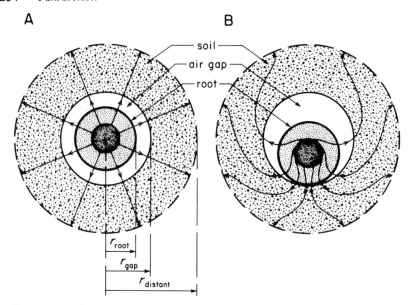

Figure 5 Pathways for radial water efflux (indicated by arrows radiating from the endodermis) for a shrunken root (A) located concentrically within the root–soil air gap and (B) touching the soil at one point on its circumference. The root radius (r_{root}) is 30% less than r_{gap}. Water is assumed to emanate uniformly from the tissues within the endodermis (heavily stippled) and then to pass across the cortex and other tissues (medium stippling), across the air gap (no stippling), and then through the soil (pebbled stippling), to reach the outer hypothetical radius of the soil to which water was lost (radius of $r_{distant}$, indicated by the dashed circle).

Besides the macroscopic aspect, root–soil contact can also be considered on a microscopic level. For instance, the size and the location of soil particles affect where soil water can be in direct contact with a root (Figure 3A). As a soil dries and water retreats into finer crevices between soil particles, the direct contact between the root surface and liquid water becomes less (Tinker, 1976; Herkelrath *et al.*, 1977b). Also, root hairs, mycorrhizal hyphae, mucilage, and soil sheaths (Newman, 1974; Tinker, 1976; Russell, 1977; Oertli, 1991) influence movement of water between soil and roots. For instance, root hairs and mucilage can form a relatively high conductance pathway (Newman, 1974) that locally bridges the gap (essentially no root hairs occurred over the region where shrinkage was examined for the three desert succulents considered here). Because of these many factors, Eq. (7) should be viewed as only a very approximate way of quantifying the influence of a root–soil air gap on water movement.

V. Overall Conductance

Now that the three parts of the water-movement pathway between the soil and roots have been considered, the overall conductance ($L_{overall}$; m s^{-1} MPa^{-1}) may be presented as

$$J_V = L_{overall} (\Psi_{distant} - \Psi_{xylem}), \qquad (8)$$

where J_V is expressed at the root surface. Constancy of water flow through the soil, across the root–soil air gap, and into the root, which is the steady-state condition, leads to

$$\begin{aligned} J_V &= L_{soil}^{eff} (\Psi_{distant} - \Psi_{gap}) \\ &= L_{gap} (\Psi_{gap} - \Psi_{surf}) \\ &= L_P (\Psi_{surf} - \Psi_{xylem}). \end{aligned} \qquad (9)$$

By Eq. (3), the effective conductance of the soil (L_{soil}^{eff}; m s^{-1} MPa^{-1}) can be expressed as

$$L_{soil}^{eff} = \frac{L_{soil}}{r_{root} \ln (r_{distant}/r_{gap})}, \qquad (10)$$

which further emphasizes the different units for L_{soil}^{eff} and L_{soil}.

Because the three parts of the water-movement pathway are in series and conductances in series add as reciprocals, $L_{overall}$ has the form

$$\frac{1}{L_{overall}} = \frac{1}{L_{soil}^{eff}} + \frac{1}{L_{gap}} + \frac{1}{L_P}. \qquad (11)$$

$L_{overall}$ can be used to predict water movement into and out of roots based on the water potential of the bulk soil and the root xylem (Ψ_{soil} and Ψ_{xylem}, respectively), which can be readily measured. Determination of the water potential in the immediate vicinity of a root (Ψ_{gap}, Ψ_{surf}) is much more difficult, although results with computer-assisted tomography applied to X-ray attenuation data (Hainsworth and Aylmore, 1986) and with nuclear magnetic resonance imaging (MacFall *et al.*, 1990) are promising.

VI. Responses to a Drying Pulse

The dimensions, L_P's (Figure 1), and shrinkage (Figure 4) for 6-week-old roots of *A. deserti*, *F. acanthodes*, and *O. ficus-indica* are similar, so the average response of the conductances for these three desert succulents will be considered. As the soil dries from -0.01 to -10 MPa, L_{soil}^{eff} decreases about 10^6-fold (Figure 6). L_{gap}, which is infinite for an unshrunken root at a Ψ_{soil} of -0.01 MPa, becomes lower than L_{soil}^{eff} at about -0.04 MPa.

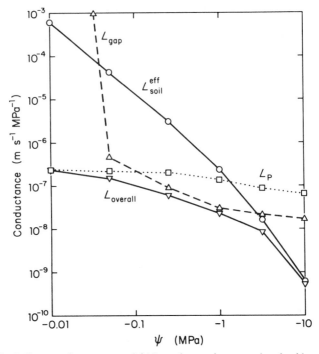

Figure 6 Influence of water potential (ψ) on the conductances involved in water movement between soil and the root xylem. Equation (10) is used to calculate $L_{\text{soil}}^{\text{eff}}$ based on L_{soil} (Figure 2), setting r_{gap} equal to the mean radius of the hydrated roots (Figure 4), and letting r_{soil} equal 30 mm, as has proved valid for studies on water uptake by *A. deserti* and *F. acanthodes* (Hunt and Nobel, 1987; Alm and Nobel, 1991). Equation (7) is used to calculate L_{gap} based on the mean shrinkage observed for 6-week-old roots of the three species (Figure 4). The mean L_P for 6-week-old roots of the three species (Figure 1) is replotted on a logarithmic scale. The three conductances are summed in series to obtain the conductance of the overall pathway, L_{overall} (Eq. (11)). Symbols indicate the values of ψ for which calculations are made.

From -0.04 to -10 MPa, L_{gap} decreases about 3×10^3-fold (Figure 6). At a Ψ_{soil} of -0.01 MPa, L_P is the lowest of the three component conductances, meaning that water movement from the bulk soil to the root xylem is then limited primarily by intrinsic properties of the roots. At the other extreme of a Ψ_{soil} of -10 MPa, L_P is the highest of the three conductances (Figure 6), meaning that the prevention of substantial water loss from a root to a dry soil results primarily from properties of the soil and the root–soil air gap. $L_{\text{soil}}^{\text{eff}}$ becomes the main contributor to L_{overall} below -1.2 to -1.5 MPa for *Triticum aestivum* in sandy loam (Hansen, 1974) and below about -2 MPa for *Gossypium hirsutum* in loamy sand

(Taylor and Klepper, 1975), similar to the results with the three desert succulents presently considered.

The kinetics and directionality of water flow during soil drying can help in understanding the rectifier like behavior of root–soil systems. In the northwestern Sonoran Desert, Ψ_{soil} can decrease from -0.01 to -10 MPa in 30 days (Young and Nobel, 1986). The consequences of the resultant changes in L_{soil}^{eff}, L_{gap}, and L_P for water exchange can be appreciated by calculating $L_{overall}$ [Eq. (11); Figure 6] and then determining J_V [Eq. (8); Figure 7]. During the first 10 d of soil drying, Ψ_{soil} remains above Ψ_{xylem} (-0.29 MPa; Nobel and Lee, 1991), so J_V is positive, meaning water uptake occurs. During this period, J_V steadily decreases because the driving force for water entry ($\Psi_{soil} - \Psi_{xylem}$) decreases and also $L_{overall}$ decreases, the latter reflecting a slight root shrinkage (Figure 4) and hence a decrease in L_{gap} (Figure 6). As the drought persists after 11 d, Ψ_{soil} becomes less than Ψ_{root} and J_V becomes negative, indicating a net water loss from the root to the soil. Now L_{gap} becomes the main limiter of water loss, remaining so until L_{soil}^{eff} becomes less than L_{gap} at a Ψ_{soil} of

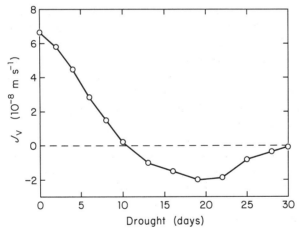

Figure 7 Changes in the volumetric flux density of water into a root (J_V) averaged for 6-week-old roots of *A. deserti*, *F. acanthodes*, and *O. ficus-indica* during a 30-d drying period ($J_V > 0$ indicates water uptake and $J_V < 0$ indicates water loss). Equation (8) is used to calculate J_V, where $L_{overall}$ at a particular $\psi_{distant}$ is obtained from Figure 6. The soil water potential decreases to -10 MPa over the 30-d period, as can occur in the center of the root zone 0.10 m below the soil surface for the sandy loam at Agave Hill (Figure 2) under conditions of relatively high temperatures and low relative humidities (Young and Nobel, 1986). The root xylem water potential (ψ_{xylem}) is assumed to be constant at -0.29 MPa (Nobel and Lee, 1991).

about -3 MPa (Figure 6), which occurs after about 20 d of drought. Indeed, the decrease in water efflux observed after 20 d (Figure 7) results primarily from the low and continually decreasing L_{soil}^{eff}.

If the drought were to last for 5 months, then the predicted 10^5-fold decrease in $L_{overall}$ (Schulte and Nobel, 1989) would mainly reflect the decrease in L_{soil} and hence in L_{soil}^{eff}. In any case, a considerable amount of water can be lost from the roots to the drying soil, as the water loss during the latter 19 d of the 30-d drought is 50% of the water taken up during the first 11 d (Figure 7; changes in root surface area during drought are taken into consideration in the calculation). Such a water loss can be reduced somewhat by shedding of roots during drought. In particular, lateral roots occurring as branches on the nodal roots of *A. deserti* are shed during soil drying, as presumably are certain lateral roots of *F. acanthodes* and *O. ficus-indica* (Nobel, 1988), although much remains to be learned concerning the drought deciduousness of roots of desert succulents.

VII. Responses to a Wetting Pulse

Exposure of 6-week-old roots of *A. deserti*, *F. acanthodes*, and *O. ficus-indica* to -10 MPa causes a nearly 20% shrinkage (Figure 4). When such roots are placed at -0.01 MPa, rehydration leads to an essentially complete reversal of the shrinkage (Figure 8). Approximately 39% of the reversal of the shrinkage occurs in 1 d, 81% in 3 d, and 96% in 6 d. The patterns are similar for the three species and for both attached and excised roots (Figure 8). Such decreases in Δx_{gap} cause L_{gap} to steadily increase during rewetting (Figure 9).

For the first 4 d after rewetting a dry soil to -0.01 MPa, the lowest conductance and hence the main limiter for $L_{overall}$ is L_{gap} (Figure 9). L_P, which is initially sevenfold higher than L_{gap}, increases about twofold by 6 d after rewetting 6-week-old roots of *A. deserti*, *F. acanthodes*, and *O. ficus-indica* (Figure 9). The rewetting is hypothesized to raise ψ_{soil} to -0.01 MPa and to maintain it there, so changes in J_V (Eq. (1)) are proportional to changes in $L_{overall}$, which increases 2.6-fold by 2 d, 5.7-fold by 4 d, and 9.1-fold by 6 d (Figure 9). Water uptake by such roots over the 6-d rewetting period amounts to 30% of the water uptake occurring during the first 11 d of drought (Figure 7). If the rewetting raised ψ_{soil} to -0.00 MPa, such as might occur after a massive rainfall, the root–soil air gap could become filled with water. In such a case, $L_{overall}$ would be raised to L_P (Figure 9), leading to a greatly enhanced water uptake.

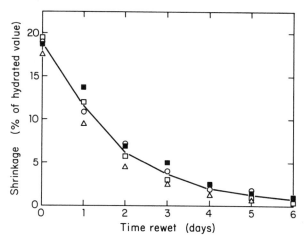

Figure 8 Reversal of root shrinkage caused by rewetting roots. Excised 6-week-old roots of *A. deserti* (○), *F. acanthodes* (△), and *O. ficus-indica* (□), as well as approximately 2-month-old attached roots of *O. ficus-indica* (■), are exposed to an atmospheric water potential of −0.01 MPa. Data are means for four to six roots with hydrated diameters of 2.15 mm for *A. deserti*, 1.40 mm for *F. acanthodes*, 2.31 mm for excised roots of *O. ficus-indica*, and 1.60 mm for attached roots of *O. ficus-indica*. (Adapted from Nobel and Cui, 1992a,b.)

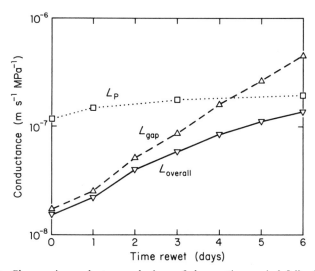

Figure 9 Changes in conductances during a 6-d rewetting period following a 30-d drought. L_P is based on values for approximately 6-week-old roots of *A. deserti*, *F. acanthodes*, and *O. ficus-indica* (North and Nobel, 1991, and unpublished observations). L_{gap} is calculated using Eq. (7) and data in Figure 8. $L_{overall}$ is calculated using Eq. (11), assuming an L_{soil}^{eff} of 5×10^{-4} m s^{-1} MPa^{-1}, as is appropriate for a ψ_{soil} of −0.01 MPa.

Water uptake can also be enhanced by the elongation of existing roots and the growth of new roots caused by soil rewetting (Lauenroth *et al.*, 1987; Klepper and Rickman, 1990). For instance, new growth in the laboratory can double water uptake by *A. deserti* in 4 d after plants, whose existing roots had been reduced by pruning, are placed in hydroponic solutions (Nobel and Sanderson, 1984). On the other hand, root growth over a 30-d period for *A. deserti* in the field increases simulated water uptake only 17 to 35% (Franco and Nobel, 1990). Root growth for desert succulents is commonly observed after soil rewetting; for example, such growth occurs in a few hours for *Opuntia decumbens* and within 24 h for *O. ficus-indica, Trichocereus bridgesii, T. pachanoi,* and *T. spachianus* (Kausch, 1965). New roots are evident in 6 h for *A. deserti* and in 8 h for *F. acanthodes* (Nobel and Sanderson, 1984; Nobel, 1988). Much remains to be learned about the induction of root growth by rainfall and its effect on water uptake for desert succulents in the field.

VIII. Conclusions

The equations presented for the conductances of the three parts of the water-movement pathway—L_{soil}^{eff} (Eq. (10)), L_{gap} (Eq. (7)), and L_P (Eq. (1))—permit a quantitative description of water movement between soils and roots. The approach presupposes that water can move in either direction at the root surface, such as is involved in "hydraulic lift" of water by roots from deep wet soil layers to more superficial drier layers for desert grasses and shrubs (Richards and Caldwell, 1987; Caldwell and Richards, 1989) and root–soil water exchange for many other species (Newman, 1974; Landsberg and Fowkes, 1978; Mooney *et al.*, 1980; Baker and von Bavel, 1988). Many simplifications, approximations, and assumptions are involved in the formulation of a conductance for the root–soil air gap, especially with regard to temperature and the eccentric location of a root in the gap. Nevertheless, L_{gap} is proposed to have a major effect on water movement to and from roots, especially for roots with living cortical cells, which needs experimental testing.

For 6-week-old roots of *A. deserti, F. acanthodes,* and *O. ficus-indica,* L_{gap} is apparently lower than L_P and L_{soil}^{eff} for ψ_{soil} from -0.12 to -2.8 MPa, meaning that L_{gap} is then the most important influence on water movement between roots and soil. Because the average ψ_{xylem} for certain desert succulents can be about -0.29 MPa (Nobel and Lee, 1991), the low L_{gap} serves primarily to limit water loss as the roots shrink away from the soil. L_{gap} may be particularly crucial for the water relations of seedlings, whose young roots are composed mainly of cortical cells that can undergo extensive and reversible shrinkage. Such roots may shrink away from the soil during the onset of drought, thus limiting plant water loss, and swell following rainfall, enabling substantial water uptake to resume.

The overall conductance, $L_{overall}$, allows quantification of water movement for the entire pathway between roots and soil, which can help settle various long-standing disputes regarding the relative water potential drops in the soil versus the root (Hansen, 1974; Newman, 1974; Faiz and Weatherley, 1978; Landsberg and Fowkes, 1978; Blizzard and Boyer, 1980; Oertli, 1991). Measured water uptake rates at intermediate soil water potentials that are much less than predicted (Herkelrath *et al.*, 1977b) could be caused by root shrinkage and hence a decrease in L_{gap}. The tenfold decrease in root plus root–soil interface conductance as ψ_{soil} for a silt loam decreases from about -0.1 to -1 MPa for *Helianthus annuus* (Bristow *et al.*, 1984) may be caused by the formation of a root–soil air gap and hence a decrease in L_{gap}. In any case, L_P is the main limiter of water uptake for very wet soils and L_{soil}^{eff} is the main limiter of water loss for dry soils, at least for desert succulents in sandy loam (Figure 6) and *Gossypium hirsutum* in loamy sand (Taylor and Klepper, 1975). The latter means for preventing water loss is widely available among plants, because the great decrease in L_{soil} (Figure 2) and hence in L_{soil}^{eff} (Eq. (10)) as a soil dries occurs regardless of whether a root shrinks or its hydraulic conductivity decreases. Thus, limitation of water loss from a root system during extensive drought depends primarily on properties of the soil and so is effective for any plant species. Understanding the factors affecting L_{soil}^{eff}, L_{gap}, and L_P is crucial for understanding how roots respond to pulses of rainfall and drought.

Acknowledgments

Financial support for the preparation of this review is gratefully acknowledged from National Science Foundation Grant DCB 90-02333 and the Environmental Research Division, Office of Health and Environmental Research, Department of Energy Contract DE-FC03-87-ER60615.

References

Alm, D. M., and Nobel, P. S. (1991). Root system water uptake and respiration for *Agave deserti*: Observations and predictions using a model based on individual roots. *Ann. Bot.* (*London*) [N.S.] **67**, 59–65.

Baker, J. M., and von Bavel, C. H. M. (1988). Water transfer through cotton plants connecting soil regions of differing water potential. *Agron. J.* **80**, 993–997.

Bernstein, L., Gardner, W. R., and Richards, L. A. (1959). Is there a vapor gap around plant roots? *Science* **129**, 1750, 1753.

Blizzard, W. E., and Boyer, J. S. (1980). Comparative resistance of the soil and the plant to water transport. *Plant Physiol.* **66**, 809–814.

Bonner, J. (1959). Water transport. *Science* **129**, 447–450.

Bristow, K. L., Campbell, G. S., and Calissendorff, C. (1984). The effects of texture on the resistance to water movement within the rhizosphere. *Soil Sci. Soc. Am. J.* **48**, 266–270.

Caldwell, M. M. (1976). Root extension and water absorption. *In* "Water and plant life: Problems and modern approaches" (O. L. Lange, L. Kappen, and E.-D. Schulze, eds.), Ecological studies, Volume **19**, pp. 63–85. Springer-Verlag, Berlin.

Caldwell, M. M., and Richards, J. H. (1989). Hydraulic lift: Water efflux from upper roots improves effectiveness of water uptake by deep roots. *Oecologia* **79**, 1–5.

Campbell, G. S. (1985). "Soil Physics with Basic: Transport Models for Soil–Plant Systems." Elsevier, Amsterdam.

Cass, A., Campbell, G. S., and Jones, L. T. (1984). Enhancement of thermal water vapor diffusion in soil. *Soil Sci. Soc. Am. J.* **48**, 25–32.

Cole, P. H., and Alston, A. M. (1974). Effect of transient dehydration on absorption of chloride by wheat roots. *Plant Soil* **40**, 243–247.

Cowan, I. R. (1965). Transport of water in the soil–plant–atmosphere system. *J. Appl. Ecol.* **2**, 221–239.

Cowan, I. R., and Milthorpe, F. L. (1968). Plant factors influencing the water status of plant tissues. *In* "Water Deficits and Plant Growth" (T. T. Kozlowski, ed.), Vol. 1, pp. 137–193. Academic Press, New York.

Dalton, F. N., Raats, P. A. C., and Gardner, W. R. (1975). Simultaneous uptake of water and solutes by plant roots. *Agron. J.* **67**, 334–339.

Dirksen, D., and Raats, P. A. C. (1985). Water uptake and release by alfalfa roots. *Agron. J.* **77**, 621–626.

Drew, M. C. (1979). Root development and activities. *In* "Arid-land Ecosystems: Structure, Functioning and Management" (R. A. Perry and D. W. Goodall, eds.), Vol. 1, pp. 573–606. Cambridge Univ. Press, Cambridge, UK.

Faiz, S. M. A., and Weatherley, P. E. (1978). Further investigations into the location and magnitude of the hydraulic resistances in the soil : plant system. *New Phytol.* **81**, 19–28.

Faiz, S. M. A., and Weatherley, P. E. (1982). Root contraction in transpiring plants. *New Phytol.* **92**, 333–343.

Fiscus, E. L. (1975). The interaction between osmotic- and pressure-induced water flow in plant roots. *Plant Physiol.* **55**, 917–922.

Fiscus, E. L. (1977). Determination of hydraulic and osmotic properties of soybean root systems. *Plant Physiol.* **59**, 1013–1020.

Fiscus, E. L. (1986). Diurnal changes in volume and solute transport coefficients of *Phaseolus* roots. *Plant Physiol.* **80**, 752–759.

Franco, A. C., and Nobel, P. S. (1990). Influences of root distribution and growth on predicted water uptake and interspecific competition. *Oecologia* **82**, 151–157.

Gardner, W. R. (1960). Dynamic aspects of water availability to plants. *Soil Sci.* **89**, 63–73.

Hainsworth, J. M., and Aylmore, L. A. G. (1986). Water extraction by single plant roots. *Soil Sci. Soc. Am. J.* **50**, 841–848.

Hansen, G. K. (1974). Resistance to water transport in soil and young wheat plants. *Acta Agric. Scand.* **24**, 37–48.

Herkelrath, W. N., Miller, E. E., and Gardner, W. R. (1977a). Water uptake by plants. I. Divided root experiments. *Soil Sci. Soc. Am. J.* **41**, 1033–1038.

Herkelrath, W. N., Miller, E. E., and Gardner, W. R. (1977b). Water uptake by plants. II. The root contact model. *Soil Sci. Soc. Am. J.* **41**, 1039–1043.

Hillel, D. (1982). "Introduction to Soil Physics." Academic Press, New York.

Huck, M. G., Klepper, B., and Taylor, H. M. (1970). Diurnal variation in root diameter. *Plant Physiol.* **45,** 529–530.

Hunt, E. R., Jr., and Nobel, P. S. (1987). Allometric root/shoot relationships and predicted water uptake for desert succulents. *Ann. Bot. (London)* [N.S.] **59,** 571–577.

Jung, Y.-S., and Taylor, H. M. (1984). Differences in water uptake rates of soybean roots associated with time and depth. *Soil Sci.* **137,** 341–350.

Katou, K., and Taura, T. (1989). Mechanism of pressure-induced water flow across plant roots. *Protoplasma* **150,** 124–130.

Kausch, W. (1965). Beziehungen zwischen Wurzelwachstum, Transpiration und CO_2— Gaswechsel bei einigen Kakteen. *Planta* **66,** 229–238.

Klepper, E., and Rickman, R. W. (1990). Modeling crop root growth and function. *Adv. Agron.* **44,** 113–132.

Landsberg, J. J., and Fowkes, N. D. (1978). Water movement through plant roots. *Ann. Bot. (London)* [N.S.] **42,** 493–508.

Lauenroth, W. K., Sala, O. E., Milchunas, D. G., and Lathrop, R. W. (1987). Root dynamics of *Bouteloua gracilis* during short-term recovery from drought. *Funct. Ecol.* **1,** 117–124.

Lopez, F. B., and Nobel, P. S. (1991). Root hydraulic conductivity of two cactus species in relation to root age, temperature, and soil water status. *J. Exp. Bot.* **42,** 143–149.

MacFall, J. S., Johnson, G. A., and Kramer, P. J. (1990). Observation of a water-depletion region surrounding loblolly pine roots by magnetic resonance imaging. *Proc. Natl. Acad. Sci. U.S.A.* **87,** 1203–1207.

Marshall, T. J., and Holmes, J. W. (1988). "Soil Physics," 2nd ed. Cambridge Univ. Press, Cambridge, UK.

Mooney, H. A., Gulmon, S. L., Rundel, P. W., and Ehleringer, J. (1980). Further observations on the water relations of *Prosopis tamarugo* of the northern Atacama Desert. *Oecologia* **44,** 177–180.

Newman, E. I. (1974). Root and soil water relations. *In* "The Plant Root and Its Environment" (E. W. Carson, ed.), pp. 363–440. Univ. Press of Virgina, Charlottesville.

Nobel, P. S. (1988). "Environmental Biology of Agaves and Cacti." Cambridge Univ. Press, New York.

Nobel, P. S. (1991). "Physicochemical and Environmental Plant Physiology." Academic Press, San Diego.

Nobel, P. S., and Cui, M. (1992a). Hydraulic conductances of the soil, the root–soil air gap, and the root: Changes for desert succulents in drying soil. *J. Exp. Bot.* **43,** 319–326.

Nobel, P. S., and Cui, M. (1992b). Shrinkage of attached roots of *Opuntia ficus-indica* in response to lowered water potentials—predicted consequences for water uptake or loss to soil. *Ann. Bot. (London)* [N.S.] **70,** 485–491.

Nobel, P. S., and Lee, C. H. (1991). Variations in root water potentials: Influence of environmental factors for two succulent species. *Ann. Bot. (London)* [N.S.] **67,** 549–554.

Nobel, P. S., and Sanderson, J. (1984). Rectifier-like activities of roots of two desert succulents. *J. Exp. Bot.* **35,** 727–737.

Nobel, P. S., Schulte, P. J., and North, G. B. (1990). Water influx characteristics and hydraulic conductivity for roots of *Agave deserti* Engelm. *J. Exp. Bot.* **41,** 409–415.

North, G. B., and Nobel, P. S. (1991). Changes in hydraulic conductivity and anatomy caused by drying and rewetting roots of *Agave deserti* (Agavaceae). *Am. J. Bot.* **78,** 906–915.

Oertli, J. J. (1991). Transport of water in the rhizosphere and in roots. *In* "Plant Roots: The Hidden Half" (Y. Waisel, A. Eshel, and U. Kafkafi, eds.), pp. 559–582. Dekker, New York.

Passioura, J. B. (1988). Water transport in and to roots. *Ann. Rev. Plant Physiol. Plant Mol. Biol.* **39,** 245–265.

Philip, J. R. (1958). The osmotic cell, solute diffusibility, and the plant water economy. *Plant Physiol.* **33**, 264–274.

Richards, J. H., and Caldwell, M. M. (1987). Hydraulic lift: Substantial nocturnal water transport between soil layers by *Artemisia tridentata* roots. *Oecologia* **73**, 486–489.

Rowse, H. R., and Goodman, D. (1981). Axial resistance to water movement in broad bean (*Vicia faba*) roots. *J. Exp. Bot.* **32**, 591–598.

Russell, R. S. (1977). "Plant Root Systems: Their Function and Interaction with the Soil." McGraw-Hill, London.

Salim, M., and Pitman, M. G. (1984). Pressure-induced water and solute flow through plant roots. *J. Exp. Bot.* **35**, 869–881.

Sanderson, J. (1983). Water uptake by different regions of the barley root. Pathways of radial flow in relation to development of the endodermis. *J. Exp. Bot.* **34**, 240–253.

Schulte, P. J., and Nobel, P. S. (1989). Responses of a CAM plant to drought and rainfall: Capacitance and osmotic pressure influences on water movement. *J. Exp. Bot.* **40**, 61–70.

Steudle, E., Oren, R., and Schulze, E.-D. (1987). Water transport of maize roots. Measurement of hydraulic conductivity, solute permeability, and of reflection coefficients of excised roots using the root pressure probe. *Plant Physiol.* **84**, 1220–1232.

Taylor, H. M., and Klepper, B. (1975). Water uptake by cotton root systems: An examination of the assumptions in the single root model. *Soil Sci.* **120**, 57–67.

Taylor, H. M., and Willatt, S. T. (1983). Shrinkage of soybean roots. *Agron. J.* **75**, 818–820.

Tinker, P. B. (1976). Roots and water: Transport of water to plant roots in soil. *Philos. Trans. R. Soc. London, Ser. B* **273**, 445–461.

Young, D. R., and Nobel, P. S. (1986). Predictions of soil water potentials in the northwestern Sonoran Desert. *J. Ecol.* **74**, 143–154.

11

Architecture and Biomass Allocation as Components of the Plastic Response of Root Systems to Soil Heterogeneity

A. H. Fitter

I. Summary

Although it is generally accepted that roots respond to locally enriched patches of soil resources by proliferation within the patch, the evidence for this comes largely from work with crop plants. Wild plant species are ecologically distinct, often slower-growing or less responsive to resource enrichment, and in those the response is not universal. Proliferation of roots within a patch seems to be an obviously adaptive response, but it carries a cost, for if the patch is short-lived or supplies only limited resources, the investment in new roots may not be repaid. Similarly, in competition with other root systems, the benefits of proliferation will depend on the ability to obtain resources rapidly and effectively as a function of the cost of the new roots. That cost depends intimately on root diameter, since coarse roots grow most rapidly but require disproportionately large amounts of resources for their construction and maintenance. Fine roots can be used to create locally high root length densities at low cost, but they have shorter lives than coarse roots and therefore a high turnover rate will be required if root length density is to be maintained. The outcome of selection for responses to patchiness therefore depends on the nature of the patches and on the types of response that root systems can make. Those responses include architectural changes that are an inevitable consequence of changes in biomass allocation, but can also occur without changes in allocation.

Architectural plasticity may represent an alternative to plasticity of biomass allocation as a response to patchiness, appropriate for different types of root system or patch. In particular, species with coarse (high diameter) roots may be poorly adapted to respond to short-term heterogeneity and may utilize architectural rather than biomass allocation plasticity, whereas fine-rooted species may be effective exploiters of small and short-lived patches through their ability to create dense networks of ephemeral rootlets rapidly. Experiments in which roots can freely explore sand containing both nutrient-rich and nutrient-poor patches reveal that architectural plasticity occurs, that fast-growing plants generally exhibit most such plasticity, and that biomass allocation plasticity and architectural plasticity are either poorly or negatively correlated. These findings demonstrate that investigation of root system architecture is a necessary part of any attempt to understand the responses of root systems to their environment.

II. Introduction: Received Wisdom

The distribution of resources in soils is never uniform, nor is that of plant roots. It is widely held that plant root systems respond to heterogeneity of soil resources by proliferation in the most nutrient- or water-rich zones, a response that intuitively appears to be highly adaptive. In this article I shall assess the evidence for this view, consider whether any a priori predictions can be made about the way in which different root systems might respond to various forms of soil heterogeneity, and present some preliminary experimental evidence. I shall not, however, consider physiological responses, which are covered by Caldwell (Chapter 12, this volume).

The most widespread, if not the strongest evidence for a link between the pattern of root and resource distribution lies in the almost invariable correlation between root densities and resource concentrations down soil profiles (Figure 1). This correlation may of course not be causal, but a reflection of the processes that lead to the two patterns: principally shoot litterfall and decomposition in the case of soil resources, and development in the case of roots. There are exceptions to the rule, notably in arid ecosystems, where water may be most abundant at depth, and where roots are typically found at a greater depth than elsewhere.

More persuasive, if less ubiquitous support for the concept comes from agronomy, where it has long been known that roots of many crop plants proliferate in fertilizer bands (Duncan and Ohlrogge, 1958; Passioura and Wetselaar, 1972). This phenomenon can clearly be seen in a nonagronomic context in the experiment displayed in Figure 2, in which *Lolium*

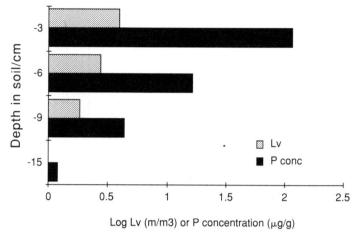

Figure 1   Relationship between water-soluble P concentration and root length density (Lv) and depth in soil at Wheldrake Ings, an ancient grassland site near York, U.K. (Data from McGonigle and Fitter, 1988.)

perenne roots were allowed to explore root compartments in two directions, that in one (C) representing a factorial combination of added nutrients and competition from roots of *Plantago lanceolata*, whereas the compartment in the other direction (A) was a standard control (Fitter, 1976). *Lolium* root density was approximately doubled in the central compartment by added nutrients, and reduced equivalently by the presence of *Plantago* roots. It was notable that *Lolium* roots almost totally failed to penetrate to the compartment beyond the *Plantago* plant (E).

Other supporting evidence comes from experiments in water culture, notably the well-known work of Drew and co-workers (1975) (Figure 3). In their experiments, part of a barley seminal axis was exposed to N or P concentrations 100 times greater than the rest of the axis, and the primary laterals arising from the enriched zone elongated farther and faster than the controls, consequently reaching a developmental stage that permitted growth of secondary laterals. Other workers have offered several significant improvements on the original experimental design (e.g., Granato and Raper, 1989), but the essential point is unchanged: Roots in the fertile zone respond by increased linear growth and consequent increases in branching. For example, in Granato and Raper's (1989) experiment, the oldest laterals were 55 mm long in $+N$ compartments but only 32 mm in $-N$, and branch densities were 12 cm^{-1} ($+N$) and 7 cm^{-1} ($-N$).

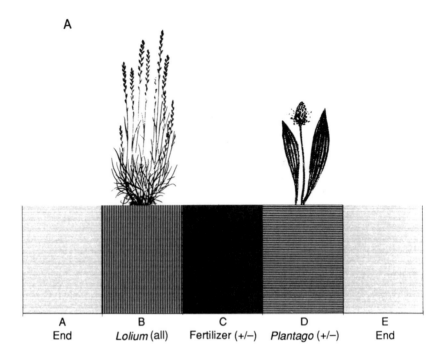

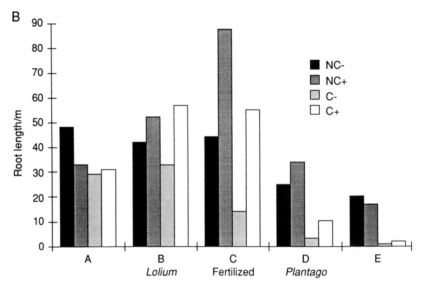

Figure 2 (A) Design of an experiment in which roots of a plant of *Lolium perenne*, planted in a compartment at B, were able to grow into four other compartments (A and C–E). The experiment comprised a factorial combination of the presence and absence of a plant of *Plantago lanceolata* in compartment D and the addition of fertilizer in compartment C. Compartments were separated by barriers containing holes through which roots could pass. (B) Root length density of *Lolium perenne* in each of the five compartments in the experiment displayed in (2a) Key: C/NC, with or without competition from a *Plantago* plant in compartment D; +/−, with or without fertilizer in compartment C.

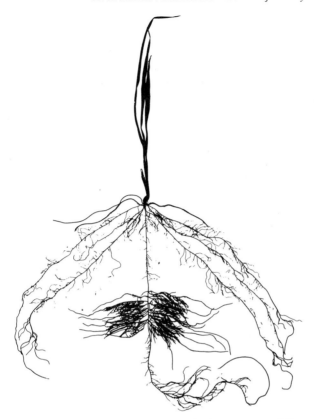

Figure 3 Root system of a barley plant grown in the apparatus of Drew *et al.* (1975), in which one seminal axis was exposed to 10 m*M* NO$_3^-$ in the central band and 0.1 m*M* NO$_3^-$ above and below the central band.

Taken together, these pieces of evidence might seem to make a convincing case for the initial proposition that roots respond to resource patches by proliferation. However, the selection of species used in them has been very limited and importantly they have mostly been crop plants (*Zea, Hordeum, Lolium*). Crop plants are a particular ecological group, characterized by rapid growth rates, high nutrient demand, and (usually) an annual habit. There is increasing evidence that some species that do not share these characteristics do not respond to soil heterogeneity in the same way. For example, Jackson and Caldwell (1989) examined root growth of two *Agropyron* species (Poaceae) to nutrient enrichment. Significantly one species exhibited the expected response of an increased root relative growth rate, but the other showed almost no short-term response at all (Figure 4), even (in another experiment) when measurements were continued for 3 weeks (Caldwell *et al.*, 1991).

This experiment clearly shows that not all species respond identically to heterogeneity. A comparative experiment by Campbell *et al.* (1991) casts light on the extent of these interspecific differences. The experiment is described in detail by Grime (Chapter 1, this volume), who shows that whereas competitively dominant species tend to have the greatest absolute amount of roots in nutrient-rich patches, they are the least precise in terms of the proportion of new root growth that they allocate to those patches. They achieve high root length densities simply by virtue of their size and ability to produce absolutely more root than smaller species. However, small (and hence often competitively inferior) species have the ability to allocate material to root growth with greater precision, as shown by the inverse relationship between the precision of allocation and a compound variable derived from a multiple regression using growth rate and seed size as independent variables (Figure 5).

It seems, therefore, that not all species do show differential proliferation to the same extent. The next step, therefore, is to ask whether there are circumstances in which it might be, counterintuitively, beneficial to a plant not to respond in this way. Exploitation of a patch can be viewed

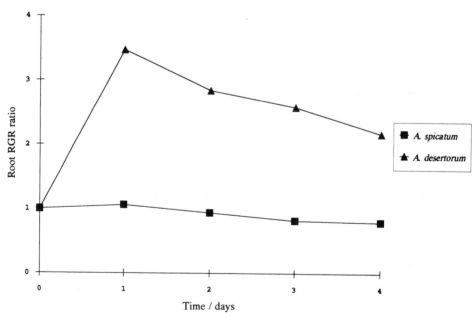

Figure 4   Ratio of root relative growth rate of two *Agropyron* species in localized patches of high nutrient availability in the experiment of Jackson and Caldwell (1989). Within the time scale of the experiment, only one species shows a growth response.

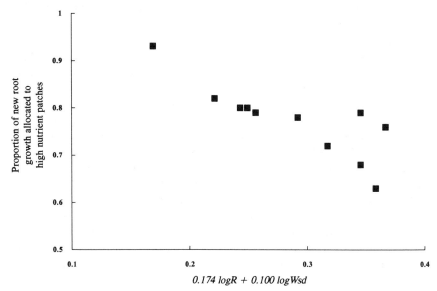

Figure 5   The precision of root foraging of 11 species in the experiments of Campbell *et al.* (1991; see also Grime, Chapter 1, this volume) as a function of plant size. Size has been expressed in terms of seed weight (*Wsd*) and growth rate (*R*) by means of a regression of root allocation on these two variables, and a new variable based on the regression coefficients constructed to display the relationship.

as a problem of resource allocation. The plant might put more resources into root growth and those resources might be concentrated in new roots within the patch. Alternatively, there might be a diversion of resources from roots in poor to those in rich patches, leading to a reduction in growth in the former. Split-root experiments (Wiersum, 1958; Drew *et al.*, 1975; Drew, 1975; De Jager, 1982; Crick and Grime, 1987; Granato and Raper, 1989) suggest that the latter typically occurs and a detailed series of experiments on peas specifically to test this proposition has recently confirmed that (Gersani and Sachs, 1992). Allocation of resources to a patch therefore involves a reduction in root growth elsewhere in the soil; responding to a patch will be beneficial, therefore, only if the gains in the patch outweigh the losses elsewhere.

III. Some Speculations

If changes in resource allocation within and between patches do occur, they will have other consequences. It is possible for root growth rates in one part of the root system to increase *without* any changes in root form or architecture, but it is unlikely since the response will only be adaptive if the new root growth occurs within the patch, which can only be achieved by a change in architecture (Figure 6). This localization of root growth can be the result of changes in any or all of four architectural variables (Table I). Indeed it would be possible for a root system to respond by such architectural plasticity with no change in resource allocation.

Increased growth in rich patches is intuitively adaptive but carries a risk. Resources have been committed and, if the patch proves to be short-lived or otherwise unprofitable to exploit, that investment may represent a net cost to the plant (or at least an opportunity cost, since the resources cannot be used elsewhere). There is likely to be benefit, therefore, from the development of fine (low diameter), more or less ephemeral roots in patches, and these will generally be laterals of high developmental order. This will minimize the potential cost, but fine roots generally have low extension rates, probably because of limited metabolic transport capacity, extend less far away from their parent root, and have a short life span, implying a high rate of turnover of resources (Table II). This will make them less effective at exploring soil, more prone to between-root (within-plant) competition which reduces their efficiency, and relatively expensive in resource cost terms.

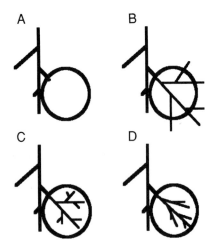

Figure 6 Topological diagrams of model root systems to indicate the distinction between biomass allocation and architectural plasticity. An initial root system (A) may respond to a patch, indicated by a circle, by increased allocation of biomass with no change in architecture (B) or a change in architecture (C and D). Architectural changes may, for example, be a change in topology, leading to a less herringbone system, with no increase in allocation (C), or a change in geometry, such as shorter link lengths and narrower branching angles (D). Biomass allocation plasticity alone (B) is ineffective at localizing new growth within the patch.

Table I The Four Architectural Variables of Root Systems and the Consequences for Root Systems of Changes in Them That Lead to Greater Root Length Density

Architectural variable	Change	Consequence
Topology (branching pattern)	Less herringbone[a]	More high-order laterals
Link length	Shorter links[b]	Greater branch density
Root diameter	Finer	High-order laterals
Branching angle	Narrower	Shorter parent–offspring link distances

[a] A herringbone root system is one that comprises a main axis and laterals only; it represents one possible topological extreme, the other being dichotomous branching.

[b] Links are segments of root between branching points (interior links) or between a branching point and a meristem (exterior links).

Against these points must be set the ability of fine-rooted species to rapidly generate high root length densities locally, simply by virtue of their ability to produce large lengths of root at relatively low cost. Eissenstat (1991), for example, showed that the greatest rate of increase in root length density among a group of *Citrus* cultivars was achieved by those with the highest specific root length (SRL). SRL is the length of root per unit mass and hence a good correlate of root diameter (Fitter, 1985). In a competitive situation, there will be great advantages to the plant that exploits a patch most rapidly.

There are, therefore, opposing benefits to be obtained from different types of roots. On the one hand, fast-growing, large-diameter roots are

Table II Generalized Values for Growth Rate, Length, and Longevity of Cereal Roots[a]

Root member	Growth rate (mm d^{-1})	Length (mm)	Life span (d)
Axis	20–30	500–1000	>100
1^y lateral	5	100	60
2^y lateral	1	40	40–50
3^y lateral	—	10	—
Hairs	—	1	2–3

[a] Data are taken from a number of sources (May *et al.*, 1967; Hackett, 1972; Russell, 1977; Fusseder, 1987) and include measurements on maize (*Zea mays*), barley (*Hordeum vulgare*), and wheat (*Triticum aestivum*).

effective at rapid exploration of soil, are unlikely to suffer within-plant competition, and can pay back construction costs over a longer period by virtue of their greater longevity; but the potential costs of these where patches prove to be of low value might be large. On the other hand, fine roots are effective at filling small volumes of soil and represent a smaller overall resource cost where a patch proves unprofitable. This paradox arises because of the disproportionate impact of root diameter on root construction and maintenance cost.

The significance of growth rate and root diameter in the exploitation of patches can be modeled simply, using the equations of Baldwin (1975) and Nye and Tinker (1977). These show that the proportion (p) of soil exploited by roots in a given volume depends on root length density, Lv (m m^{-3}), the diffusion coefficient of the resource under consideration, D (m^2 s^{-1}), and time, t (s):

$$p = 1 - e^{-D \cdot t \cdot Lv}.$$

We are considering, however, a dynamic situation where Lv is continually changing. Therefore

$$Lv_t = Lv_0 \cdot e^{g \cdot t},$$

where Lv_0 is Lv at time 0 and g is the root specific growth rate (m m^{-1} s^{-1}). It is a simple matter then to calculate the proportion of a given soil volume exploited at any given time by one root system or by a combination of two or more root systems, which may differ in g. One can set some arbitrary value of p (say, 95%) at which the soil volume is held to be fully exploited and determine what proportion of the soil has been utilized by each species at that point. For $p = 0.95$, the time to saturation (t_{sat}) is approximately $3/D \cdot Lv$.

Until detailed data on the relationships between root specific growth rate and root diameter are available for a number of species, it is not especially fruitful to perform such simulations. When such data emerge, however, simulations could clearly be used to investigate the costs and benefits of exploiting patches of different type by various kinds of root systems.

IV. On Patches

There is therefore a conflict between the results achievable by coarse and fine roots, whose resolution might be sought in a consideration of the nature of patches. Regrettably few data exist on soil spatial and temporal heterogeneity, though it is clear that both are extensive. Gupta and Rorison (1975), for example, showed that both spatial and temporal variation occurred in the availability of a wide range of nutrients down a podzolic soil profile, and confirmed the significance of this by means of a seedling bioassay using soil taken from different horizons at different times of the year. Seedlings of *Rumex acetosa* grown for 5 weeks in soil from A and B horizons always grew poorly, at whatever time of year the sampling was made; in contrast, *R. acetosa* grew about seven times larger in soil taken from the A_0 horizon in early summer as in soil sampled at the end of the summer (Figure 7). Similarly, Veresoglou and Fitter (1984) found large seasonal variations in P and K availability in three adjacent grassland soils separated by only a few meters, and also extensive variation in both vertical and horizontal spatial dimensions. Jackson and Caldwell (1992, 1993) have used geostatistical techniques to reveal extensive pattern in soil nutrient availability at scales from 12.5 cm to 1 m. Such pattern is well known to soil scientists, who advise careful sampling of field soils for chemical analysis to overcome the problem of local variability, and it is obvious to all field ecologists. It is, of course, an inevitable consequence of the nature of nutrient supply and nutrient cycles in soil and of the low rates of transport processes in most soils (Nye and Tinker,

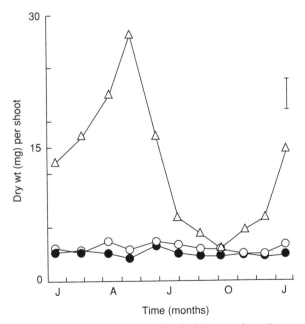

Figure 7 Seasonal variation in shoot growth of a bioassay plant (*Rumex acetosa*) on soil taken at different times of year from three horizons of a podzolic soil ($\triangle$, A_0; $\bigcirc$, A; $\bullet$, B). (Reproduced from Gupta and Rorison, 1975, with permission.)

1977). The message from all these sources is therefore clear: soil is a patchy environment.

These patches have three important attributes, namely, extent, frequency, and distribution, in both time and space (Table IIIa). Different types of response by roots would seem to be appropriate in relation to patches possessing various attributes. The distribution of patches (Table IIIb) may select for either plastic responses (where patches are unpredict-

Table III A Simple Classification of Patches

(a) Basic Attributes of Patches

Scale	Attribute		
	Distribution	Extent	Number
Spatial	Pattern	Size	Abundance
Temporal	Predictability	Duration	Frequency

(b) Root Responses to Patch Distribution in Space or Time

Range of variation of patch:	Random	Clumped	Regular (horizons) (seasons)
Type of response of roots:	Plasticity (response by differential growth)		Genotypic differentiation (rooting depth, time of growth)

(c) Root Characteristics in Relation to Patch Extent in Space or Time

Root characteristic	Patch size or duration	
	Small/short	Large/long
Diameter	Small	Large
Longevity	Short	Long

(d) Root Responses to Patch Number in Space or Time

	Patch abundance or frequency	
	Few	Many
Ease of location	Difficult	Bound to be found
Type of response	Speed of response less critical?	Fast growth, high density

able through being randomly distributed) or genotypically fixed responses (e.g., plants with shallow root patterns growing in soils where deeper soil layers are in some way unfavorable). The size or duration of patches (Table IIIc) may select for roots with distinct characteristics: fine (low diameter) or short-lived roots would be expected to be best suited to small or short-lived patches. Finally, patches that are difficult to locate because they are rare in either space or time (Table IIId) will not impose on roots a need for rapid exploitation, since other roots are less likely to encounter them.

These features of patches raise an important conflict: root systems will maximize acquisition of resources in a patchy environment by preferential allocation of root growth to rich patches. This, however, involves two stages, namely, locating (and possibly recognizing) and exploiting the patch. It is apparent that root characteristics that increase the probability of locating patches are not those that improve the ability to exploit it. The former will be achieved by fast-growing, long-lived root systems, which can maintain a high length density throughout the soil; these are likely to be roots of high diameter. Exploitation of patches, however, requires generation of a high root length density in a short time, something that can only be achieved by fine-rooted species. These fine roots will, however, be short-lived. The trade-off between these two extremes may help to explain the diversity of root diameter found between plant species, even those that coexist in a community (Fitter, 1985).

Generally, it seems likely that small and short-lived patches will most favor fine-rooted species with a high degree of allocation plasticity, whereas long-lived patches in particular may be exploitable by thick-rooted species with similarly long-lived roots. Since increasing root length density by increased biomass allocation is particularly expensive for coarse-rooted species, one might predict that they would use changes in architecture, which are known to alter exploitation efficiency of a root system, as an alternative response to patchiness.

These considerations lead to a number of predictions.

(i) There will be much variation among species in the degree of root system plasticity that they exhibit, depending on the types of patches they have evolved to exploit. This prediction is supported by the discovery that there is a wide range of biomass allocation plasticity (Figure 5).

(ii) Biomass allocation plasticity and architectural plasticity will not necessarily be related. Indeed they may be viewed as different ways of achieving proliferation, and hence may be appropriate to different types of patch. Short-lived patches should encourage reduced root diameter, and small patches more dichotomous branching, shorter links, and possibly wider angles, all of which tend to ensure that root density within the patch is increased. Some evidence to test this prediction is given in Section V.

(iii) Since large-diameter roots are expensive to produce and may have greater potential life spans than finer roots, one can predict that they will be less able to react to short-term heterogeneity but be more suited to long-lived patches. Plants with coarse roots, therefore, may well exhibit less biomass allocation plasticity and more architectural plasticity.

V. Experimental Tests

As yet, few experiments have been performed explicitly to test these ideas. The experiments of Grime and his co-workers mentioned earlier and described in Chapter 1 (this volume), however, give useful insights. The negative relationship between plant size (defined as a function of seed weight and relative growth rate) and allocation plasticity (Figure 5) has already been shown to conform to prediction (i). We have analyzed the roots from some of these experiments architecturally, using the methods described in Fitter and Stickland (1991). We used several architectural parameters, including exterior and interior link lengths and the topological index given by the ratio of altitude to its expected value assuming random growth $(a/E(a))$; altitude is the number of links in the longest path from any exterior link to the shoot base. We measured these parameters in both high and low patches and in the treatment in which a uniform high nutrient supply was given to all quadrants of the root system. This requires the architectural analysis of root system fragments rather than of whole root systems as in earlier work (e.g., Fitter and Stickland, 1991), but this approach is supported by our own data from field-grown plants (Fitter and Stickland, 1992a) and by the work of Van Pelt and Verwer (1984), who demonstrated that "cut trees" (i.e., those from which extremities have been pruned at random) do not differ topologically from the original tree. This result follows from the fractal nature of the trees (Fitter and Stickland, 1992b): They exhibit a high degree of self-similarity.

There were large differences among the 11 species tested for all architectural variables, but treatment differences occurred only for exterior link length and topology, not for interior link length. Both types of exterior link (exterior/exterior—those that join other exterior links—and exterior/interior—those that join interior links) were longer in high-nutrient regions, whether these were high patches or uniformly high (Figure 8). The topological index $a/E(a)$ was lowest in high patches, but similar in both low patches and the high uniform treatment (Figure 8). A high value for $a/E(a)$ implies a more herringbone branching pattern (fewer high-order laterals) and has been shown to be more efficient at

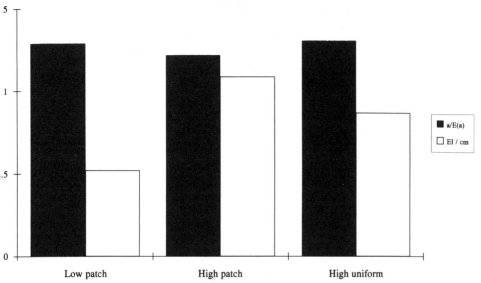

Figure 8 Mean value for 11 species from the Sheffield foraging experiments of the topological index $a/E(a)$, the ratio of altitude to its expected value assuming random branching, and of the length of exterior–interior links. The topological index is highest, implying a more nearly herringbone branching pattern, in the low-nutrient patches and in the high uniform treatment. Treatment differences are significant in both cases as determined by analysis of covariance: for $a/E(a)$, $F_{2.247} = 5.90$, $P = 0.003$; for EI link length $F_{2.247} = 31.2$, $P < 0.0001$.

exploiting soil in nutrient-poor conditions (Fitter *et al.,* 1991); the higher value in low-nutrient than in high-nutrient patches, therefore, conforms to that prediction, but the high value in the uniform nutrient treatment cannot be interpreted in the absence of a uniform low-nutrient treatment.

To rank the species for their responsiveness to patchiness, the variables that responded to nutrient patchiness—exterior link length and $a/E(a)$—were used to calculate an index of architectural plasticity. This index represents the mean for these two variables of the absolute value of the difference of the low patch: high patch ratio from unity. With the exception of *Lolium perenne,* which is the least plastic species in terms of root architecture, the remaining ten species form a recognizable ecological series (Table IV). The most striking feature, however, is the relationship between architectural and allocation plasticity (Figure 9). For the nine species as a whole, there is no relationship. One species (*Lolium perenne*), however, represents an apparent outlier to a negative relationship for the other eight. *Lolium perenne* has already been noted as an exception to the ecological pattern of architectural plasticity (Table IV). Figure 9 therefore confirms prediction (ii)—that biomass allocation

Table IV Architectural Plasticity in Eleven Species from the Sheffield Foraging Experiments

Species	Index[a]	Category
Poa annua	0.52	ruderals and weeds
Chenopodium album	0.40	
Urtica dioica	0.37	competitive dominants
Epilobium hirsutum	0.36	
Arrhenatherum elatius	0.35	
Festuca ovina	0.31	abundant, subdominant
Koeleria macrantha	0.29	species
Cerastium fontanum	0.27	sparsely distributed species
Hypericum perforatum	0.20	
Campanula rotundifolia	0.15	
Lolium perenne	0.11	

[a] The index is a measure of the difference in root system architecture between high- and low-nutrient patches: high values mean high plasticity (see text for details)

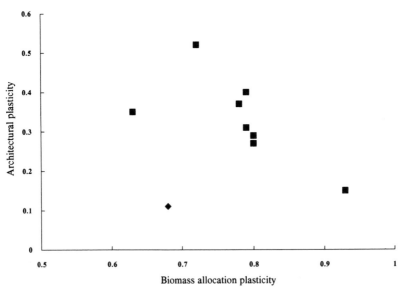

Figure 9 Relationship between architectural and biomass allocation plasticity in a group of nine herbaceous species grown in the foraging experiments of Campbell *et al.* (1991). Biomass allocation plasticity is the proportion of new root growth that develops in nutrient-rich patches and architectural plasticity is an index based on topological and geometrical variables of the root systems (see text). The point marked with a diamond represents *Lolium perenne*.

plasticity and architectural plasticity will not be related—and suggests that there may be a high degree of control of the response of root systems for patchiness.

I have tested prediction (iii), that coarse-rooted species will be less reactive to patches in terms of root growth, by comparing root growth of four closely related species in the family Caryophyllaceae grown between two glass sheets separated by 4 mm of soil. One patch in the soil was pure potting compost and the remainder a mixture of 10% compost with sand. Root growth was traced at regular intervals and relative extension rates (RER, m m^{-1} d^{-1}) were calculated (A. H. Fitter, unpublished data). There was a close relationship between the ratio of the RER values in the rich patch to that in the poor patch and the mean root diameter of terminal roots across the four species (Figure 10). The more coarse-rooted species were much less responsive, as predicted.

These data represent a preliminary attempt to investigate the architectural components of root system plasticity. Other variables await proper investigation, including root diameter and branching angle, but it is already apparent that architectural changes are at least as important as

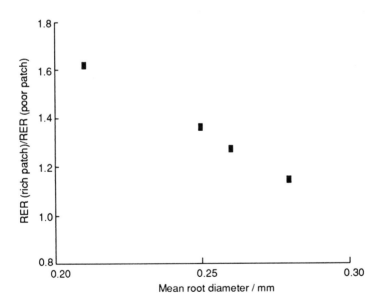

Figure 10 Relationship between root diameter and root growth response to nutrients, measured as the ratio of the relative extension rate (RER, m m^{-1} d^{-1}) in pure potting compost to that in a mixture of 10% potting compost and 90% sand (see text). Root diameter was measured optically under a dissecting microscope. Species used were *Silene vulgaris*, *S. noctiflora*, *S. alba*, and *Agrostemma githago*.

alterations in biomass allocation in determining the response of root systems to heterogeneity in soil. Future studies should include a consideration of architecture if a proper understanding of the responses of roots is to be achieved.

Acknowledgments

This chapter has profited from discussion with Philip Grime and Martyn Caldwell. Some of the work discussed here was carried out in collaboration with Mary Harvey, Tim Stickland, and James Merryweather in York, and with Bruce Campbell and Joanna Mackey in Sheffield.

References

Baldwin, J. P. (1975). A quantitative analysis of the factors affecting plant nutrient uptake from some soils. *J. Soil Sci.* **26,** 617–626.

Caldwell, M. M., Manwaring, J. H., and Durham, S. L. (1991). The microscale distribution of neighbouring plant roots in fertile soil microsites. *Funct. Ecol.* **5,** 765–772.

Campbell, B. D., Grime, J. P., and Mackey, J. M. L. (1991). A trade-off between scale and precision in resource foraging. *Oecologia* **87,** 532–538.

Crick, J. C., and Grime, J. P. (1987). Morphological plasticity and mineral nutrient capture in two herbaceous species of contrasted ecology. *New Phytol.* **107,** 403–414.

De Jager, A. (1982). Effects of a localised supply of $H_2PO_4^-$, NO_3^-, SO_4^{2-}, Ca^{2+} and K^+ on the production and distribution of dry matter in young maize plants. *Neth. J. Agric. Sci.* **30,** 193–203.

Drew, M. C. (1975). Comparison of the effect of a localized supply of phosphate, nitrate, ammonium and potassium on the growth of the seminal root system and the shoot in barley. *New Phytol.* **75,** 479–490.

Drew, M. C., Saker, L. R., and Ashley, T. W. (1975). Nutrient supply and the growth of the seminal root system in barley. I. The effect of nitrate concentration on the growth of axes and laterals. *J. Exp. Bot.* **24,** 1189–1202.

Duncan, W. G., and Ohlrogge, A. J. (1958). Principles of nutrient uptake from fertilizer bands. II. *Agron. J.* **50,** 605–608.

Eissenstat, D. M. (1991). On the relationship between specific root length and the rate of root proliferation: a field study using citrus rootstocks. *New Phytol.* **118,** 63–68.

Fitter, A. H. (1976). Effects of nutrient supply and competition from other species on root growth of *Lolium perenne* in soil. *Plant Soil* **45,** 177–189.

Fitter, A. H. (1985). Functional significance of root morphology and root system architecture. *In* "Ecological Interactions in Soil" (A. H. Fitter *et al.*, eds.), pp 87–106. Blackwell, Oxford.

Fitter, A. H. (1987). An architectural approach to the comparative ecology of plant root systems. *New Phytol.* **106,** (Suppl) 61–77.

Fitter, A. H., and Stickland, T. R. (1991). Architectural analysis of plant root systems. II. Influence of nutrient supply on architecture in contrasting species. *New Phytol.* **119,** 383–389.

Fitter, A. H., and Stickland, T. R. (1992a). Architectural analysis of plant root systems. III. Studies on plants under field conditions. *New Phytol.* **121,** 243–248.

Fitter, A. H., and Stickland, T. R. (1992b). Fractal characterisation of root system architecture. *Funct. Ecol.* **6,** 632–635.

Fitter, A. H., Stickland, T. R., Harvey, M. L., and Wilson, G. W. (1991). Architectural analysis of plant root systems. I. Architectural correlates of exploitation efficiency. *New Phytol.* **119,** 375–382.

Fusseder, A. (1987). The longevity and activity of the primary root of maize. *Plant Soil* **101,** 257–265.

Gersani, M., and Sachs, T. (1992). Developmental correlations between roots in heterogeneous environments. *Plant, Cell Environ.* **15,** 463–469.

Granato, T. C., and Raper, C. D. (1989). Proliferation of maize (*Zea mays* L.) roots in response to localized supply of nitrate. *J. Exp. Bot.* **40,** 263–275.

Gupta, P. L., and Rorison, I. H. (1975). Seasonal differences in the availability of nutrients down a podzolic profile. *J. Ecol.* **63,** 521–534.

Hackett, C. (1972). A study of the root system of barley. II. Relationships between root dimensions and nutrient uptake. *New Phytol.* **68,** 1023–1030.

Jackson, R. B., and Caldwell, M. M. (1989). The timing and degree of root proliferation in fertile-soil microsites for three cold desert perennials. *Oecologia* **81,** 149–153.

Jackson, R. B., and Caldwell, M. M. (1992). The scale of nutrient heterogeneity around individual plants and its quantification with geostatistics. *Ecology* **74,** 612–614.

Jackson, R. B., and Caldwell, M. M. (1993). Soil heterogeneity around individual perennial plants as quantified by geostatistics. *J. Ecol.* **81** (in press).

May, L. H., Randles, F. H., Aspinall, D., and Paleg, L. G. (1967). Quantitative studies of root development. II. Growth in the early stages of development. *Aust. J. Biol. Sci.* **20,** 273–283.

McGonigle, T. P., and Fitter, A. H. (1988). Ecological consequences of arthropod grazing on VA mycorrhizal fungi. *Proc. R. Soc. Edinburgh, Sect. B: Biol. Sci.* **94,** 25–32.

Nye, P. H., and Tinker, P. B. H. (1977). "Solute Movement in the Soil–Root System." Blackwell, Oxford.

Passioura, J. B., and Wetselaar, R. (1972). Consequences of banding nitrogen fertilizers in soil. II. Effects on the growth of wheat roots. *Plant Soil* **36,** 467–473.

Russell, R. S. (1977). "Plant Root Systems." McGraw-Hill, London.

Van Pelt, J., and Verwer, R. W. H. (1984). Cut trees in the topological analysis of branching patterns. *Bull. Math. Biol.* **45,** 269–285.

Veresoglou, D. S., and Fitter, A. H. (1984). Spatial and temporal patterns of growth and nutrient uptake of five co-existing grasses. *J. Ecol.* **72,** 259–272.

Wiersum, L. K. (1958). Density of root branching as affected by substrate and separate ions. *Acta Bot. Neerl.* **7,** 174–190.

12

Exploiting Nutrients in
Fertile Soil Microsites

Martyn M. Caldwell

I. Introduction

It is well known that roots have the proclivity to proliferate in regions
of local nutrient enrichment (e.g., Passioura and Wetselaar, 1972; Drew
and Saker, 1975, 1978) and this activity likely plays a role in exploiting
fertile soil microsites. Grime (Chapter 1, this volume) considers several
aspects of root foraging, including trade-offs between the precision and
speed of root proliferation, and Fitter (Chapter 11, this volume) ad-
dresses several theoretical issues, such as importance of root architectural
flexibility in exploiting different types of patches. This chapter will
consider a few questions of root distribution and proliferation in rich
microsites. It will also examine relationships between actual nutrient
acquisition from fertile soil patches, elevated physiological uptake capacity of roots, altered partitioning of soil phosphate, and the influence of
other plant roots in fertile patches. These findings and generalizations
are based primarily on field studies. Since most soil resource acquisition
is by the fine, actively absorbing elements of root systems, these fine
roots are emphasized in this treatment.

II. Root Distributions and "Foraging" Responses

A. Irregular Root Distributions

At scales from decimeters to meters, roots tend to be very unevenly
distributed in soils (Soriano *et al.*, 1987; Tardieu and Manichon, 1986;
Tardieu, 1988; Kinsbursky and Steinberger, 1989). At scales of millime-

ters to centimeters, the fine root distribution in nature is difficult to observe without the use of transparent barriers against which roots are forced to grow. One approach is to capture roots from the field in their original positions in three-dimensional soil space using a freezing and slicing technique (Caldwell *et al.*, 1987, 1991b). The procedure is necessarily destructive, but provides the opportunity to see the actual positions of roots in the soil as they are naturally before the sudden freezing. Additionally, it is possible to determine the species of the roots in two-way mixtures of some species (e.g., shrub and grass species) by a simple chemical procedure (Caldwell *et al.*, 1987, 1991b). Studies with this approach also indicate a very uneven distribution of roots (Figure 1). Although it is tempting to expect that aggregations of roots correspond with fertile soil microsites, this is not always the case. For example, there was little correlation between root density and local soil phosphate (P) or NO_3^- in a centimeter-scale analysis from field samples (Figure 2).

The trajectory that root growth assumes and the ultimate distribution of roots will be greatly influenced by soil physical and chemical properties (Passioura and Wetselaar, 1972; Whiteley and Dexter, 1983; Wang *et al.*, 1986), by root branching patterns (Tardieu and Pellerin, 1990), and by biotic factors such as mycorrhizal infection (Hetrick, 1991) and the presence of other roots (Mahall and Callaway, 1991; Caldwell *et al.*, 1991b). So apart from active "foraging" behavior of roots, there may be numerous reasons to expect roots to be very unevenly distributed in soils. Thus, root foraging for nutrient opportunities in patches or pulses must be considered against the background of very irregular root distributions in nature.

The irregular distribution of roots violates the common assumptions of most theoretical treatments of resource acquisition by roots in homogeneous soils (Nye and Tinker, 1977; Barber, 1984). Such models normally assume roots to be evenly distributed in soil. The implications of uneven root distribution for nutrient acquisition by roots were explored by Baldwin *et al.* (1972), who used a simple electrical analog to show the reduced effectiveness of root system nutrient acquisition from homogeneous soil when roots have an aggregated distribution. Few root/soil nutrient or water uptake models have incorporated irregular root distributions. A notable exception is the recent model of Lafolie *et al.* (1991) in which actual distributions of maize roots mapped in the field served as initial information for the model of water uptake by the crop. When roots are not regularly distributed in the soil, the model predicts that roots access the soil moisture reserve much less effectively and that moisture diffusion resistances in the soil are more important than if roots are regularly distributed, as is assumed in most models (Tardieu *et al.*, 1992). Thus, when soils are assumed to be uniform in moisture and nutrient

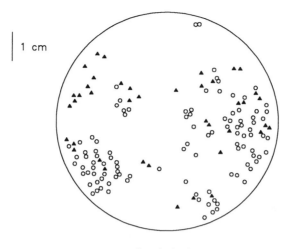

○ Crested wheatgrass
▲ Sagebrush

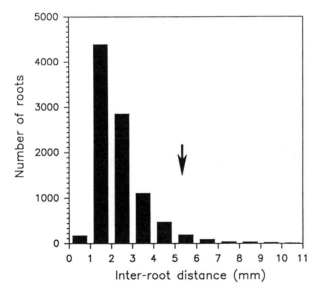

Figure 1 Soil core cross-section map (top) of root distribution of perennial grass (crested wheatgrass, *Agropyron desertorum*) and shrub (sagebrush, *Artemisia tridentata*) roots. The microscale mapping was conducted following the procedures described in Caldwell *et al.* (1991b). Frequency distribution (bottom) of nearest-neighbor distances between fine roots in soil core sections collected in field studies. Over 9300 target roots were involved. The arrow indicates the average inter-root distance if the roots were evenly spaced. Roots are decidedly aggregated in distribution. (From M. M. Caldwell and J. H. Manwaring, unpublished.)

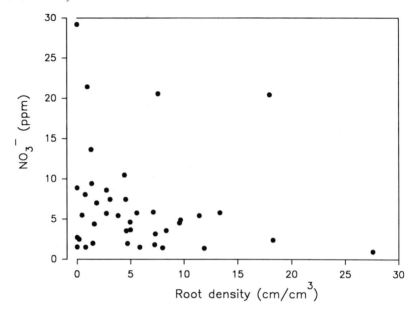

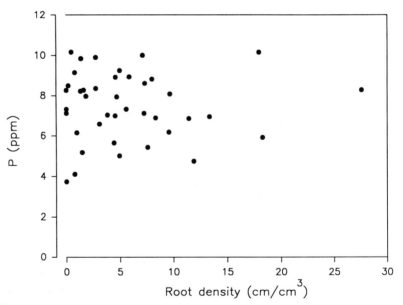

Figure 2   Relationships between root length density and concentrations of soil $Ca(OH)_2$-extractable nitrate and $NaHCO_3$-extractable phosphate in soil samples of approximately $0.5 \ cm^3$ volume. (From M. M. Caldwell and J. H. Manwaring, unpublished.)

content within vertical layers before roots begin depleting these resources, uneven root distributions appear to be theoretically disadvantageous. However, if moisture and nutrients are distributed in a patchy fashion, higher root densities in resource-rich sites could be of advantage.

B. Root Proliferation in Fertile Microsites

When roots do proliferate in fertile microsites, the amount and speed of the proliferation response can vary among species (Grime, Chapter 1, this volume). For example, a burst of root proliferation was seen for the perennial tussock grass *Agropyron desertorum* within 24 hr after injecting nutrient solution into soil to create a fertile patch, whereas roots of another perennial tussock grass species, *Pseudoroegneria spicata,* did not respond within a 2-week period (Jackson and Caldwell, 1989).

The degree of proliferation appears to be modulated by the nutrient concentration in the patch; at lower nutrient concentration the pattern and speed of proliferation are similar to those at higher nutrient concentration, but the amount of proliferation is damped (Jackson and Caldwell, 1989). The type of nutrient also plays a role; for example, proliferation responses have been shown for NO_3^-, NH_4^+, or P, but not for K in studies of Drew (1975) and Philipson and Coutts (1977). And nutrient demand of the plant plays a role. Friend *et al.* (1990) showed that the overall nitrogen nutrition of conifer seedlings influenced the degree of the proliferation response in a localized fertile substrate compartment. Seedlings deficient in N exhibited a greater proliferation response than did nitrogen-sufficient plants. Apart from the intensity or speed of proliferation response, qualitative features can also change such as the branching patterns of roots in patches as discussed by Fitter (Chapter 11, this volume).

At a smaller scale, responses of root hairs and mycorrhizae to soil conditions have been shown. Stimulation of root hair growth with low nutrient availability has been shown by several workers (see Robinson and Rorison, 1987). Under some circumstances this should contribute to root acquisition of nutrients. A distinct foraging response of mycorrhizal hyphae similar in behavior to that of roots has been demonstrated (St. John *et al.*, 1983). As hyphae encountered small organic matter patches, they branched prolifically in the patches. Friese and Allen (1991) described categories of VAM hyphae that differed in architecture and behavior even though the hyphal strands were similar in morphology. For example, "runner hyphae" appeared to be specialized for exploring the soil and infecting roots whereas "absorption hyphal networks" were specialized for proliferation and nutrient acquisition.

Moving from laboratory and container-grown plant studies to the field, further complications become apparent, especially when roots of more than one species occur in patches. Several experiments have been conducted in field plots containing mixtures of the prominent shrub of the U.S. Great Basin, *Artemisia tridentata,* and two prevalent perennial tussock grass species of the region, *Agropyron desertorum* and *Pseudoroegneria spicata.* These examined both the root microdistribution of competing plants and the effectiveness of exploiting P from fertile microsites created by injecting liquid fertilizer between neighboring plants (Caldwell *et al.,* 1991a,b). Several features became apparent in these studies: (1) Root proliferation in fertile patches is not always apparent. (2) The amount of root proliferation in soil microsites can be influenced by the presence of other species' roots. (3) There can be a tendency for individual roots of different species to avoid one another. (4) In either control or fertile microsites, there is not necessarily a clear relationship between root length density and nutrient acquisition. Elaboration of these points follows.

Root densities in soil patches 3 weeks following fertilization were often not greater than in control patches given only distilled water for three of the species (Figure 3). (Studies showing negative results often do not appear in the literature, but in our experience, even with repeated trials, root proliferation in patches where it is expected is not always found,

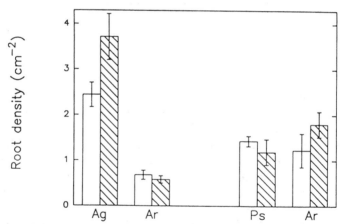

Figure 3   Mean root length densities of core sections from enriched soil patches (hatched bars) and similar control (water only) patches (open bars) 3 weeks following fertilization. This is represented for patches containing *Agropyron desertorum* (Ag) and *Artemisia tridentata* (Ar) and for patches containing *Pseudoroegneria spicata* (Ps) with *Artemisia tridentata.* Only *Agropyron* exhibited a significant proliferation in the enriched patches compared with the controls. The abundance of *Artemisia* roots was considerably greater when with *Pseudoroegneria* than when with *Agropyron* in both control and enriched patches. Error bars represent one standard error of the mean. (Adapted from Caldwell *et al.,* 1991b.)

especially in the field.) The amount of root proliferation in both the fertile and control microsites varied among species and for the shrub *Artemisia,* the degree of root proliferation was very dependent on the species of grass roots occurring in the microsites (Figure 3). When with *Agropyron,* root densities of *Artemisia* were much lower than when *Artemisia* roots shared the microsites with roots of *Pseudoroegneria.* Furthermore, even the tendency for roots of *Artemisia* to proliferate in response to the added nutrients (compared with distilled-water control patches) depended on the presence of the species of grass roots in the patches. When *Pseudoroegneria* roots were present, *Artemisia* roots tended to proliferate in the fertile microsites but did not when *Agropyron* roots were present.

Even at the scale of individual roots, there was indication that the distribution of roots was influenced by the presence of other species' roots. A nearest-neighbor analysis revealed that roots of *Artemisia* and the grasses tended to segregate (avoid one another) which may in part involve interference at the individual root level (Caldwell *et al.,* 1991b; M. M. Caldwell and J. H. Manwaring, unpublished). With other species, Mahall and Callaway (1991) presented evidence at the level of individual roots that the presence of one plant's roots could cause considerable reduction in growth of another plant's roots. With one species, *Ambrosia dumosa,* this only happened if roots of another plant made contact with the *Ambrosia* roots and it was very specific to the genotype of the plants involved. With *Larrea tridentata,* if roots of another plant were within several millimeters of the *Larrea* roots, this was sufficient to reduce root growth and the effect was not specific to plant genotype. These studies conducted in a sand culture system periodically flushed with nutrient solution indicated that the influence of individual roots on growth of other roots was not due to resource depletion but rather to an allelochemic effect. Subsequent studies showed that the presence of activated charcoal could essentially quench this interference effect (Mahall and Callaway, 1992).

In fertile soil patches, greater local root length density should theoretically contribute to greater acquisition of nutrients in these patches. Such relationships have not often been tested in soil, especially in the field. As addressed later (Section V,C,3), a clear relationship between root length density and nutrient procurement was not apparent.

III. Root Uptake Kinetics and Local Soil Fertility

In addition to the amount of active root length in a soil patch, physiological uptake capacity of the root tissue for specific ions can play an important role in nutrient acquisition. The physiological nutrient uptake capacity, commonly termed uptake kinetics, is determined by the genetic constitution of a plant, the soil nutrient status, and the demand for a particular nutrient in the plant (Clarkson, 1985). If different portions of a plant's root system encounter soil patches of different nutrient content, there is evidence that roots adjust their uptake kinetics corresponding to the local soil nutrient levels (Jackson *et al.*, 1990; Jackson and Caldwell, 1991). These field studies have indicated that the elevation of uptake kinetics was both very sizeable, as much as 80% at some external assay nutrient concentrations, and also rapid, occurring within 3 days (Jackson *et al.*, 1990; Jackson and Caldwell, 1991) (Figure 4). Since, these changes are among roots of the same plant, demand for nutrients would be the same. Thus, changes in uptake kinetics of different portions of a plants' root system should be solely in response to soil conditions.

In laboratory studies conducted with nutrient solution or sand culture systems, small proportions of the root system exposed to elevated nutrient concentrations exhibited increased uptake capacity, even within 24 hr (Drew and Saker, 1978). Although such rapid increases of uptake capacity have not been documented often (usually because of the measurement schedule of the experiments), most studies report increased uptake capacity within a matter of days and this usually preceded root proliferation (Drew and Saker, 1978; Burns, 1991). Under circumstances

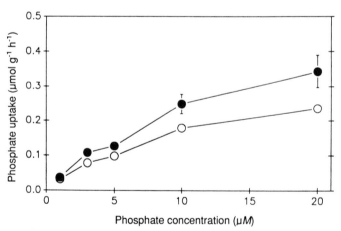

Figure 4   The rate of phosphate uptake for *Pseudoroegneria spicata* roots extracted from enriched (●) and control (○) soil patches in the field as a function of the solution P concentration in the uptake assays. These tests were conducted 3 days following patch enrichment. Error bars indicate one standard error of the mean. (From Jackson *et al.*, 1990.)

of high demand, small portions of the root system can increase nutrient inflow rates as much as four- to five-fold (Drew and Saker, 1978; Robinson and Rorison, 1983).

The importance of elevated uptake kinetics for nutrient acquisition is theoretically much greater for nutrients with relatively high diffusivity such as NO_3^- than for very immobile nutrients such as P (Barber, 1984). Uptake of immobile nutrients tends to be limited by diffusion in soil rather than by uptake at the root surface. However, even for P, a theoretical analysis indicates that elevated kinetics can contribute significantly to P acquisition in very fertile microsites (Caldwell *et al.*, 1992), as will be discussed later.

IV. Root Exudates in Fertile Patches

Much is known about root exudates and their role in promoting mineralization and mobilization of certain nutrients. Particular emphasis has been directed to exudation of protons, phosphatases, and chelating agents, called phytosiderophores (e.g., Hedley *et al.*, 1982; Kroehler and Linkins, 1988; Treeby *et al.*, 1989; Gahoonia and Nielsen, 1992). Deficiency of P stimulates release of phosphatases (Boutin *et al.*, 1981) and iron deficiency enhances the exudation of phytosiderophores (Treeby *et al.*, 1989). Although there is considerable evidence for the facilitation of nutrient acquisition by these exudates, it is not well understood if exudation rates of roots change when roots encounter fertile soil patches.

In a split-root experiment with half of a tomato plant root system under high-P and half under low-P conditions, the two halves of the root system differed considerably (Boutin *et al.*, 1981). Roots in the low-P substrate had 50% greater root surface phosphatase activity (per root mass) than did roots in the high-P substrate. The control of exudate release by the local nutrient status of the roots was thought to be the primary factor resulting in these differences. If one extends these results to heterogeneous soil conditions, one might expect lower rather than greater exudate release in fertile soil patches. Preliminary experiments with plant roots in soil with patches of applied organic P in the form of lecitin or phytin did not result in consistent differences in root surface phosphatase activity of roots in the organic P patches compared with roots in control patches (J. H. Manwaring and M. M. Caldwell, unpublished).

Although exudates are important in facilitating acquisition of certain nutrients under some conditions, whether plants release greater quantities of these exudates to more effectively garner resources from fertile patches remains an issue in need of study.

V. Acquisition of Nutrients from Patches and Pulses

Increased root uptake kinetics combined with root proliferation in nutrient-rich portions of the soil provide considerable potential for plants to exploit rich patches when encountered by roots. Such root behavior also has important implications for temporal variability in nutrient availability. Pulses of nutrients may become available when soils are wetted following dry periods or in freeze/thaw activity (Lee *et al.*, 1983; Chapin, 1988). Leaching of an ash layer from previous fires may also provide pulses of nutrients in the soil beneath the ash (Grier and Cole, 1971; Grier, 1975). Apart from nutrient opportunities presented in patches or pulses, situations may arise in which much of the root system is deprived of nutrients. For example, strong leaching may remove much of the NO_3^- to lower soil depths where only a small proportion of the root system is exposed to relatively high concentrations of NO_3^- (Burns, 1980, 1991). Therefore, flexibility in the root system should be beneficial in many circumstances. The dynamic nature of both root proliferation and changes in uptake kinetics indicates the potential for effective and very rapid exploitation of nutrients in fertile patches. Yet, do these characteristics actually translate into effective procurement of nutrients from enriched patches?

A. Laboratory Experiments

Several laboratory studies using liquid or sand culture have investigated acquisition of nutrients and plant growth when portions of root systems have been exposed to compartments with high nutrient levels (e.g., Drew and Saker, 1975, 1978; Robinson and Rorison, 1983; Burns, 1991). These and other studies indicate that the combination of enhanced root uptake kinetics and root proliferation results in effective nutrient acquisition. For example, if only one-third of the root system is supplied with a macronutrient such as N or P, after a period of adjustment this fraction is sufficient to provide the plant with this nutrient at the same rate as if the entire root system was supplied with the same quantity of the nutrient (Drew and Saker, 1975; Edwards and Barber, 1976). A review of several other estimates and experiments suggests that as little as 10% of the root system can be capable of supplying nutrients to maintain growth at the same rate as plants with entire root systems (Burns, 1980).

B. Complications under Field Conditions

In the field, assessing nutrients acquired from a patchy soil environment is necessarily more complicated. Diffusion of nutrients in the soil solution slows nutrient uptake especially if soil moisture is less than field capacity. As discussed earlier, roots are not uniformly distributed in the soil either within patches or in the bulk soil. Mycorrhizae and other rhizosphere microorganisms combined with root exudates may play important roles in facilitating nutrient acquisition. Furthermore, usually roots of more than one plant occur in a soil patch. Although these complications generally apply to all nutrients, for P they may be more important because of its low soil solution concentrations, its very low diffusivity, and the importance of mycorrhizae for its acquisition in many situations. One approach to assessing P uptake from soil patches in the field involves the use of a double-isotope-labeling technique. Fortunately, for P two radioisotopes, ^{32}P and ^{33}P, are available. A brief account of such experiments follows.

C. A Case Study of Patch Exploitation

1. Experiments with Dual-Isotope Labeling To determine how effectively plants were acquiring P from a nutrient-rich patch under field conditions, uptake of P from artifically created nutrient-enriched soil patches was compared with uptake from soil volumes of similar size that received only water (Figure 5). The plants involved were in a competitive setting and roots of neighboring plants were present in both types of patches. The enriched and control patches were each labeled by injecting different radioisotopes of P, ^{32}P or ^{33}P (designated generically as *P), into the soil along with the solutions used to create the patches (Caldwell *et al.*, 1991a). The ratio of radioisotopes appearing in the shoot tissues of the plant in question provides information on the relative acquisition of P, both that added to the soil (P_a) and the indigenous P (P_i) already present before the creation of the patches. However, it is necessary to know how P in the soil is partitioned between that in the soil solution (l) and that adsorbed on soil particles (s).

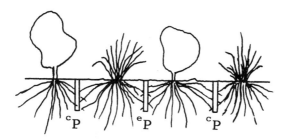

Figure 5 Replicate plant set used in dual-isotope-labeling experiments. Two species were alternated in a row with even spacing. The enriched-nutrient patch was located in the center (^{e}P) and given a phosphate isotope label that was either ^{32}P or ^{33}P. The control patches (^{c}P) were between the neighbors and center plants. These received the other phosphate isotope label. The two center plants were evaluated for isotope content. (Adapted from Caldwell *et al.*, 1991a.)

When *P and $^{31}P_a$ are placed in the soil, they are distributed among P_l, P_s, and a large component of strongly absorbed and precipitated P that is not available to the plant (Barber, 1984). The exchange between P_l and P_s is rapid (1 to 2 days in well-mixed soil in the laboratory and somewhat longer if simply injected into the soil in the field as in these isotope experiments). The *P and $^{31}P_a$ should also exchange rapidly with the P_i in both solution (P_l) and absorbed (P_s) phases (Mattingly, 1957). Thus, at equilibrium

$$\frac{*P_l}{*P_s} = \frac{P_l}{P_s}.$$ (1)

Roots remove P_l, but this is replaced by the comparatively large P_s in these soils with high buffering capacity. This replacement is assumed to be rapid relative to the diffusion of P, an effective equilibrium between P_s and P_l is in effect, and P_s is not depleted, at least in the short term. Given such an equilibrium, the specific activities of P_l of the control (c) and enriched (e) patches are $^{c,*}P_l/^cP_{i,l}$ and $^{e,*}P_l/(^eP_{i,l} + ^eP_{a,l})$, respectively. These can be used to estimate the short-term relative P acquisition from control and enriched patches. Therefore, the relative uptake from the enriched and control patches corrected for P concentration of the soil solution would be

$$\frac{^eP_{uptake}}{^cP_{uptake}} = \frac{^{e,*}P_{plant}}{^{c,*}P_{plant}} \times \frac{^eP_{i,l} + ^eP_{a,l}}{^cP_{i,l}},$$ (2)

where $^{e,*}P_{plant}/^{c,*}P_{plant}$ is the ratio of radioisotopes from the enriched and control patches appearing in the shoot tissues of the plant. The relationship between solution P in the enriched and control patches in Eq. (2) is, of course, specific to the soil in question and must be determined experimentally as in Kovar and Barber (1988) or by some other suitable technique.

When P is added to soil, P_l, P_s, and the unavailable P component all increase, but the ratio P_l/P_s also often changes (Figure 6). The P_l can increase proportionately much more than P_s when P is added to soil and implications of this will be discussed later. Estimates of P_l in the control and enriched patches are taken from relationships as in Figure 6A and are used in Eq (2).

The P acquisition from control and enriched patches is shown in Figure 7 for paired combinations of three perennial species in the configurations shown in Figure 5. The amount of P acquired by each species from the patches was conservatively estimated from the amount of *P appearing in the shoot tissues of each plant and the specific activities of P_l in the two types of patches. Acquisition of P from both control and enriched

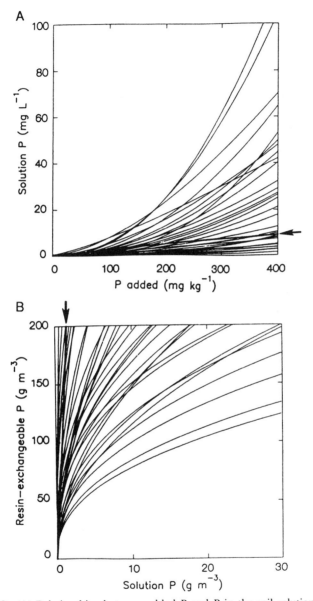

Figure 6 (A) Relationships between added P and P in the soil solution (P_l) for 33 soils surveyed by Kovar and Barber (1988) and the soil used in the dual-isotope-labeling experiments (indicated by the arrow). (B) Isotherm relationships between P in soil solution (P_l) and readily exchangeable phosphate adsorbed on soil particles as reflected by resin-exchangeable P (P_s) for 33 soils surveyed by Kovar and Barber (1988) and the soil used in the dual-isotope-labeling experiments (indicated by the arrow).

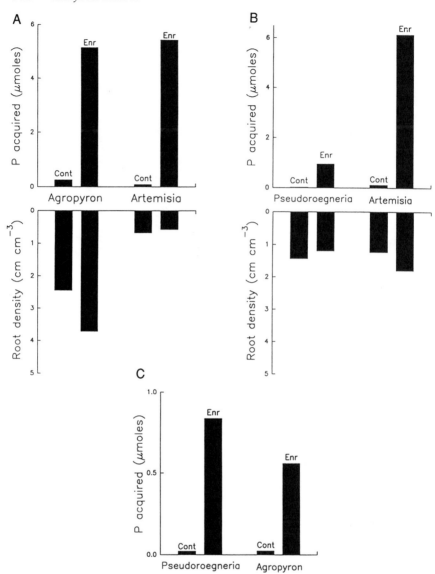

Figure 7 Acquisition of P from enriched and control patches in three experiments with paired combinations of three perennial plants in the configurations shown in Figure 5. The P acquired includes both P from that added in the patches and that already in the soil. For combinations of the shrub (*Artemisia*) with the two grass species, root densities in the patches from a parallel experiment are shown (as in Figure 3). The experiments with the different species combinations were conducted in different plots where there were inevitably some differences in soil depth and moisture. Therefore, absolute ranges of P acquisition by the species cannot be directly compared when they were in combination with different species. (Adapted from information in Caldwell *et al.*, 1991a,b, 1992.)

patches in these experiments was always in the presence of neighboring plant roots. Determinations of root length density for the competing species in the patches from a parallel study (Figure 3) are also shown for situations in which the shrub *Artemisia* was competing with either tussock grass species. Such determinations were not possible when the two grasses were together since the roots of the these species could not be distinguished in the patches. In each situation, all three species very effectively exploited P from the enriched patches—the ratio $^e P_{uptake}/ ^c P_{uptake}$ was at least 20 for all species and could exceed 70. In all cases, P was procured from the patches at least in proportion to the relative enrichment of solution P in the fertilized patches, that is, $^e P_{uptake}/ ^c P_{uptake} \geq {}^e P_l/ {}^c P_l$. (In this soil, the added P increased total available P, i.e., $P_s + P_l$, by a factor of 4, but P_l by a factor of 20.)

2. Factors Contributing to Nutrient Exploitation

Model simulations are useful in assessing the relative importance of root uptake kinetics, soil equilibrium chemistry, and root proliferation in nutrient acquisition from enriched soil patches, and these factors will vary in their importance according to the nutrient element, species, and soil type.

Simulations using the model of Barber and Cushman (Barber, 1984) were in general agreement with the experimental findings from the dual-isotope studies (Caldwell *et al.*, 1992). Almost half of the increased P gain from enriched patches could be attributed to the increased P_l in these fertilized microsites (Figure 8) even though the calcareous soil in which these experiments were performed released relatively little P into solution when compared with many other soils (Figure 6A).

Calculations by Kovar and Barber (1989) help explain previous experiments of the Barber research group (e.g., Borkert and Barber, 1985) showing that applying P fertilizer in greater concentrations to a small fraction of the soil volume was more effective than a more dilute application to a larger fraction of the soil volume. Their simulations revealed that without any root proliferation or elevated root uptake kinetics, concentrating the added P to as little as 4% of the rooted soil volume represented the optimal plant P for the soil in question, even though only 4% of the roots were in contact with the added P.

The local increase of P_l also increased the relative effectiveness of enhanced root uptake capacity in the simulation of Caldwell *et al.* (1992). As mentioned earlier, the marginal benefit from elevated P root uptake kinetics is very small in most soils because diffusion tends to limit uptake of P and our simulations produced the same result for the control patches. In the enriched patches, the local increase of P_l altered this situation appreciably and contributed more to P acquisition than did root proliferation for *Agropyron* (Figure 8). There were also substantial

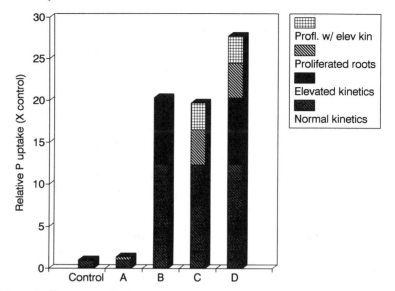

Figure 8 Factors contributing to P acquisition from enriched patches relative to that from the control patches for *Agropyron desertorum* based on model stimulations for four theoretical cases. Case A is a control patch with root proliferation as found in the enriched patch. Case B is uptake from an enriched patch without root proliferation, showing uptake with normal (control patch) kinetics and with elevated kinetics. Case C shows uptake from an enriched patch where antecedent roots do not have elevated kinetics, but root proliferation with and without elevated kinetics contributes to P acquisition. Case D shows uptake from an enriched patch showing contributions of both elevated kinetics and root proliferation. (From Caldwell *et al.*, 1992.)

differences in the manner that roots of the three species reacted to enriched as opposed to control patches. Some of the species exhibited no root proliferation in the enriched patches (Figures 3 and 7) and yet still effectively exploited enriched patches. For example, *Pseudoroegneria* when grown with *Artemisia* had the same root density in control and enriched patches, but its ratio $^eP_{uptake}/^cP_{uptake}$ was 74. In soils that release more P into solution, the efficacy of elevated root uptake kinetics in small fertile patches would be even greater.

Similar experiments have not been conducted with nitrate patches, in part because two suitable isotopes are not available. However, the importance of elevated root NO_3^- uptake kinetics should be even more pronounced. Unlike P, NO_3^- is not absorbed by soil, which results in both high diffusivity in soil and high soil solution NO_3^-—generally two to three orders of magnitude greater than P_l (Barber, 1984).

3. Acquisition of Phosphate in Competitive Settings Fertile soil microsites may represent important resources for plants, but competition for nutrients in these patches may be intense. In both control and enriched patches, *Agropyron* and *Artemisia* acquired the same amount of P when they were grown together, while *Artemisia* procured about six to seven times more P than did *Pseudoroegneria* (Figure 7). This corresponds with an earlier experiment (Caldwell *et al.*, 1985), in which *Artemisia* was located between neighbors of *Agropyron* and *Pseudoroegneria* and labeled, unfertilized patches were placed in the interspaces. The shrub acquired six times more *P from the soil interspace shared with *Pseudoroegneria* than from the interspace shared with *Agropyron* even though the shrub had the same investment in root length and mycorrhizal infection in both soil interspaces. Even though *Agropyron* fared much better in competition with *Artemisia* than did *Pseudoroegneria*, in the experiments with the two grasses in competition, they were equivalent in their acquisition of P from both control and enriched patches (Figure 7).

These differences in ability of the two grass species to acquire P in competition with *Artemisia* correspond to some extent with the grass root length density in the patches. *Agropyron* was much more effective in acquiring P than was *Pseudoroegneria* when with *Artemisia* and *Agropyron* also had greater root density in patches than did *Pseudoroegneria*. However, differences in P acquisition were not proportional to the differences in grass root density in the patches. *Agropyron* root densities were never more than threefold greater than those of *Pseudoroegneria* in patches shared with *Artemisia* and usually the difference was less. Thus, in competition with *Artemisia*, *Agropyron* is at least twice as effective in P acquisition per unit root length as *Pseudoroegneria*.

When the shrub and grasses are compared, the relationship between P procurement and root density was also not proportional. Although *Agropyron* had eight- to tenfold more roots at 1 week and four- to sixfold more roots than the shrub at 3 weeks (Caldwell *et al.*, 1991b), acquisition of P by the two species for the entire 5-week experiment was equivalent (Figure 7). *Artemisia* acquired six to eight times more isotope than did *Pseudoroegneria*, yet in most cases *Pseudoroegneria* had greater root densities than the shrub. The shrub appears to be four to ten times more effective in P acquisition per unit root length than the grasses under the circumstances of these experiments. The basis of the greater effectiveness of P acquisition of the *Artemisia* roots is not clear. Uptake kinetic experiments indicated that the capacity for P uptake was similar for the three species (Jackson *et al.*, 1990). All three species have vesicular-arbuscular mycorrhizal associations of the genus *Glomus*. Earlier studies at this time of year indicated that the species have comparable infection rates—if anything, the shrub had somewhat lower infection rates than the grasses (Caldwell *et al.*, 1985).

Other factors may have been involved in the different abilities of these species to acquire P from the patches. At the time of year when these experiments were conducted (July and early August), all three species remain active in photosynthesis and transpiration (Caldwell *et al.*, 1981; Caldwell and Richards, 1989). However, the grasses are less active than the shrub, which might explain some of the difference in P uptake between *Artemisia* and the grasses. Nevertheless, P uptake kinetics of the grass species in July remained nearly as great as in May and June (Jackson *et al.*, 1990). Roots of these species may also differ in exudates, which might either facilitate P mobilization in the soil or cause some interference in the acquisition of P by neighboring plant roots. Throughout the growing season, *Artemisia* exhibited greater root-associated phosphatase activity than the two grasses (J. H. Manwaring, unpublished), which might contribute to the difference in P acquisition between the shrub and the grasses. Such factors that might have contributed to differences among the species are only speculative. In any case, it is clear that simple relationships between root investment in soil microsites and acquisition of resources should not necessarily be expected.

VI. Interactions between Shoots and Roots in a Variable Environment

Just as resource acquisition by roots has significant effects on photosynthesis and light harvesting by plant shoots, reducing the light that is available to plants has important implications for root function. This operates at different time scales. On the scale of weeks to months, reduced light can lead to a decreased demand for nutrients, increased shoot/root ratios, and a general scaling down of root function (e.g., Lambers *et al.*, 1990). On the scale of hours to days, reduced light causes reductions in root functions such as root growth, respiration and ion uptake (Massimino *et al.*, 1980, 1981; Crapo and Ketellapper, 1981). Though reductions in root function of shaded plants are expected, it is less clear if reduced light might have a particularly strong effect on the ability of plants to exploit nutrients from pulses and patches compared with the effects of shading on acquisition of nutrients from a uniform soil environment.

As light reductions lead to curtailment of root function, root systems appear to place priorities on maintaining different root functions. For example, with shading, roots of wheat, barley, and tomato first reduced root growth, then potassium uptake, and finally respiration (Crapo and Ketellapper, 1981). A clear priority seemed to lie with maintenance, as respiration of the root was the last process to be affected by the shading.

In a shading experiment conducted in the field (Jackson and Caldwell, 1992), shading appeared to have no general effect on nutrient uptake capacity of roots, however, the ability of the roots to elevate uptake capacity in fertile microsites was limited by the shading (Figure 9). Again, this suggests a priority of the root system when shading occurs. In this case, the plant maintained current uptake capacity but sacrificed physiological changes that would help to exploit the fertile microsites. Since ion uptake represents a major ATP sink for roots (van der Werf *et al.*, 1988), the costs of elevating uptake capacity of roots may be quite significant.

Although generalization from only a few studies is hardly warranted, they suggest that plants might be less influenced by short-term shading when roots are acquiring nutrients from a homogeneous soil than from a variable soil environment.

VII. Summary and Conclusions

The soil environment in nature can be very heterogeneous in both space and time, which greatly complicates assessment of how effectively roots might be able to exploit soil resources. Root proliferation and elevation of uptake capacity in fertile soil regions are intuitively appealing characteristics of root systems that should contribute to resource acquisition

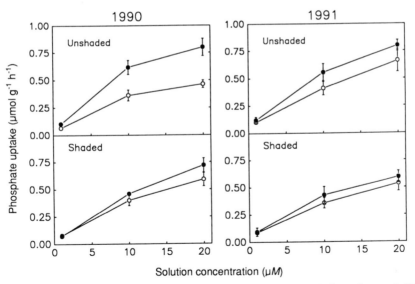

Figure 9 Effect of shading on root uptake capacity in enriched (●) and control (○) soil patches for *Agropyron desertorum*. Rates of phosphate uptake as a function of test solution concentrations are shown for shaded and unshaded plants from which roots were extracted in the field 4 days after the patches were applied. (From Jackson and Caldwell, 1992.)

without undue complication from other soil or plant factors. However, this may be overshadowed by several factors associated with natural soil environments. First, roots exist in a complicated biological environment and root behavior can be influenced by biotic factors such as the presence of neighboring plant roots. Second, soil equilibrium chemistry through its effects on the abundance of nutrients in solution, especially in the case of phosphate, can greatly influence the relative importance of root proliferation or uptake kinetics for nutrient acquisition. Third, even with the presence of roots, effective nutrient acquisition depends on the ability of the plant to supply the energy needed for root uptake and assimilation under nutrient-enriched conditions.

Field experiments using dual-isotope techniques indicate that, despite the complications, plants effectively acquire phosphate from fertilized microsites relative to control microsites—in ratios of 20 to >70. This is greatly influenced by the partitioning of phosphate in soil solution and surface-adsorbed components. Model simulations indicate the contribution of plant and soil characteristics to effective nutrient exploitation, but do not fully explain differences among species when in competition.

Acknowledgments

Some of this work stems from research supported by the National Science Foundation (BSR 8705492 and DEB 9208212) and the Utah Agricultural Experiment Station. Helpful comments on the manuscript were provided by R. B. Jackson and A. L. Friend.

References

Baldwin, J. P., Tinker, P. B., and Nye, P. H. (1972). Uptake of solutes by multiple root systems from soil. II. The theoretical effects of rooting density and pattern on uptake of nutrients from soil. *Plant Soil* **63,** 693–708.

Barber, S. A. (1984). "Soil Nutrient Bioavailability." Wiley, New York.

Borkert, C. M., and Barber, S. A. (1985). Predicting the most efficient phosphorus placement for soybeans. *Soil Sci. Soc. Am. J.* **49,** 901–904.

Boutin, J. P., Provot, M., and Roux, L. (1981). Effect of cycloheximide and renewal of phosphorus supply on surface acid phosphatase activity of phosphorus deficient tomato roots. *Physiol. Plant.* **51,** 353–360.

Burns, I. G. (1980). Influence of the spatial distribution of nitrate on the uptake of N by plants: A review and a model for rooting depth. *J. Soil Sci.* **31,** 155–173.

Burns, I. G. (1991). Short- and long-term effects of a change in the spatial distribution of nitrate in the root zone on N uptake, growth and root development of young lettuce plants. *Plant Cell Environ.* **14,** 21–33.

Caldwell, M. M., and Richards, J. H. (1989). Hydraulic lift: Water efflux from upper roots improves effectiveness of water uptake by deep roots. *Oecologia* **79,** 1–5.

Caldwell, M. M., Richards, J. H., Johnson, D. A., Nowak, R. S., and Dzurec, R. S. (1981). Coping with herbivory: Photosynthetic capacity and resource allocation in two semiarid *Agropyron* bunchgrasses. *Oecologia* **50,** 14–24.

Caldwell, M. M., Eissenstat, D. M., Richards, J. H., and Allen, M. F. (1985). Competition for phosphorus: Differential uptake from dual-isotope-labeled interspaces between shrub and grass. *Science* **229**, 384–386.

Caldwell, M. M., Richards, J. H., Manwaring, J. H., and Eissenstat, D. M. (1987). Rapid shifts in phosphate acquisition show direct competition between neighbouring plants. *Nature (London)* **327**, 615–616.

Caldwell, M. M., Manwaring, J. H., and Jackson, R. B. (1991a). Exploitation of phosphate from fertile soil microsites by three Great Basin perennials when in competition. *Funct. Ecol.* **5**, 757–764.

Caldwell, M. M., Manwaring, J. H., and Durham, S. L. (1991b). The microscale distribution of neighbouring plant roots in fertile soil microsites. *Funct. Ecol.* **5**, 765–772.

Caldwell, M. M., Dudley, L. M., and Lilieholm, B. (1992). Soil solution phosphate, root uptake kinetics and nutrient acquisition: Implications for a patchy soil environment. *Oecologia* **89**, 305–309.

Chapin, F. S. (1988). Ecological aspects of plant mineral nutrition. *Adv. Mineral Nutr.* **3**, 161–191.

Clarkson, D. T. (1985). Factors affecting mineral nutrient acquisition by plants. *Annu. Rev. Plant Physiol.* **36**, 77–115.

Crapo, N. L., and Ketellapper, H. J. (1981). Metabolic priorities with respect to growth and mineral uptake in roots of *Hordeum, Triticum* and *Lycopersicon. Am. J. Bot.* **68**, 10–16.

Drew, M. C. (1975). Comparison of the effects of a localized supply of phosphate, nitrate, ammonium and potassium on the growth of the seminal root system, and the shoot, in barley. *New Phytol.* **75**, 479–490.

Drew, M. C., and Saker, L. R. (1975). Nutrient supply and the growth of the seminal root system in barley. II. Localized, compensatory increases in lateral root growth and rates of nitrate uptake when nitrate supply is restricted to only part of the root system. *J. Exp. Bot.* **26**, 79–90.

Drew, M. C., and Saker, L. R. (1978). Nutrient supply and the growth of the seminal root system in barley. III. Compensatory increases in growth of lateral roots, and in rates of phosphate uptake in response to a localized supply of phosphate. *J. Exp. Bot.* **29**, 435–451.

Edwards, J. H., and Barber, S. A. (1976). Nitrogen flux into corn roots as influenced by shoot requirement. *Agron. J.* **68**, 471–473.

Friend, A. L., Eide, M. R., and Hinckley, T. M. (1990). Nitrogen stress alters root proliferation in Douglas-fir seedlings. *Can. J. For. Res.* **20**, 1524–1529.

Friese, C. F., and Allen, M. F. (1991). The spread of VA mycorrhizal fungal hyphae in the soil: Inoculum types and external hyphal architecture. *Mycologia* **83**, 409–418.

Gahoonia, T. S., and Nielsen, N. E. (1992). The effects of root-induced pH changes on the depletion on inorganic and organic phosphorus in the rhizosphere. *Plant Soil* **143**, 185–191.

Grier, C. C. (1975). Wildfire effects on nutrient distribution and leaching in a coniferous ecosystem. *Can. J. For. Res.* **5**, 599–607.

Grier, C. C., and Cole, D. W. (1971). Influence of slash burning on ion transport in a forest soil. *Northwest Sci.* **45**, 100–106.

Hedley, M. J., White, R. E., and Nye, P. H. (1982). Plant-induced changes in the rhizosphere of rape (*Brassica napus* var. emerald) seedlings. III. Changes in *L* value, soil phosphate fractions and phosphatase activity. *New Phytol.* **91**, 45–56.

Hetrick, B. A. D. (1991). Mycorrhizas and root architecture. *Experientia* **47**, 355–362.

Jackson, R. B., and Caldwell, M. M. (1989). The timing and degree of root proliferation in fertile-soil microsites for three cold-desert perennials. *Oecologia* **81**, 149–153.

Jackson, R. B., and Caldwell, M. M. (1991). Kinetic responses of *Pseudoroegneria* roots to localized soil enrichment. *Plant Soil* **138**, 231–238.

Jackson, R. B., and Caldwell, M. M. (1992). Shading and the capture of localized soil nutrients: Nutrient contents, carbohydrates, and root uptake kinetics of a perennial tussock grass. *Oecologia* **91**, 457–462.

Jackson, R. B., Manwaring, J. H., and Caldwell, M. M. (1990). Rapid physiological adjustment of roots to localized soil enrichment. *Nature (London)* **344**, 58–60.

Kinsbursky, R. S., and Steinberger, Y. (1989). Root and microbial biomass dynamics under the canopy of the desert shrub *Zygophyllum dumosum*. *Oecologia* **80**, 498–500.

Kovar, J. L., and Barber, S. A. (1988). Phosphorus supply characteristics of 33 soils as influenced by seven rates of phosphorus addition. *Soil Sci. Soc. Am. J.* **52**, 160–165.

Kovar, J. L., and Barber, S. A. (1989). Reasons for differences among soils in placement of phosphorus for maximum predicted uptake. *Soil Sci. Soc. Am. J.* **53**, 1733–1736.

Kroehler, C. J., and Linkins, A. E. (1988). The root surface phosphatases of *Eriophorum vaginatum:* Effects of temperature, pH, substrate concentration and inorganic phosphorus. *Plant Soil* 3–10.

Lafolie, F., Bruckler, L., and Tardieu, F. (1991). Modeling root water potential and soil–root water transport. 1. Model presentation. *Soil Sci. Am. J.* **55**, 1203–1212.

Lambers, H., Cambridge, M. L., Konings, H., and Pons, T. L., eds. (1990). "Causes and Consequences of Variation in Growth Rate and Productivity of Higher Plants." SPB Academic Publishing, The Hague, The Netherlands.

Lee, J. A., Harner, R., and Ignaciuk, R. (1983). Nitrogen as a limiting factor in plant communities. *In* "Nitrogen as an Ecological Factor" (J. A. Lee, S. McNeill, and I. H. Rorison, eds.), pp. 95–112. Blackwell, Oxford.

Mahall, B. E., and Callaway, R. M. (1991). Root communication among desert shrubs. *Proc. Natl. Acad. Sci. U.S.A.* **88**, 874–876.

Mahall, B. E., and Callaway, R. M. (1992). Root communication mechanisms and intracommunity distributions of two Mojave Desert shrubs. *Ecology* **73**, 2145–2151.

Massimino, D., André, M., Richaud, C., Daguenet, A., Massimino, J., and Vivoli, J. (1980). Évolution horaire au cours d'une journée normale de la photosynthèse, de la transpiration, de la respiration foliaire et racinaire et de la nutrition N.P.K. chez *Zea mays*. *Physiol. Plant.* **48**, 512–518.

Massimino, D., André, M., Richaud, C., Daguenet, A., Massimino, J., and Vivoli, J. (1981). The effect of a day at low irradiance of a maize crop. I. Root respiration and uptake of N, P and K. *Physiol. Plant.* **51**, 150–155.

Mattingly, G. E. G. (1957). The use of the isotope ^{32}P in recent work on soil and fertilizer phosphorus. *Soils Fert.* **20**, 59–68.

Nye, P. H., and Tinker, P. B. (1977). "Solute Movement in the Soil–root System." Univ. of California Press, Berkeley.

Passiouria, J. B., and Wetselaar, R. (1972). Consequences of banding nitrogen fertilizers in soil. II. Effects on the growth of wheat roots. *Plant Soil* **36**, 461–473.

Philipson, J. J., and Coutts, M. P. (1977). The influence of mineral nutrition on the root development of trees. II. The effect of specific nutrient elements on the growth of individual roots of Sitka spruce. *J. Exp. Bot.* **28**, 864–871.

Robinson, D., and Rorison, I. H. (1983). A comparison of the responses of *Lolium perenne* L., *Holcus lanatus* L. and *Deschampsia flexuosa* (L.) Trin. to a localized supply of nitrogen. *New Phytol.* **94**, 263–273.

Robinson, D., and Rorison, I. H. (1987). Root hairs and plant growth at low nitrogen availabilities. *New Phytol.* **107**, 681–693.

St. John, T. V., Coleman, D. C., and Reid, C. P. P. (1983). Growth and spatial distribution of nutrient-absorbing organs: Selective exploitation of soil heterogeneity. *Plant Soil* **71**, 487–493.

Soriano, A., Golluscio, R. A., and Satorre, E. (1987). Spatial heterogeneity of the root system of grasses in the Patagonian arid steppe. *Bull. Torrey Bot. Club* **114**, 103–108.

Tardieu, F. (1988). Analysis of the spatial variability of maize root density. II. Distances between roots. *Plant Soil* **107**, 267–272.

Tardieu, F., and Manichon, H. (1986). Caractérisation en tant que capteur d'eau de l'enracinement due mais en parcelle cultivée. II. Une méthode d'étude de la répartition verticale et horizontale des racines. *Agronomie* **6**, 415–425.

Tardieu, F., and Pellerin, S. (1990). Trajectory of the nodal roots of maize in fields with low mechanical constraints. *Plant Soil* **124**, 39–45.

Tardieu, F., Bruckler, L., and Lafolie, F. (1992). Root clumping may affect the root water potential and the resistance to soil–root water transport. *Plant Soil* **140**, 291–301.

Treeby, M., Marschner, H., and Römheld, V. (1989). Mobilization of iron and other micronutrient cations from a calcareous soil by plant-borne, microbial, and synthetic metal chelators. *Plant Soil* **114**, 217–226.

van der Werf, A., Kooijman, A., Welschen, R., and Lambers, H. (1988). Respiratory energy costs for the maintenance of biomass, for growth and for ion uptake in roots of *Carex diandra* and *Carex acutiformis*. *Physiol. Plant.* **72**, 483–491.

Wang, J., Hesketh, J. D., and Woolley, J. T. (1986). Preexisting channels and soybean rooting patterns. *Soil Sci.* **141**, 432–437.

Whiteley, G. M., and Dexter, A. R. (1983). Behaviour of roots in cracks between soil peds. *Plant Soil* **74**, 153–162.

13

Coping with Environmental Heterogeneity: The Physiological Ecology of Tree Seedling Regeneration across the Gap– Understory Continuum

F. A. Bazzaz and P. M. Wayne

I. Introduction

Forest environments can be extremely heterogeneous at the scale perceived by regenerating tree seedlings. The amount and types of environmental heterogeneity in forest ecosystems are influenced by tree- and branch-fall disturbances that create gaps in the forest canopy. Canopy gaps greatly modify local microclimatic conditions and the availability of resources, and thus pose a very different set of conditions for regenerating tree seedlings—relative to the intact understory (e.g., Chazdon and Fetcher, 1984; Fetcher *et al.*, 1985). These disturbance-induced modifications to environmental heterogeneity are believed to play a critical role in regulating the dynamics and coexistence of forest species, and also to have been important in the evolution of tree physiological, architectural, and life history traits (Ricklefs, 1977; Pickett, 1983; Bazzaz, 1983; Canham and Marks, 1985; Hubbell and Foster, 1986; Denslow, 1987; Bazzaz and Sipe, 1987; Koike, 1988). Yet, only limited progress has been made in assessing finer-scale patterns of spatial and temporal environmental heterogeneity across the gap–understory continuum, and the responses of tree seedlings to these patterns.

In 1986 our lab set out to investigate experimentally the consequences of small- and large-scale forest disturbances to the physiological ecology and growth of tree seedlings in New England mixed-hardwood forests. Beginning with a view of disturbance as a relatively sudden change in physical and chemical fluxes (Bazzaz, 1983; Bazzaz and Sipe, 1987), we set out to characterize critical aspects of microenvironmental conditions experienced by seedlings within and between small- ($75 m^2$) and medium-sized ($300 m^2$) tree-fall gaps, compared with the intact understory. Largely focusing on two groups of congeners varying in shade tolerance and successional status—intolerant pioneer birches (*Betula*) and tolerant later-successional maples (*Acer*)—we also set out to characterize the physiological, architectural, and demographic responses of these species to the mosaic of microenvironmental conditions following different disturbance events. Underlying this research has been the conviction that an understanding of community-level issues, such as species coexistence and succession, can be achieved through the study of microsite physical-chemical fluxes and the physiological ecology of resource processing by individual organisms.

In this chapter, we first present a conceptual framework for describing the spatiotemporal complexes of multiple resources and environmental factors across the gap–understory continuum. We then describe some of the patterns of variation in environmental factors in gaps and adjacent understories, focusing on our studies in the Harvard Forest. Finally, we present data describing some of the physiological and demographic responses of established birch and maple seedlings to the spatiotemporal microclimatic complexes across the gap–understory continuum. The processes of seed production, dispersal, and germination are not discussed here, despite their critical role in forest regeneration (Murray, 1988; Houle and Payette, 1991). The primary questions we focus on are: What are the ecologically relevant quantitative and/or qualitative differences in the microenvironmental conditions at different points across the gap–understory continuum, and what are the physiological and demographic characteristics of species' seedlings that occupy different portions of the continuum?

II. Spatial and Temporal Aspects of Environmental Heterogeneity

Characterizing the "relevant" qualitative and quantitative patterns of environmental factors within and between microsites of regenerating tree seedlings is a formidable challenge because there are many abiotic and biotic factors to consider, and most of these vary simultaneously on many temporal and spatial scales. Recent developments in hierarchy theory have begun to develop a framework for describing this complexity (e.g., Allen and Starr, 1982; O'Neil *et al.*, 1986; Pickett *et al.*, 1989; Kolasa and Rollo, 1991). In the following we briefly define some of the terminology and concepts we have employed in our research, and more clearly define the specific questions and scales we have focused on.

A. Resources, Conditions, and Signals

Environmental factors influencing regenerating seedlings can be classified as either resources, conditions, or signals. *Resources* are defined here as depletable or consumable substances required by plants for normal maintenance, growth, and reproduction (e.g., nutrients, light, water) (Ricklefs, 1990; Begon *et al.*, 1990). Other physical, chemical, or biological attributes of the environment that, while not being consumable, also influence biological processes are referred to as *conditions* (Begon *et al.*, 1990); conditions such as soil pH or air temperature modify or control the rate or efficiency of resource use by seedlings. A particular set of conditions, which also influence seedling development, but not necessarily directly via resource utilization, are *signals* or *cues*. For example, both blue light and the ratios of red to far-red light provide information to seedlings regarding their local environment via specific receptors (e.g., cryptochrome or phytochrome) (Taiz and Zeigler, 1991), but do not necessarily act directly via resource utilization. While it may be heuristic for ecologists to distinguish these three 'anthropocentric" categories, it is important to bear in mind that from a seedling's "phytocentric" perspective, these categories may be less differentiable. For example, inherently depauperate soils, competing neighbors, or low soil temperature may all have the same functional effect, that is, seedlings perceive low levels of soil resources.

B. Scales and Perception of Environmental Heterogeneity

Any meaningful discussion of heterogeneity must begin with the awareness that heterogeneity is scale dependent (Kolasa and Rollo, 1991). The quantity and qualitative nature of heterogeneity within an ecological "entity" (e.g., an ecosystem or leaf) will depend on the spatial and temporal scales used to describe the entity. For example, patterns of spatial heterogeneity following forest disturbances can be described at the landscape, community, seedling, or cellular scale, and temporal heterogeneity of a given forest patch can be described over centuries, years, hours, or seconds. The most relevant scale(s) and unit(s) used to characterize environmental heterogeneity will depend on the questions one is asking. In our work in the Harvard Forest, we have largely focused on three scales of heterogeneity, each scale being investigated to address different sets of questions: between-"patch" (β-diversity), within-"patch" (α-

diversity), and within-seedling microsite environmental heterogeneity. Between-patch heterogeneity (e.g., average differences between intact understories versus large gaps) is investigated to address landscape/community-scale issues such as ecosystem productivity and species coarse-scale distributions, dynamics, and coexistence. Within-patch heterogeneity (e.g., variation within gaps) is studied to assess finer-scale patterns of species distributions and issues related to species' resource partitioning, coexistence, and niche characteristics. Within-seedling environmental heterogeneity (e.g., patchiness of resources within the rhizosphere or canopy of a seedling) is investigated to discover the challenges and constraints that individual plants face, particularly in foraging and integrating environmental variation across and within modules, that is, within-individual plasticity in physiology, morphology, and architecture.

It is important to make the distinction between what ecologists perceive and measure and what tree seedlings perceive and respond to. Kolasa and Rollo (1991) suggested the distinction between measured versus functional heterogeneity. *Measured heterogeneity* is what ecologists measure: It is a product of their informed perspective and the instruments they use to assay the environment. In contrast, *functional heterogeneity* is the heterogeneity that ecological entities (e.g., ecosystems or seedlings) actually perceive and respond to. However, a population of regenerating seedlings may perceive many different "functional" heterogeneities. Seedlings of different species, genotypes, or phenotypes will likely perceive and experience the same "measured" heterogeneity quite differently. Thus, while two ecologists can go to the same unit of landscape and, using similar instruments and protocols, arrive at similar measures of heterogeneity, this is likely for two randomly chosen seedlings.

C. Spatiotemporal Components of Environmental Heterogeneity

Although environmental heterogeneity clearly plays an important role at all scales of research in forest ecology, detailed descriptions of how most environmental factors vary in space and time are not well characterized; even the basic components or categories of heterogeneity have not yet been well outlined (Bell, 1992). The aspects of environmental heterogeneity that have been described have mainly focused on spatial variability, generally for one factor at a time, and mainly on the between-patch scale (see Kolasa and Rollo, 1991). In the following and in Figure 1, we outline some potentially important spatial and temporal aspects of single and multiple environmental factors relevant to the three scales we investigate in our research at the Harvard Forest. It is important to recognize that each of these components of environmental heterogeneity describes just one aspect of the heterogeneity within a given entity or defined patch in space-time, and that variation of all these aspects can occur simultaneously.

1. Cumulative Quantity of a Factor Integrated across Space or Time   By far the most often considered factor for characterizing the environment of a seedling or patch is the total cumulative or average level of a resource or factor, integrated across units of space or through periods of time (Figure 1a). Average or total values of most environmental factors are often highly correlated with ecological processes at many hierarchical levels. For example, both the productivity of forest patches and the growth of individual tree seedlings show clear patterns with increasing total or average N availability (Figure 1a) (Safford and Czapowski, 1986; Thompson *et al.*, 1988; Walters and Reich, 1989). However, though similar correlations of population, individual, or module level responses exist with average or total quantities of many other environmental factors, this does not necessarily mean that it is specifically the same aspects of environmental factors (i.e., total or average) that elicit such responses. For many factors, when integrated total values increase, so do spatial and temporal patterns of availability. For example, as total daily integrated PFD increases in tropical forest patches, peak daily values, overall variability of daily values, and the durations of values above or below critical carbon gain thresholds also vary (Chazdon and Fetcher, 1984; Chazdon and Field, 1987).

2. Frequency Distribution of Factor Levels Sampled across Space or through Time   Frequency distributions of factor levels within a seedling or patch across space or time are also a potentially important component of heterogeneity (Figure 1b). Similar total or average values within an entity can be composed of distributions of subsamples with very different minimum values, peak values, degrees of skewness, and overall variability (e.g., % CV). For example, the total PFD incident upon the canopy of seedlings can be distributed across the leaves within a seedling quite differently (Figure 1b); monolayer canopies of understory seedlings with horizontal leaf angles may be expected to exhibit very skewed distribution of PFD toward the lower PFD classes in contrast to more uniform distributions of PFD in seedlings growing orthotropically in gaps with much leaf overlap and variation in leaf angles (e.g., Chazdon *et al.*, 1988; Oberbauer *et al.*, 1988). Similar differences in distributions of PFD might also be expected at the patch level, such as gaps with and without the presence of competing herbs. Some modeling evidence suggests that leaf maximum photosynthetic rate is affected by the frequency distribution of PFD through time, as well as total amount of incident light (Takenaka, 1989). At the patch scale, the frequency distribution of light microenvironments

Single Resources, Controllers, or Signals

Examples

a. Cumulative Quantity of a Factor Integrated across Space or Time

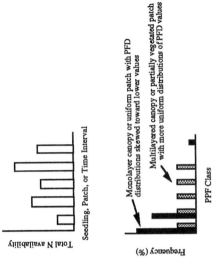

Spatial variation in total nitrogen available to adjacent tree seedlings or patches per month or year

or

Monthly or annual variation in nitrogen availability to an individual seedling or patch

b. Aspects of Frequency Distribution of Factor Levels Sampled across Space or through Time [e.g. Minimum, Maximum (Peak), Skewness, Variability (% CV).]

Frequency distribution of PFD levels across leaves within an individual seedling's canopy or across patches on the forest floor at noon

or

Frequency distribution of PFD levels available to an individual leaf or patch across the day or season

c. Proportion of Patch or Duration of Time with Factor Levels above (or below) Critical Biological Thresholds

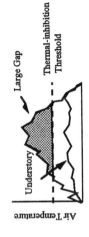

Minutes/day or days/season during which air temperatures exceed thresholds for thermal-inhibition (e.g. gap vs understory)

or

Proportion of patch exposed to temperatures above thermal-inhibition thresholds in a gap vs understory

d. Spatial Location or Timing of Critical Factor Values

Seasonal timing of critically low soil water potential in two contrasting patches or seedling microsites

or

Depth within two soil profiles or seedling rhizospheres at which soil water potential drops below critical damage thresholds

e. Spatial and Temporal Pattern of
Different Factor Levels

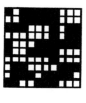

Uniform Variability   Patchy Variability

Spatial distribution of litter depth
across patches in the forest with
uniform and patchy canopies

or

Temporal patterns of litter depth
within a subsection of a patch

Multiple Resources, Controllers, or Signals

f. Congruency of Multiple Factors in Time
or Space

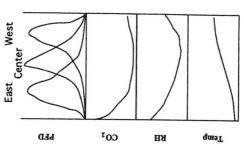

Temporal congruency of 3 diurnal
PFD regimes with more optimal levels
of CO_2 availability, relative humidity,
and air temperature for seedlings
established in the center, or east or
west edges of a gap

or

Spatial congruency within a
seedling or patch of peak values of PFD
with more optimal values of CO_2,
relative humidity, and air temperature
at noon

Figure 1   Some potentially relevant spatial and temporal aspects of single and multiple environmental factors.

355

in space or time can have significant effects on the relative abundances and distributions of regenerants (Grubb, 1977).

3. Duration of Time or Proportion of Patch with Factor Levels above or below Critical Biological Thresholds   Many ecological processes may be significantly regulated by the duration of time or spatial proportion of seedlings or patches that experience extreme, critical values (Figure 1c). For example, short periods of very high temperatures can have significant physiological or demographic consequences to seedlings (Tubbs, 1969; Turner and Kramer, 1980; Weis and Berry, 1988). Similarly, the proportion of patches exposed to unusually high temperatures likely influences spatial patterns of seedling survivorship. Thus for many environmental factors, identifying critical thresholds such as inhibitory levels, compensation points, or optima and characterizing the spatial and temporal distributions of these thresholds may be as or more important than just characterizing the effects of averages and total levels.

4. Timing or Spatial Location of Critical Values of an Environmental Factor   The specific timing or spatial locations of threshold events may also be ecologically important (Figure 1d). For example, the occurrence of a drought early in the growing season when first-year seedlings are small and shallow rooted may have more severe consequences than a drought later in the season. Similarly, drying of only the upper layers of the soil may have very different consequences than soil moisture deficits throughout a tree's rooting profile (Figure 1d). In nutritional studies with herbaceous plants, nitrogen pulses early in the growing season generally result in greater reproduction than equivalent pulses later in the season (Benner and Bazzaz, 1988; Miao and Bazzaz, 1990).

5. Spatial and Temporal Pattern of Different Factor Levels   Assuming that the distribution of an environmental factor across parts of a seedling or patch is not homogeneous, the spatial or temporal patterns of subsamples may also be an ecologically important component of heterogeneity (Figure 1e). Variability within an entity can be arranged regularly across space with a checkerboard pattern, or irregularly, with clumped spatial arrangements. For example, the distribution of leaf litter might be expected to exhibit more clumped distributions in gaps than in the understory because of the greater wind in gaps (Figure 1e) (Miller *et al.,*1991). Spatial and temporal patterns of litter fall are likely to significantly influence the establishment of individual seedlings, as well as the density and species composition of regenerating communities (Collins, 1990; Facelli and Pickett, 1991; Vázquez-Yanes and Orozoco-Segovia, 1992).

6. Congruency of Multiple Factors in Time or Space Most work on environmental heterogeneity has focused on single factors. However, the performance of individuals and populations is dependent on multiple environmental factors. Patterns of spatial and temporal variations of single factors may overlay with one another and result in higher or emergent levels of heterogeneity. For some combinations of environmental factors, patterns may vary in concert, as the status of one factor biophysically influences the status of another factor (e.g., air temperature and relative humidity). In other cases, variations between two factors may be quite orthogonal (e.g., wind speed and soil nutrition). Figure 1f shows three hypothetical patterns of daily PFD availability in different locations in gaps, and the resulting temporal congruence (i.e., simultaneous occurrence) of PFD with favorable levels of other factors critical for carbon gain. Even though seedlings receiving morning, midday, and afternoon light regimes all receive the same total amount of PFD, with equivalent peaks and frequency distributions, the temporal congruency of light availability relative to other critical factors for carbon gain (e.g., CO_2 availability and relative humidity and air temperature) differs substantially. Commonly observed midday depressions in photosynthesis are generally due to such temporal incongruencies of PFD with one or more of the foregoing factors (Schultze, 1986; Tenhunen *et al.*, 1987). The spatial congruency of multiple factors, for example, light, water, and nitrogen, is also of critical importance to tree seedling performance, as the status of one factor can influence the response to the other (e.g., Thompson *et al.*, 1988; Walters and Reich, 1989; Schulze, 1991).

III. Environmental Variation across the Gap–Understory Continuum in the Harvard Forest

Our field research on the effects of canopy gaps on environmental heterogeneity has been conducted at the Harvard Forest, Petersham, Massachusetts, U.S.A. Studies were conducted within a 4-hectare stand that developed following a clear-cutting in 1890. The stand is currently dominated by *Quercus rubra, Acer rubrum, Fagus grandifolia, Betula alleghaniensis,* and *Betula papyrifera,* with scattered individuals of *Betula lenta, Prunus serotina, Fraxinus americana, Pinus strobus,* and *Tsuga canadensis.* Soils within the study area are primarily (Gloucester) stony loams and are well-drained, shallow, prone to drought, and relatively infertile (Sipe and Bazzaz, 1993a). During the fall of 1987, twelve gaps were created by felling carefully selected canopy trees. Trees were chosen such that six of the resulting gaps were approximately 30 m^2 and six gaps were 75 m^2. All twelve gaps were oval in shape with their long axes oriented east–west. Following felling, boles, crowns, and brush were removed from gaps with minimal disruption to the soil. All above- and belowground vegetation was also removed from plots to eliminate shoot competition. Though downed and regrowing vegetation are critical components of heterogeneity naturally experienced by regenerating tree seedlings, we eliminated this variability in order to focus on the more predictable microclimatic patterns generated by canopy gap sizes, shapes, and diurnal and seasonal patterns. To investigate microclimatic variation within gaps, 1.5×2.0-m subplots were located in the northeast, northwest, southeast, southwest, and center regions. Plots were arranged in an identical spatial pattern in the understory.

Fifteen portable microclimate stations were designed and constructed to quantify physical aspects of the gap–understory continuum, including photosynthetic photon flux (PFD), air temperature, soil temperature at two depths (1 and 15 cm), vapor pressure, and wind speed (Sipe and Bazzaz, 1994a). All stations were connected to Campbell Scientific 21X field data loggers that enabled the simultaneous sampling of all fifteen stations. The stations were deployed such that each of the five plots in a large gap, a small gap, and an understory patch within a given block were simultaneously recorded. The standard sampling program scanned all sensors at 10-sec intervals and stored 10-min means. Microclimate was recorded for 3 to 5 days every other week throughout the growing seasons of 1988 and 1989. In the following we highlight some of the general patterns we have observed, focusing primarily on light and air and soil temperature, and also on some recent measures of CO_2 variation obtained in the same forest (Bazzaz and Williams, 1991).

A. Light (PFD)

An important feature distinguishing microenvironments along the gap–understory continuum are the seasonal and diurnal timing and movements of direct beam radiation. Seasonal maps of direct beam radiation isoclines measured both empirically and using simple trigonometric models reveal that north plots in large gaps at this latitude receive direct beam radiation for approximately 7 months of the year, including the entire growing season, whereas in small gaps, north plots receive only 3–4 weeks of direct beam radiation centered around the summer solstice (Figure 2). In contrast, the south edges of both large and small gaps, as well as all understory microsites, receive no direct beam radiation except for occasional sunflecks (Sipe and Bazzaz, 1994a). Differences in the duration of direct beam radiation along the north–south axes of gaps results from the daily course of the sun through the southern sky at this latitude; seasonal direct beam patterns, and therefore the amount of regular north–south variation in direct beam radiation within gaps, differ markedly with latitude (e.g., Canham, 1988; Poulson and Platt, 1989;

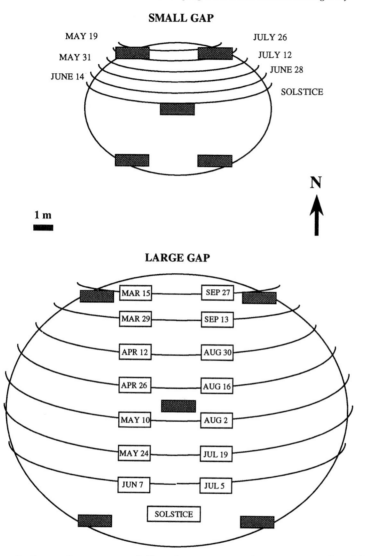

Figure 2 Seasonal movement of direct beam arcs at solar noon in small and large gaps, based on trigonometric calculations and field mapping. The drawings are idealized somewhat to represent the average patterns for gaps of each size. (From Sipe and Bazzaz, 1994a.)

Canham *et al.*, 1990). Superimposed on seasonal patterns are differences within and between patches in the diurnal movements of direct beam radiation (Figure 3). For example, on clear days near the summer solstice, north plots in large gaps receive direct beam radiation two to three times longer during the day than those in small gaps. The diurnal east–west

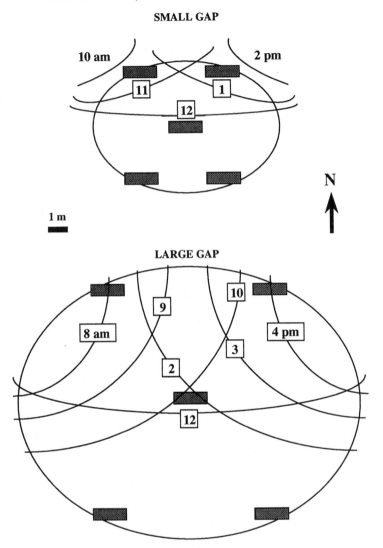

Figure 3 Diurnal movement of direct beam arcs near summer solstice for small and large gaps. Drawings are idealized to represent the average for gaps of each size. Hours are solar time. (From Sipe and Bazzaz, 1994a.)

movement of the sun also results in a phase shift in the timing of direct beam patches of light. In large gaps, northeast plots receive direct beam radiation 2.0–2.5 hr later than northwest plots; this phase shift is reduced to 1.5–2.0 hr in small gaps. While northeast and northwest plots within large and small gaps receive approximately similar daily total PFD, the differences in the absolute timing of high light result in different degrees of temporal congruencies of PFD with other critical resources that also vary diurnally (e.g., temperature, CO_2, and humidity; see the following discussion).

The interaction of seasonal and diurnal spatial and temporal variation in direct beam radiation, in combination with a smaller amount of variation between microsites in diffuse light, results in an extremely broad gradient of total daily PFD across the gap–understory continuum (Canham, 1988; Canham *et al.*, 1990). On clear days near the solstice, total daily PFD varies more than 30-fold, ranging from 0.9 mol day-1 in some understory microsites to 29.0 mol day-1 along the northern edges of large gaps (Figure 4). Much of this spatial variation occurs within gaps, with values within large gaps ranging from 29.0 mol day-1 in the north to 6.29 mol day-1 in the south. While the ranges and (absolute) variance in total PFD are greatest in large gaps and least in understory, when this spatial variation between plots is scaled by the average (e.g., % CV), relative variation (% CV) of small gaps > understory > large gaps. The average time courses of PFD availability across the day were also more variable in gaps. Absolute variances of 10-min averages on clear days were 40-fold greater, on average, in large gaps than in the understory; however, relative variation (% CV) of diurnal 10-min means was greater in the understory than in gaps. On overcast days, the total amount PFD, the overall breadth of the total PFD gradient, and the overall spatial and temporal variability were all significantly reduced.

B. Air and Soil Temperature

In many respects, spatial and temporal variation in air and soil temperatures is qualitatively similar to PFD variation. Mean air temperature, for example, is highly correlated with mean PFD across the gap–understory continuum (Figure 5). Across the continuum, air temperatures are generally greater than soil surface temperatures (1 cm) which are in turn greater than deep soil temperatures for the entire 24-hr period (Figure 6). Soil surface temperatures show more spatial variation across the gradient than either air or deep soil temperatures. As with PFD, much of the spatial variation in temperature occurred within gaps, with little spatial variation in air and soil temperature in the understory. However, unlike PFD, both the absolute variation (diurnal range and variance of 10-min averages) and the % CV in soil temperatures are greater in gaps than in the understory.

C. Carbon Dioxide

Although global change issues have focused attention on the importance of carbon dioxide as a critical plant resource, research on plant responses

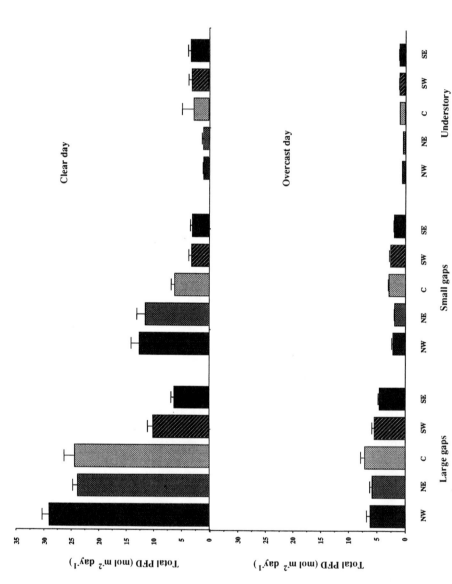

Figure 4 Variation in total photosynthetic photon flux density across the gap–understory continuum in the Harvard Forest on a representative clear (mid-June) and overcast (mid-August) day. Error bars represent one standard error. (Data from Sipe and Bazzaz, 1994a.)

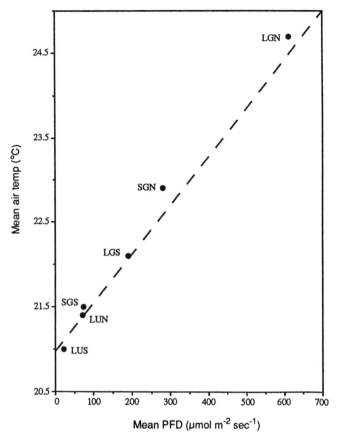

Figure 5 Relationship between mean daily irradiance (PFD) and air temperature on clear days across the gap–understory continuum. Microsite codes are: LGN = large gap north, SGN = small gap north, LGS = large gap south, SGS = small gap south, LUN and LUS = understory. (Data from Sipe and Bazzaz, 1994a.)

to CO_2 has largely focused on predicted increases in average global atmospheric levels (Bazzaz, 1990). Our research at the Harvard Forest has demonstrated that finer-scale spatial and temporal variation in CO_2 concentrations within forest stands can be quite substantial, and also quite different from the well-mixed atmospheric concentrations above the forest canopy (Wofsy *et al.*, 1993), or as measured in remote areas such as Mauna Loa (Keeling, 1986). Working largely in understory microsites within a maple–oak-dominated forest, Bazzaz and Williams (1991) sampled seasonal variation in CO_2 concentrations along vertical profiles. Average CO_2 concentrations show a marked seasonal pattern (Figure 7). At or near the soil surface, CO_2 concentrations were highest in midsum-

Large Gap

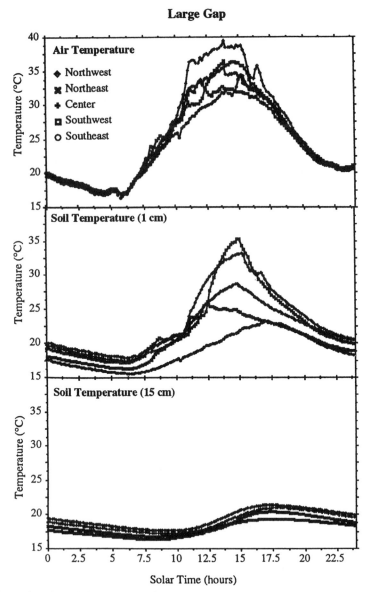

Figure 6 Diurnal time courses of air and soil temperature for a representative clear day near the solstice in a large gap and in the adjacent understory in the Harvard Forest. Air temperatures were measured at 25 cm above ground. (Data from Sipe and Bazzaz, 1994a.)

Understory

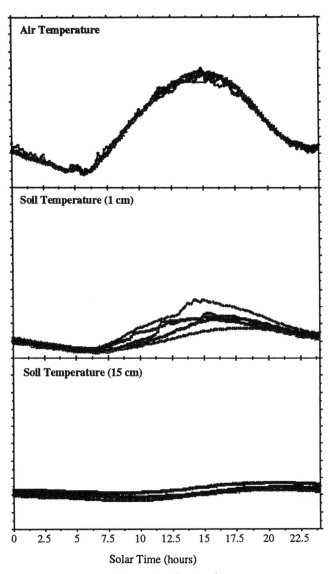

Figure 6 *(Continued)*

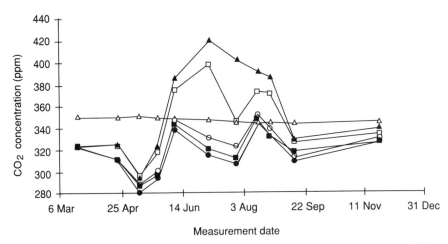

Figure 7 CO_2 concentrations in 1985 at Harvard Forest, average over 3- to 7-day periods at various times during the growing season and at five indicated heights above ground: ▲, 0.05 m; □, 0.20 m; ■, 3.00 m; ○, 6.00 m; ●, 12.0 m. Data from Mauna Loa CO_2 observatory (△) are given on the same axis for comparison. The date given is the midpoint of the sampling interval. (From Bazzaz and Williams, 1991.)

mer, probably due to microbial respiration associated with rapid decomposition and root activity at higher temperatures. With increasing height, average CO_2 concentrations decreased and exhibited a curious two-peaked curve during the growing season. In another study, Thomas and Bazzaz (1993) compared diurnal variation in CO_2 concentrations in both understory and gap environments at three heights (average gap size = 175 m²). At both 0.1 and 0.5 m, CO_2 concentrations in gaps were significantly higher across most of the day than in adjacent understory plots (Figure 8). However, at 2.0 m above the soil surface, these microsite differences disappeared, probably due to the greater mixing of air. The magnitudes of these horizontal and vertical differences in CO_2 concentrations are substantial, and within the range reported to have significant effects on tree seedling performance (Eamus and Jarvis, 1989; Bazzaz, 1990; also see the following). The consequences to tree seedling regeneration of this spatial (vertical and horizontal) and temporal (seasonal and diurnal) variation in the CO_2 availability, and CO_2's spatial and temporal congruency with other environmental factors across the gap–understory continuum, is an exciting area of research that has not yet received sufficient attention.

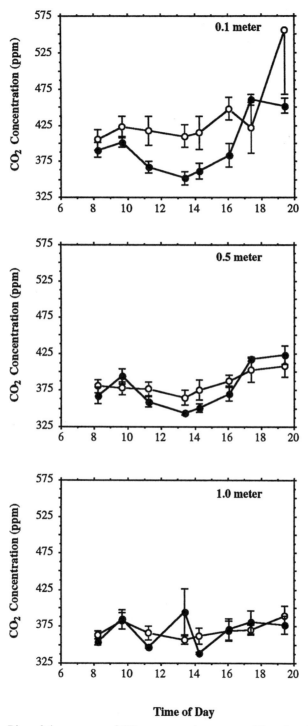

Figure 8 Diurnal time courses of CO_2 concentrations on gap (○) and understory (●) environments in the Harvard Forest, measured at 0.1, 0.5, and 1.0 m above the soil surface. Error bars represent one standard error. (Data from Thomas and Bazzaz, 1993.)

IV. Responses of Seedlings to the Gap–Understory Continuum

The Birch–Maple System: Our research on the responses of tree seedlings and saplings to environmental heterogeneity across the gap–understory continuum has largely focused on seven co-occurring birch (*Betula*) and maple (*Acer*) species. We chose these species as a model system for a number of reasons. First, the seven species vary in successional status and shade tolerance, ranging from the pioneer, very intolerant gray birch (*B. populifolia*) to the late-successional, shade-tolerant sugar maple (*A. saccharum*). Second, the two genera also represent a contrast in their dependence on advance regeneration (*Acer*) versus postdisturbance regeneration (*Betula*). Finally, the use of closely related species allowed us to assess the presence of evolutionary constraints and opportunities within both genera while simultaneously minimizing the potential sources of variation. Within and between these genera there are significant differences in the physiological, architectural, and life history characteristics, including size, longevity, potential growth rates, and reproductive traits. Some of these and other characteristics of the species are summarized in Table I. Next we compare the responses of birch and maple seedlings to aspects of the environmental variation across the gap–understory continuum, namely, responses to (a) cumulative quantities of environmental factors; (b) aspects of the time courses of environmental factors; (c) critical thresholds of environmental factors; and finally (d) the responses to multiple interacting factors.

A. Growth Responses to Cumulative Quantities of Environmental Factors

1. Responses to Total PFD   As described earlier, the gradient in total PFD across the gap–understory continuum is very broad. In two related experiments, the growth of first-year birch and maple seedlings at two extreme and one intermediate point along the total PFD gradient was compared to assess niche partitioning in species responses to total PFD (Sipe and Bazzaz, 1994a,b). While all seven species showed the least amount of growth in the extreme low light of the understory, the two most tolerant species, sugar and striped maple, exhibited more growth in this microsite than the other five species (Figure 9). Furthermore, species differed substantially in their response to light levels within large gaps. The relatively intolerant birches consistently exhibited greatest growth in the exposed north edges of large gaps, particularly the least tolerant gray birch, whereas the more tolerant maples grew largest in the intermediate light levels of the south sides of large gaps (Figure 9). The relatively small size of red maple seedlings across the entire gradient

Table I Some Life History and Autecological Characters of Birch and Maple Species Studied at Harvard Forest

	Gray birch	White birch	Black birch	Yellow Birch	Red maple	Striped maple	Sugar maple
Common names	Gray birch	White birch	Black birch	Yellow Birch	Red maple	Striped maple	Sugar maple
Scientific names	*Betula populifolia*	*Betula papyrifera*	*Betula lenta*	*Betula alleghaniensis*	*Acer rubrum*	*Acer pensylvanicum*	*Acer saccharina*
Generic subgroup	Albae	Albae	Costatae	Costatae	Rubra	Macrantha	Saccharina
Shade tolerance	Very intolerant	Intolerant	Intermediate	Intermediate-tolerant	Tolerant	Very tolerant	Very tolerant
Successional status	Very early	Early	Mid	Mid-late	Mid	Mid-late	Late
Water requirement	Broad	Broad-dry	Broad-dry	Moderate-high	Broad	Moderate	High
Longevity (yrs: Max, Mean)	50+, 30+	200+, 75+	265+, 75+	300+, 90+	150+, 90+	60+, 25+	400+, 150+
Mean mature size							
Height (m)	8+	15+	15–20	15–20	18+	8+	20+
Diameter (cm)	20–25	25–50	30–60	60+	45–75	15–25	50–90
Minimum fruiting age (yrs)	3	15	40	40	4	10	40–60
Mean seed weight (mg)	0.11	0.26	0.46	0.80	40.0	20.0	74.0
Dispersal time	Sept-December	August-Spring	Sept-December	Sept-Spring	April-July	Oct-November	Oct-December
Frequency of good fruiting years	1	1–2	1–2	1–2	1–2	?	3–7
Sprout occurrence	Profuse	Profuse	Common	Rare	Profuse	Profuse	Common

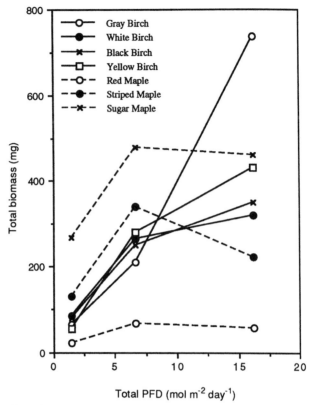

Figure 9 Growth responses of 1-year-old birch and maple seedlings to PFD variation along the gap–understory continuum in the Harvard Forest. (Data from Wayne and Bazzaz, 1994c, and Sipe and Bazzaz, 1994d.)

largely results from the fact that seedlings of this species do not get established until late spring when their seeds mature (Canham and Marks, 1985). These results suggest that across the broad gradient of total PFD in temperate forests, sympatric species (even cogeners) show some evidence for partitioning the environment, a result that differs from recently published studies in the tropics (e.g., Denslow *et al.*, 1990).

2. Responses to Light Quality Coincident with variation in total PFD along the gap–understory continuum are changes in the ratio of red : far-red light (e.g., Lee, 1987). To test whether the very intolerant gray birch, which rarely establishes successfully in the understory, and the intermediate tolerant yellow birch, which regularly establishes in the understory, exhibited differing responses to the ratio of red : far-red light, we grew seedlings of both species in a controlled glasshouse environment at approx 50% full sun, filtered by either neutral shadecloth (R : Fr = 1.0) or deep-dyed polyester (R : Fr = 0.6; GramColor®, Hollywood, Calif.). As has been reported in other studies (e.g., Morgan and Smith, 1979; Kwesiga and Grace, 1986; Turnbull, 1991), the species that commonly experiences understory conditions (i.e., yellow birch) showed much less photomorphogenetic response to low R : Fr than gray birch (Figure 10). In contrast, gray birch seedlings growing under reduced R : Fr regimes showed significant responses in many sun–shade characteristics, including increased stem height, internode length, petiole length, and reduced leaf area ratio. These responses are critical in gap environments where severe competition puts a premium on height

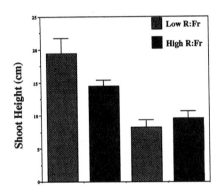

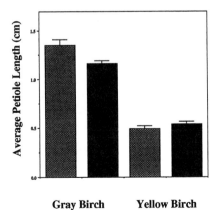

Gray Birch Yellow Birch

Figure 10 Responses of pioneer gray birch (*Betula populifola*) and later-successional yellow birch (*B. alleghaniensis*) seedlings to high (1.0) and reduced (0.6) red : far-red ratios. Error bars represent one standard error. (Data from Ackerly and Wayne, 1993.)

growth and light preemption (Canham and Marks, 1985). Thus, in addition to different responses to variation in light quantity, co-occurring closely related species can also show very different sensitivities to light quality, with one species responding to it and another not.

3. Responses to a Soil Moisture Gradient   Although soil moisture availability was not measured across the gap–understory continuum in the Harvard Forest, personal observations and soil moisture data from other studies in forest gaps (e.g., Geiger, 1965; Becker *et al.*, 1988; Ashton, 1992) suggest that regular gradients exist, with decreasing amounts of available surface soil moisture ranging from understory to exposed gap microsites. Controlled environment studies of the responses of gray birch and red maple, two species with broad moisture requirements, to an experimental soil moisture gradient ranging from chronically dry to flooded revealed a very different growth response pattern than we observed on light quantity and quality gradients (Miao *et al.*, 1991). Gray birch and red maple showed similar total biomass responses to this gradient, both peaking in the mesic condition (Figure 11). However, the allocational and architectural patterns underlying these responses dif-

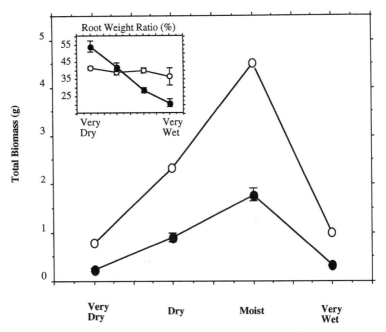

Figure 11   Growth and allocational responses of gray birch (○) and red maple (●) seedlings to an experimental soil moisture gradient. (Data from Miao *et al.*, 1991.)

fered markedly between species. Gray birch exhibited few allocational or architectural changes across the entire gradient, whereas red maple greatly varied both root weight ratio and the ratio of leaf area to root mass (Figure 11). Some data suggest that the lack of allocation changes in gray birch may have been compensated for by alterations in tissue water relations (Morse *et al.*, 1994). At the dry end of gradient, plants showed a significant decrease in osmotic potential and a significant increase in tissue elastic moduli. The reduction in tissue osmotic potential may have helped shift leaf zero turgor point to lower water potentials, while decreases in elasticity may have enabled plants to improve the soil–plant water potential gradients for smaller changes in leaf water content. These results suggest that species can occupy similar positions along the continuum by deploying very different suites of traits. They also caution against assuming that growth or survivorship responses parallel physiological or architectural differences.

4. Growth Responses of Seedlings to Elevated CO_2 Growth responses of seedlings of the four birch species to two atmospheric CO_2 concentrations (380 versus 690 μL L^{-1}) were compared in a glasshouse study conducted at high light (900 μmol m^{-2} s^{-1}). While all species showed increases in response to elevated CO_2, gray birch, the most shade intolerant, exhibited the greatest responsiveness (Rochefort and Bazzaz, 1993), much as it did in response to total PFD and light quality. In another study conducted at lower light levels (PFD = 520 ± 30), seedlings of seven co-occurring tree species (beech, white birch, black cherry, white pine, red maple, sugar maple, and hemlock) also showed variability in their responsiveness to CO_2, however, in this study, shade-tolerant species such as beech, sugar maple, and hemlock showed a greater responsiveness than did intolerants (Bazzaz *et al.*, 1990). Species differences in responsiveness to CO_2 suggest that their relative regenerative/competitive ability along the gap–understory continuum may be partially determined by CO_2 concentrations near the soil surface, however, it is clear that the status of other factors influences responses to CO_2.

B. Aspects of the Time Courses of Environmental Factors

Because of the complex temporal and spatial patterns in resource availability across the gap–understory continuum, seedlings located in different positions in gaps can receive similar total amounts of a resource, but in potentially very different temporal or spatial patterns of availability. In two related studies, we examined the effects of changes in diurnal time courses of PFD, independent of total PFD, on the physiology, growth, and architecture of birch seedlings (Wayne and Bazzaz, 1993a,b). In the first experiment, gray and yellow birch were grown in two sets of contrasting diurnal light regimes, "gaps" and "shadehouses," and at four levels of total integrated PFD (approximately 12, 27, 50, and 70% of full sun) in an experimental garden. In gaps, seedlings received relatively more heterogeneous diurnal light regimes, with midday full-sun peaks lasting between 40 and 280 min on sunny days, depending on gap size. In shadehouses, seedlings received similar daily total and average PFD as in gaps, but received no midday direct sunlight, and an overall more uniform temporal distribution of light. Differences in the daily time courses of light availability, independent of total PFD, significantly affected growth of seedlings. Both species grew significantly larger in shadehouse environments, but gray birch was more responsive to diurnal time courses, especially at lower total PFD (Figure 12). Diurnal time courses of light availability also significantly influenced seedling sun–shade responses. For most physiological and morphological characters, responses to increasing total PFD were generally more sensitive in gap than in shadehouse regimes. These results suggest that time courses and distributions of PFD availability, independent of cumulative totals, can significantly affect the performance of tree seedlings, and that experimental studies employing uniform light regimes that incorporate little of the temporal variability experienced by seedlings in natural gaps may underestimate plasticity within tree species, and also the potential for species niche differences and coexistence (Wayne and Bazzaz, 1993a).

In a comparison study, we attempted to further decouple components of gap light regimes and investigated whether the diurnal timing of high light availability (sun patches), independent of totals, peaks, and

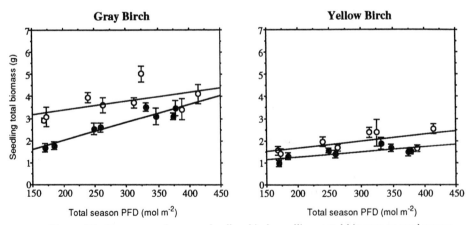

Figure 12 Responses of gray and yellow birch seedlings total biomass to total season integrated PFD in experimental gap (●) and shadehouse (○) light regimes. Regressions were fit to all data points and superimposed over means for eight light levels. Bars represent one standard error. (From Wayne and Bazzaz, 1993a.)

frequency distributions of light, influences the physiology and growth of four birch species (Wayne and Bazzaz, 1993b). In this study, seedlings were grown for 2 years along the east or west sides of experimental gaps and at two moisture levels. Seedlings positioned in the west received sun patches earlier in the day than those in the east, when environmental conditions for carbon gain were generally more favorable (e.g., air and leaf temperatures were lower, and relative humidity higher during morning sun patches). Diurnal time courses of leaf-level carbon gain reveal that seedlings positioned along the west edges of gaps fixed more carbon earlier in the day than those in the east and, in many cases, peak net photosynthetic rates were greater for west-positioned seedlings (Figure 13). Leaf-level daily carbon gain (integrated across the day) in Year 2 was also greater for west- than east-positioned plants, and for the less shade tolerant gray birch, differences between west and east seedlings were greatest at lower soil moisture levels. However, despite these (small) effects on leaf gas exchange, the timing of high light availability, and its temporal congruence with other factors critical to carbon gain, had no significant effects on first- or second-year seedling biomass. This lack of integrated growth responses to the timing of high light availability in garden studies parallels patterns of birch and maple seedling growth along the east and west edge of gaps in the Harvard Forest (Wayne and Bazzaz, 1993c; Sipe and Bazzaz, 1993c). Together, these results suggest that responses of birch and maple seedlings to controlled variations in the timing of high light availability, and its congruency with other diurnally varying factors critical to carbon gain, are generally much smaller than responses to variations in other components of daily light regimes such as total integrated PFD or light quality.

C. Threshold Effects

Threshold phenomena that occur chronically can have significant effects on the dynamics of forest regeneration. Figure 14 shows the relationship between seedling survivorship of three species of maple and the diurnal duration of supraoptimal temperatures (Sipe and Bazzaz, 1994c). Survivorship of all species dropped markedly in microsites that were exposed to high temperatures for more than 30% of the day, and these survivorship patterns were clearer than those of survivorship and mean daily temperature. Species differed in their responses to high-temperature exposure, with red maple being least sensitive to chronic high temperature. This observation may partially explain the overall greater survivorship of advanced regenerant red maple seedlings in the exposed north sides of large gaps (Sipe and Bazzaz, 1994c).

The exposure of tree seedlings to extreme environmental events, even for short periods of time, can also have very significant and long-lasting

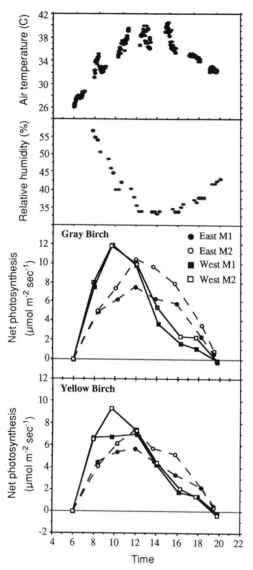

Figure 13 Diurnal time course of environmental conditions and second-year seedlings gas exchange of gray and yellow birch seedlings growing along the east and west sides of experimental gaps. M1 and M2 refer to low- and high-moisture treatments, respectively. Each point represents the average of four seedlings. Bars are one standard error. (From Wayne and Bazzaz, 1993b.)

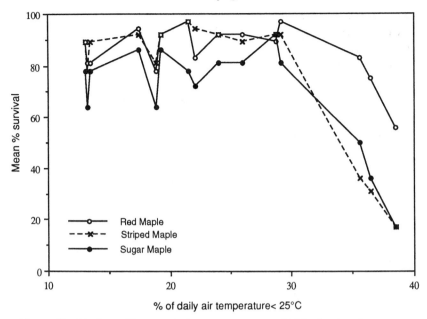

Figure 14   Maple seedlings survival versus percentage of 12-hr day that air temperature exceeded 25°C across the gap–understory continuum in Harvard Forest. Each of the 15 points represents a unique location along the gap–understory continuum. (Data from Sipe and Bazzaz, 1994c.)

consequences for performance. Across the gap–understory continuum, air temperatures in some microsites regularly reach near 40°C for short periods of the day (Figure 6). This important component of variation is overlooked when only average temperatures are considered. In a controlled environment study, we specifically investigated the consequences of short-term exposures to extreme temperature events on both the short-term physiological and longer-term carbon gain responses of two birch and one maple species (Bassow *et al.,* 1994). Seedlings of gray birch, yellow birch, and striped maple were raised at high light in a day/night temperature regime of 28/22 ± 2°C. During midsummer, a subset of seedlings was exposed to just one hot day (45°C). Stomatal conductance rates were substantially depressed in the heat-stressed plants the day following the heat shock event and showed no signs of recovery 2 days after the shock. There were also substantial amounts of leaf necrosis following heat shocks, further reducing carbon gain potential. Seedling total biomass 105 days after the extreme heat event was substantially lower in shocked versus unshocked plants, and these differences were greater for gray birch and striped maple than for yellow birch. Thus transient extreme events can have long-lasting consequences not only to individual seedling performance, but also to species' relative performances, therefore potentially influencing species' relative competitive abilities in regenerating communities.

D. Interactions between Multiple Factors

Though it is useful to isolate individual environmental factors and study their effects on tree seedling performance, seedlings are simultaneously exposed to many environmental factors each with a complex spatial and temporal pattern of variation. Increasing evidence suggests that the status of one factor greatly influences seedling responses to other factors, and in order to make significant progress in our understanding of the physiological ecology of regeneration, we must acknowledge and face this complexity (Chapin *et al.*, 1987; Mooney *et al.*, 1991). For example, Crabtree and Bazzaz (1993) studied the responses of birch seedlings to different forms of nitrogen, which are currently changing in the Harvard Forest as a result of atmospheric nitrogen deposition (Aber *et al.*, 1989). Whereas birch seedlings showed no response to nitrogen form at low light levels, black birch showed a clear preference for ammonium at high light (Figure 15).

In a fully crossed design varying CO_2, light, and nutrients, Bazzaz and Miao (1993) investigated the growth responses of six Harvard Forest species: gray and yellow birch, red and striped maple, red oak, and white ash. While elevated CO_2 significantly stimulated the growth of all seedlings, the degree of CO_2-induced growth enhancement was largely contingent on the availability of light and nutrients and on species identity

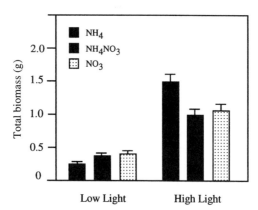

Figure 15   The response of 1-year-old black birch seedlings to form of nitrogen addition at two light levels. Error bars are one standard error. (From Crabtree and Bazzaz, 1993.)

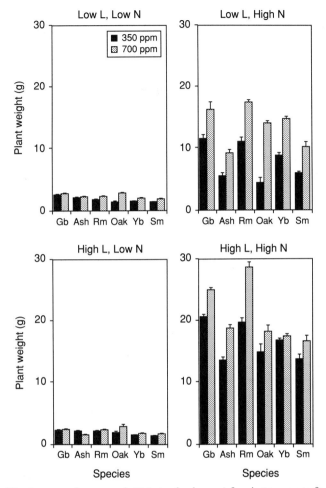

Figure 16 Average plant growth (±1 standard error) for six temperate forest species grown under 350 or 700 ppm CO_2 with various light and nutrient treatments for 175 days. Species are depicted from left to right in order of increasing shade tolerance. Gb, gray birch; Ash, white ash; Rm, red maple, Oak, red oak; Yb, yellow birch; Sm, striped maple. (Data from Bazzaz and Miao, 1993.)

(Figure 16). For example, relatively shade-intolerant species showed a greater response to CO_2 at high nutrients, whereas more shade-tolerant species showed greater CO_2 enhancements at low light availability. More multifactor experiments are needed to develop a better understanding of how the status of one factor alters the responses of a seedling to other factors. Such controlled environment experiments that simultaneously vary multiple factors should begin to give us better insight into the responses of naturally established seedlings to the complex patterns of light, CO_2, nutrients, and other factors across the gap–understory continuum, and the differential responses of species to this heterogeneity.

V. Environmental Heterogeneity and Tree Seedling Plasticity

Our descriptions of both spatial and temporal environmental variability across the gap–understory continuum suggest that pioneer species that regularly occupy disturbed gap environments will experience more environmental heterogeneity on average than later-successional species that regularly occupy understory microsites. It has been hypothesized that pioneer species that occupy disturbed, heterogeneous sites should be expected to exhibit greater phenotypic flexibility and broader niches than those regularly occupying less variable, later-successional environments (Bazzaz, 1979, 1983, 1987). It has also been suggested that the types of phenotypic plasticity, in addition to the quantity, may also vary regularly with habitat preferences of species. For example, Grime and colleagues (1986) suggest that species of productive environments that experience high levels of competition respond to environmental heterogeneity primarily through plasticity in the production and development of root and shoot modules (i.e., morphological plasticity), whereas species common to relatively unproductive environments generally acclimate to variation via physiological plasticity within existing modules.

Numerous studies with herbaceous species (reviewed in Bazzaz, 1987) and some studies with tree seedlings have reported greater physiological and morphological plasticity in species common to recently disturbed environments (e.g., Bazzaz and Carlson, 1982; Chazdon and Field, 1987; Bongers and Popma, 1988; Koike, 1988; Ramos and Grace, 1990; Strauss DeBenedetti and Bazzaz, 1991), though there are exceptions (e.g., Walters and Field, 1987; Sims and Pearcy, 1989; Turnbull, 1991). Detailed studies of the responses of four co-occurring birch species varying in shade tolerance and successional status to an experimental gradient of three gap sizes have allowed us to compare both the quantity and types of plasticity exhibited by species varying in successional status within a genus (Table II) (Wayne, 1991). Of the 18 physiological, morphological, and architectural characters investigated, the shade-intolerant pioneer gray birch was most plastic for 8 traits, whereas the later-successional yellow birch was most plastic for only 2 characters. When the earlier-successional gray and white birch were grouped together and compared to the two later-successional black and yellow birch, the pattern is more dramatic, with pioneers exhibiting greater plasticity in 15 of the 18 traits.

However, our data on birch plasticity do not suggest clear differences in the types of traits varied by pioneer and nonpioneer species. For example, three of the four species were very plastic with respect to branch production and dark respiration, and all species exhibited large variation in average leaf size across the three gap sizes. In contrast, all species showed relatively little variation in chlorophyll a/b ratios, leaf weight ratios, and maximum photosynthetic rates. Thus for birches, there do not seem to be clear differences in the types (i.e., physiological versus morphological) of traits varied by pioneer and later-successional species, though individual species do vary to some degree with respect to the traits that are most plastic. Additional studies with other groups of species and other forest ecosystems are needed to better evaluate whether species from early- and later-successional environments regularly adapt to environmental heterogeneity with different suites of traits.

VI. Conclusions

Disturbance results in increased heterogeneity at many scales of organization: Our work in Harvard Forest suggests that canopy gap disturbances generate significant amounts of ecologically relevant heterogeneity at the three ecological scales considered. At the landscape level, higher resources in disturbed patches (e.g., PFD and CO_2) result in both greater productivity and differences in species dominance, a result consistent with other studies conducted in temperate forests (e.g., Borman and Likens, 1979; Phillips and Shure, 1990). Incorporation of this patchiness into landscape-level models of ecosystem processes, such as carbon or water flux, may improve predictability (Forman and Godron, 1986; Shugart *et al.*, 1986; Bazzaz, 1993). Our studies also reveal a great deal of variation within patches, suggesting that for many ecological questions, classification of regeneration environments as either gaps or understory may be an undesirable oversimplification (e.g., Lieberman *et al.*, 1989). For most of the environmental variables we measured in this study, much of the total variation occurring across the entire gap–understory continuum exists within gaps, particularly in large ones, and this within-gap heterogeneity is sufficient to allow differentiation of species responses. Thus even though gaps make up a small percentage of most temperate forest ecosystems (e.g., Runkle, 1985), they contribute a disproportionate amount of environmental heterogeneity to the system, and understanding the finer-scale variation within gaps may be critical for understanding issues related to species coexistence and community organization. Finally, at the scale perceived by individual seedlings, there also appears to be regular variation across the gap–understory continuum in the amount

Table II Comparative Physiological and Morphological Plasticity of Seedlings of Four Sympatric Birch Species Varying in Successional Status and Shade Tolerance and Grown in Experimental Forest Gaps of Three Sizes.[a]

Character	All species averaged	Relative plasticity ranking	Gray birch	White birch	Black birch	Yellow birch	Intolerant average	Tolerant average
Leaf area ratio	22.6	5	24.6	23.8	19.6	22.6	24.2	21.1
Leaf weight ratio	2.30	17	3.2	0.9	3.3	1.6	2.1	2.4
Support weight ratio	15.4	11	15.0	21.9	14.6	10.0	18.5	12.3
Root weight ratio	12.0	12	12.3	12.9	14.2	8.8	12.6	11.5
Total leaf area	21.9	6	17.1	31.1	17.1	22.3	24.1	19.7
Specific leaf weight	20.6	7	26.6	20.0	15.7	20.0	23.3	17.9
Height	18.2	8	16.6	23.5	15.4	17.2	20.1	16.3
Leaf number	16.86	9	37.23	4.98	7.69	17.53	21.1	12.6
Branch number	50.3	1	97.44	42.74	15.18	45.83	70.9	30.5

Leaf dark respiration	38.4	2	57.2	35.3	56.3	5.0	46.3	30.6
Initial slopes	6.4	14	9.9	6.9	5.5	3.3	8.4	4.4
Maximum net photosynthesis	6.3	15	6.6	9.8	5.5	3.2	8.2	4.4
Nitrogen/g	15.4	10	22.3	14.0	13.1	12.3	18.1	12.7
Nitrogen/area	5.8	16	7.0	10.1	5.2	7.1	8.5	6.2
Chlorophyll/g	24.0	4	17.7	39.2	24.1	14.8	28.5	19.5
Chlorophyll/area	8.1	13	7.9	8.2	3.8	12.6	8.0	8.2
Chlorophyll a/b ratio	2.0	18	2.0	1.1	2.0	2.8	1.5	2.4
Average leaf size	33.6	3	43.0	30.4	25.1	35.7	36.7	30.4
Number of traits in which species are ranked most plastic			8	6	2	2	16	3

[a] Plasticity was quantified using the coefficient of variation (%) of mean performance across the three gap sizes. Characters with the lowest ranking exhibited the greatest plasticity.

of heterogeneity. Most factors show more diurnal and seasonal variation within gaps, particularly in the exposed north microsites, than in the understory. Although data on within-seedling spatial heterogeneity are not yet available, it is probably likely that it is also more variable in disturbed habitats, particularly if tree falls are accompanied by uprooting, which often drastically alters variation in soil depth, soil profile organization, and moisture and nutrient availability (e.g., Orians, 1983; Beatty and Stone, 1986; Brandani *et al.*, 1987). As discussed earlier, these patterns in the amounts of within-seedling variation may be related to the amounts and types of plasticity exhibited by species that dominate in these different microsites.

Heterogeneity itself is heterogeneous: The nature of environmental heterogeneity across the gap–understory continuum is itself complex. Although some resources, conditions, and signals apparently vary independently (e.g., wind and soil mineral nutrition), many factors vary as complexes (e.g., PFD, air and soil temperature, humidity). The presence of environmental complexes suggests caution in attributing particular demographic or physiological responses of tree seedlings to particular environmental factors. For example, the typically observed increase in specific leaf weight (SLW) in exposed environments is often considered a response to higher levels of PFD. However, SLW is also known to vary with air temperature, soil and atmospheric moisture availability, nutrient availability, and wind, all of which may vary simultaneously, and therefore potentially confound responses to light. Furthermore, any particular environmental factor can vary spatially and temporally in a variety of ways in addition to average or total cumulative levels. Focusing solely on correlations of seedling responses with cumulative totals of environmental factors may give a limited view of the heterogeneity that species have regularly experienced throughout their evolution, and the types of heterogeneity that ecologically differentiate their responses and distributions across the gap–understory continuum.

Species both do and do not differ at various levels of organization in their responses to heterogeneity across the gap–understory continuum. As discussed in the preceding, early-successional species were generally physiologically and morphologically more responsive (plastic) to measured heterogeneity than were later-successional species. Some species also differed significantly in their patterns of growth and survivorship across the gap–understory continuum, supporting the theory that disturbance-induced heterogeneity facilitates species coexistence and the maintenance of diversity. However, some species exhibited very similar growth and survivorship responses to environmental variation, but differed greatly in the physiological and morphological traits deployed to achieve equivalent responses. Thus different species may be able to occupy similar regions of the gap–understory continuum, but do so by deploying different suites and combinations of traits. Because individual traits generally do not vary independently of one another, studies attempting to interpret the adaptive value of particular physiological or morphological traits will be most useful when multiple traits are investigated simultaneously, and when species comparisons are made within phylogenetic lineages, where on average closely related species share common constraints (e.g., Givnish, 1987; Chazdon, 1991). Combining studies of species phylogeny, patterns of character evolution, and ecological responses to disturbance-induced heterogeneity is an exciting and potentially very fruitful area of future research.

Acknowledgments

We thank David Ackerly, Glenn Berntson, Charlie Canham, Beth Farnsworth, Elizabeth Newell, and Timothy Sipe for very helpful comments on earlier drafts of this manuscript. This research was partially funded by grants from the National Science Foundation (BSR-86-11308 and LTER 525337591/2), Department of Energy (DE-FGO2-84-ER60257), and NIGEC 901214-HAR.

References

Aber, J. D., Nadelhoffer, K. J., Steudler, P., and Mellilo, J. M. (1989). Nitrogen saturation in northern forest ecosystems. *BioScience* **39**, 378–386.

Ackerly, D. D., and Wayne, P. M. (1993). The response of tropical and temperate tree seedlings to moderate changes in light quality. In preparation.

Allen, T. F. H., and Starr, T. B. (1982). "Hierarchy: Perspectives for Ecological Complexity." Univ of Chicago Press, Chicago.

Ashton, P. M. S. (1992). Some measurements of the microclimate with a Sri Lankan tropical rainforest. *Agric. For. Meteorol.* **59**, 217–235.

Bassow, S. L., McConnaughay, K. D. M., and Bazzaz, F. A. (1994). How will extreme temperature fluctuations affect temperate trees in an elevated CO_2 atmosphere? *Ecol. Appl.* (in press).

Bazzaz, F. A. (1979). The physiological ecology of plant succession. *Annu. Rev. Ecol. Syst.* **10**, 351–371.

Bazzaz, F. A. (1983). Characteristics of populations in relation to disturbance in natural and man-modified ecosystems. *In* "Disturbance and Ecosystems: Components of Response" (H. A. Mooney and M. Gordon, eds.), pp. 259–275. Springer-Verlag, Berlin.

Bazzaz, F. A. (1987). Experimental studies on the evolution of niche in successional plant populations. *In* "Colonization, Succession, and Stability" (A. J. Gray, M. J. Crawley, and P. J. Edwards, eds.), pp. 245–271. Blackwell, Oxford.

Bazzaz, F. A. (1990). The response of natural ecosystems to the rising global CO_2 levels. *Annu. Rev. Ecol. Syst.* **21**, 167–196.

Bazzaz, F. A. (1993). Scaling in biological systems: Population and community perspectives. *In* "Scaling Physiological Processes: Leaf to Globe" (J. R. Ehlringer and C. B Field, eds.), pp. 233–254. Academic Press, San Diego.

Bazzaz, F. A., and Carlson, R. W. (1982). Photosynthetic acclimation to the variability of the light environment of early and late successional plants. *Oecologia* **54**, 313–316.

Bazzaz, F. A., and Miao, S. L. (1993). Successional status, seed size, and responses of tree seedlings to CO_2, light, and nutrients. *Ecology* **74**, 104–112.

Bazzaz, F. A., and Sipe, T. W. (1987). Physiological ecology, disturbance, and ecosystem recovery. *In* "Potentials and Limitations of Ecosystem Analysis" (E. D. Schulze and H. Zwolfer, eds.), pp. 203–227. Springer-Verlag, Berlin.

Bazzaz, F. A., and Williams, W. E. (1991). Atmospheric CO_2 concentrations within a mixed forest: Implications for seedling growth. *Ecology* **72**, 12–16.

Bazzaz, F. A., Coleman, J. S., and Morse, S. R. (1990). Growth responses of seven major co-occurring tree species of the northeastern United States to elevated CO_2. *Can. J. For. Res.* **20**, 1479–1484.

Beatty, S. W., and Stone, E. L. (1986). The variety of soil microsites created by tree falls. *Can. J. For. Res.* **16**, 539–548.

Becker, P. E., Rabenold, E., Indol, J. R., and Smith, A. P. (1988). Water potential gradients for gaps and slopes in a Panamanian tropical moist forest's dry season. *J. Trop. Ecol.* **4**, 173–184.

Begon, M., Harper, J. L., and Townsend, C. R. (1990). "Ecology: Individuals, Populations, and Communities." Blackwell, Oxford.

Bell, G. (1992). Five properties of environment. *In* "Molds, Molecules, and Metazoa: Growing Points in Evolutionary Biology" (P. Grant and H. Horn, eds.), pp. 33–56. Princeton Univ Press, Princeton, NJ.

Benner, B. L., and Bazzaz, F. A. (1988). Carbon and mineral accumulation and allocation in two annual plant species in response to timing of nutrient addition. *J. Ecol.* **76**, 19–40.

Bongers, F., and Popma, J. (1988). Is exposure-related variation in leaf characteristic of tropical rain forest species adaptive? *In* "Plant Form and Vegetation Structure" (M. J. Werger, P. J. M. van der Aart, H. J. During, and J. T. A. Verhoeven, eds.), pp. 191–200. SPB Academic Publishing, The Hague, The Netherlands.

Borman, F. H., and Likens, G. E. (1979). "Patterns and Process in a Forested Ecosystem." Springer-Verlag, New York.

Brandani, A., Hartshorn, G. S., and Orians, G. H. (1987). Internal heterogeneity of gaps and species richness in Costa Rican tropical wet forest. *J. Trop. Ecol.* **4**, 99–119.

Canham, C. D. (1988). An index for understory light levels in and around canopy gaps. *Ecology* **69**, 1634–1637.

Canham, C. D., and Marks, P. L. (1985). The responses of woody plants to disturbance: Patterns of establishment and growth. *In* "The Ecology of Natural Disturbance and Patch Dynamics" (S. T. A. Pickett and P. S. White, eds.), pp. 197–217. Academic Press, Orlando, FL.

Canham, C. D., Denslow, J. S., Platt, W. J., Runkle, J. R., Spies, T. A., and White, P. S. (1990). Light regimes beneath closed canopies and tree-fall gaps in temperate and tropical forests. *Can. J. For. Res.* **20**, 620–631.

Chapin, F. S., Bloom, A. J., Field, C. B., and Waring, R. H. (1987). Plant responses to multiple environmental factors. *BioScience* **37**, 49–57.

Chazdon, R. L. (1991). Plant size and form in the understory palm. *Geonoma:* Are species a variation on a theme. *Am. J. Bot.* **78**, 680–694.

Chazdon, R. L., and Fetcher, N. (1984). Photosynthetic light environments in a lowland tropical rain forest in Costa Rica. *J. Ecol.* **72**, 553–564.

Chazdon, R. L., and Field, C. B. (1987). Determinants of photosynthetic light environments in six rainforest *Piper* species. *Oecologia* **73**, 222–230.

Chazdon, R. L., Williams, K., and Field, C. B. (1988). Interactions between crown structure and light environment in five rain forest *Piper* species. *Am. J. Bot.* **75**, 1459–1471.

Collins, S. L. (1990). Habitat relationships and survivorship of tree seedlings in hemlock-hardwood forest. *Can. J. Bot.* **68**, 790–797.

Crabtree, R. C., and Bazzaz, F. A. (1993). Seedling response of four birch species to simulated nitrogen deposition: Ammonium versus nitrate. *Ecol. Appl.* **3**, 315–321.

Denslow, J. S. (1980). Gap partitioning among tropical rainforest trees. *Biotropica* **12**, Suppl., 47–55.

Denslow, J. S. (1987). Tropical rainforest gaps and tree species diversity. *Annu. Rev. Ecol. Syst.* **18**, 431–451.

Denslow, J. S., Schultz, J. C., Vitousek, P. M., and Strain, B. R. (1990). Growth responses of tropical shrubs to treefall gap environments. *Ecology* **71**, 165–179.

Eamus, D., and Jarvis, P. G. (1989). The direct effects of increases in the global atmospheric concentrations of CO_2 on natural and commercial temperate trees and forests. *Adv. Ecol. Res.* **19**, 1–53.

Facelli, J. M., and Pickett, S. T. A. (1991). Plant litter: Its dynamics and effects on plant community structure. *Bot. Rev.* **57**, 1–32.

Fetcher, N., Oberbauer, S. F., and Strain, B. R. (1985). Vegetation effects on microclimate in lowland tropical forests of Costa Rica. *Int. J. Biometeor.* **29**, 145–155.

Forman, R. T. T., and Godron, M. (1986). "Landscape Ecology." Wiley, New York.

Geiger, R. (1965). "The Climate Near the Ground," rev. ed. Harvard Univ. Press, Cambridge, MA.

Givnish, T. J. (1987). Comparative studies of leaf form: Assessing the relative roles of selective pressures and phylogenetic constraints. *New Phytol.* **106**, Suppl., 131–160.

Grime, J. P., Crick, J. C., and Rincon, J. E. (1986). The ecological significance of plasticity. *Symp. Soc. Exp. Biol.* **40**, 5–19.

Grubb, P. J. (1977). The maintenance of species richness in plant communities: The importance of the regeneration niche. *Biol. Rev. Cambridge Philos. Soc.* **52**, 107–145.

Houle, G., and Payette, S. (1991). Seed dynamics of *Abies balsamea* and *Acer saccharum* in a deciduous forest of northeastern North America. *Am. J. Bot.* **78**, 895–905.

Hubbell, S. P., and Foster, R. B. (1986). Biology, chance, and history and the structure of tropical rainforest tree communities. *In* "Community Ecology" (J. Diamond and T. J. Case, eds.), pp. 314–329. Harper & Row, New York.

Keeling, C. D. (1986). Atmospheric CO_2 concentrations. Mauna Loa Observatory, Hawaii, 1958–1986. *Oak Ridge Natl. Lab. [Rep.] ORNL-NDP (U.S.)* **NDP-001/R1**.

Koike, T. (1988). Leaf structure and photosynthetic performance as related to the forest successional of deciduous broad-leaved trees. *Plant Species Biol.* **3**, 77–87.

Kolasa, J., and Rollo, C. D. (1991). Introduction: The heterogeneity of heterogeneity: A glossary. *In* "Ecological Heterogeneity" (J. Kolosa and S. T. A. Pickett, eds.), pp. 1–23. Springer-Verlag, New York.

Kwesiga, F. K., and Grace, J. (1986). The role of the red/far-red ratio in the response of tropical tree seedlings to shade. *Ann. Bot. (London)* [N.S.] **57**, 283–290.

Lee, D. W. (1987). The spectral distribution of radiation in two neotropical rainforests. *Biotropica* **19**, 161–166.

Lieberman, M., Lieberman, D., and Peralta, R. (1989). Forests are not just Swiss cheese: Canopy stereogeometry of non-gaps in tropical forests. *Ecology* **70**, 550–552.

Miao, S. L., and Bazzaz, F. A. (1990). Response to nutrient pulses of two colonizers requiring different disturbance frequencies. *Ecology* **71**, 2166–2178.

Miao, S. L., Wayne, P. M., and Bazzaz, F. A. (1991). Elevated CO_2 differentially alters the responses of cooccurring birch and maple seedlings to a moisture gradient. *Oecologia* **90**, 300–304.

Miller, D. R., Lin, J. D., and Lu, Z. N. (1991). Some effects of surrounding forest canopy architecture on the wind field in small clearings. *For. Ecol. Manage.* **45**, 79–91.

Mooney, H. A., Winner, W. E., and Pell, E. J. (1991). "Responses of Plants to Multiple Stresses." Academic Press, San Diego.

Morgan, D. C., and Smith, H. (1979). A systematic relationship between phytochrome-controlled development and species habitat, for plants grown in simulated natural radiation. *Planta* **145,** 253–258.

Morse, S. R., Wayne, P. M., Miao, S., Bazzaz, F. A. (1994). Elevated CO_2 and drought alter tissue water relations of birch (*Betula populifolia*) seedlings. *Oecologia* (in press).

Murray, K. G. (1988). Avian seed dispersal of three noetropical gap-dependent plants. *Ecol. Mongr.* **58,** 271–298.

Oberbauer, S. F., Clark, D. B., Clark, D. A., and Quesada, M. (1988). Crown light environments of saplings of two species of rain forest emergent trees. *Oecologia* **75,** 207–212.

O'Neil, R. V., DeAngelis, D. L., Wade, J. B., and Allen, T. F. H. (1986). "A Hierarchical Concept of Ecosystems." Princeton Univ Press, Princeton, NJ.

Orians, G. H. (1983). The influence of treefalls in tropical forests on tree species richness. *Trop. Ecol.* **23,** 255–279.

Phillips, D. L., and Shure, D. J. (1990). Patch-size effects on early succession in southern Appalachian forests. *Ecology* **71,** 204–212.

Pickett, S. T. A. (1983). Differential adaptations of tropical species to canopy gaps and its role in community dynamics. *Trop. Ecol.* **24,** 68–84.

Pickett, S. T. A., Kolasa, J., Armesto, J. J., and Collins, S. (1989). The ecological concept of disturbance and its expression at various hierarchical levels. *Oikos* **54,** 129–136.

Poulson, T. L., and Platt, W. J. (1989). Gap light regimes influence canopy tree diversity. *Ecology* **70,** 553–555.

Ramos, J., and Grace, J. (1990). The effects of shade on the gas exchange of seedlings of four tropical trees from Mexico. *Funct. Ecol.* **4,** 667–677.

Ricklefs, R. E. (1977). Environmental heterogeneity and plant species diversity: A hypothesis. *Am. Nat.* **111,** 376–381.

Ricklefs, R. E. (1990). "Ecology," 3rd ed. Freeman, New York.

Rochefort, L., and Bazzaz, F. A. (1993). Growth responses to elevated CO_2 in seedlings of four co-occurring birch species. *Can. J. For. Res.* (in press).

Runkle, J. R. (1985). Disturbance regimes in temperate forests. *In* "The Ecology of Natural Disturbance and Patch Dynamics" (S. T. A. Pickett and P. S. White, eds.), pp. 17–33. Academic Press, Orlando, FL.

Safford, L. O., and Czapowski, M. M. (1986). Fertilizer stimulates growth and mortality in a young Populus–Betula stand: 10-year results. *Can. J. For. Res.* **16,** 807–811.

Schulze, E. D. (1986). Carbon dioxide and water vapor exchange in response to drought in the atmosphere and the soil. *Annu. Rev. Plant Physiol.* **37,** 247–274.

Schulze, E. D. (1991). Water and nutrient interactions with plant water stress. *In* "Responses of Plants to Multiple Stresses" (H. A. Mooney, W. E. Winner, and E. J. Pell, eds.), pp. 90–103. Academic Press, San Diego.

Shugart, H. H., Antonovsky, M. Y., Jarvis, P. G., and Sandford, A. P. (1986). CO_2, climatic change and forest ecosystems: Assessing the response of global forests to the direct effects of increasing CO_2 and climatic change. *In* "The Greenhouse Effect, Climatic Change, and Ecosystems" (B. Bolin, B. R. Doos, J. Jager, and R. A. Warrick, eds.), pp. 475–521. Wiley, New York.

Sims, D. A., and Pearcy, R. W. (1989). Photosynthetic characteristics of a tropical forest understory herb, *Alocasia macrorrhiza*, and a related crop species, *Colocasia esculenta*, grown in contrasting light environments. *Oecologia* **79,** 53–59.

Sipe, T. W. (1990). Gap partitioning among maples (*Acer*) in the forests of central New England. Ph.D. Thesis, Harvard University, Cambridge, MA.

Sipe, T. W., and Bazzaz, F. A. (1994a). Gap partitioning in the forests of central New England: Microclimatic patterns. In press.

Sipe, T. W., and Bazzaz, F. A. (1994b). Gap partitioning among maples (*Acer*) in the forests of central New England: Shoot architecture and photosynthesis. In press.

Sipe, T. W., and Bazzaz, F. A. (1994c). Gap partitioning among maples (*Acer*) in the forests of central New England: Survival, growth and study synthesis. In press.

Sipe, T. W., and Bazzaz, F. A. (1994d). Establishment of maple seedlings (*Acer*) across the gap–understory mosaic in central New England forests. In preparation.

Spurr, S. H. (1956). Forest associations of the Harvard Forest. *Ecol. Monogr.* **26,** 245–256.

Strauss-DeBenedetti, S., and Bazzaz, F. A. (1991). Plasticity and acclimation to light in tropical Moraceae of different successional positions. *Oecologia* **87,** 377–387.

Taiz, L., and Zeigler, E. (1991). "Plant Physiology." Benjamin/Cummings, Redwood City, CA.

Takenaka, A. (1989). Optimal photosynthetic capacity in terms of utilizing a natural light environment. *J. Theor Biol.* **139,** 517–529.

Tenhunen, J. D., Pearcy, R. W., and Lange, O. L. (1987). Diurnal variation in leaf conductance and gas exchange in natural environments. *In* "Stomatal Function" (E. Ziegler, G. D. Farquar, and I. R. Cowan, eds.), pp. 323–351. Standford Univ. Press, Standford, CA.

Thomas, S., and Bazzaz, F. A. (1993). Spatial heterogeneity in understory CO_2 levels in a tropical and temperate forest. Unpublished data.

Thompson, W. A., Stocker, G. C., and Kriedemann, P. E. (1988). Growth and photosynthetic response to light and nutrients of *Flindersia brayleyana,* a rainforest tree with broad tolerance to sun and shade. *Aust. J. Plant Physiol.* **15,** 299–315.

Tubbs, C. H. (1969). The influence of light, moisture, and seedbed on yellow birch regeneration. *U. S., For. Ser., Res. Pap. NC* **NC-27.**

Turnbull, M. H. (1991). The effect of light quantity and quality during development on the photosynthetic characteristics of six Australian rainforest tree species. *Oecologia* **87,** 110–117.

Turner, N. C., and Kramer, P. J. (1980). "Adaptations of Plants to Water and High Temperature Stress." Wiley, New York.

Vázques-Yanes, C., and Orozoco-Segovia, A. (1992). Effects of litter from a tropical rainforest on tree seed germination and establishment under controlled conditions. *Tree Physiol.* **11,** 391–400.

Walters, M. B., and Field, C. B. (1987). Photosynthetic light acclimation in two rainforest Piper species with different ecological amplitudes. *Oecologia* **72,** 449–456.

Walters, M. B., and Reich, P. B. (1989). Responses of *Ulmus americana* seedlings to varying nitrogen and water stress. I. Photosynthesis and growth. *Tree Physiol.* **5,** 159–172.

Waring, R. H., and Schlesinger, W. H. (1985). "Forest Ecosystems: Concepts and Management." Academic Press, New York.

Wayne, P. M. (1991). Effects of the daily timecourse of light availability on the sun–shade responses and regeneration of birch seedlings. Ph. D. Thesis. Harvard University.

Wayne, P. M., and Bazzaz, F. A. (1993a). Effects of the daily timecourse of light availability on the sun–shade responses and regeneration of birch seedlings. *Ecology* **74,** 1500–1515.

Wayne, P. M., and Bazzaz, F. A. (1993b). Morning vs afternoon sun patches in experimental forest gaps: Consequences of temporal incongruency of resources to birch regeneration. *Oecologia* **94,** 235–243.

Wayne, P. M., and Bazzaz, F. A. (1994). Radiation, ectomycorrhizae, and the responses of four sympatric birches (*Betula*) to the gap–understory continuum. *J. Ecol.* (submitted for publication).

Weis, E., and Berry, J. A. (1988). Plants and high temperature stress. *Symp. Soc. Exp. Biol.* **42,** 329–346.

Wofsy, S. C., Gouldon, M. L., Munger, J. W., Fan, S. M., Bawkin, P. S., Daube, B. C., Bassow, S. L., and Bazzaz, F. A. (1993). Net exchange of CO_2 in mid-latitude forests. *Science* **260,** 1314–1317.

Wylie, R. B. (1951). Principles of foliar organization shown by sun-shade leaves from ten species of deciduous dicotyledonous trees. *Am. J. Bot.* **38,** 355–361.

14

Spatial Heterogeneity at Small Scales and How Plants Respond to It

Graham Bell and Martin J. Lechowicz

I. Introduction

The lives of plants are strongly influenced by their being rooted in place and having to endure their local situation without being able to seek more favorable conditions (Bradshaw, 1965). Plants must respond to minute-to-minute fluxes in insolation, change in temperature and humidity from day to night, day-to-day variation in precipitation, seasonal cycles of nutrient availability, annual variation in the length of the growing season, and countless similar elements of environmental variation. Both the variety and the functional significance of such responses to temporal variation in the environment are amply documented in this book. Despite being rooted in place, plants experience not only temporal but also spatial variation in their environment. As a plant grows and extends into adjacent areas (Caldwell, Chapter 12, and Fitter, Chapter 11, this volume), it is likely to encounter different conditions—in other words, the plant's "perception" of environmental variation extends beyond the scale of the individual. Nor is the extent of these spatial effects limited to the scale of roots and rhizomes. As pollen and seeds disperse, the effects of spatial variance at even greater distances are "perceived" by the plant as variation in its success in transmitting genes through progeny to subsequent generations in the local population. This makes spatial structure in the local environment, as well as temporal variation, important to the study of plant responses to environmental heterogeneity. We might estimate environmental heterogeneity with instrumental measures of the physical and chemical environment, by quantitative description of the biota, or by the response of bioassays using standardized test plants. Such data on environmental heterogeneity may be interpreted from various points of view (Addicott *et al.*, 1987; Wiens, 1989; Kolasa and Rollo, 1991), but we will be concerned here with the spatial pattern of microenvironmental patches at scales ranging from that of individual plants to that of plant populations. One of our goals in this chapter is to illustrate how plants respond to environmental heterogeneity at the scale of the population. We shall argue that it is necessary to judge plant responses to environmental heterogeneity not only at the scale of the individual plant, but also at the scale of the plant population.

The most straightforward response of a plant to adversity is to die or, less dramatically, to set little or no seed. If this were generally the case, most populations of plants would comprise a large number of specialized types, each flourishing within a narrow range of environmental conditions and perishing elsewhere. At large spatial scales, this is often true: We find different species of plants in different habitats. Whether comparable specialization to a locale mosaic of different environments can also account for genetic diversity within populations (references in Bell 1990a,b, 1991a,b) and species diversity within communities (references in Bazzaz, 1991) is less certain. There is some evidence for such genetic specialization from field experiments that compare the performance of plants from the same genetic stock in their native site and in more or less distant sites. Such reciprocal-transplant experiments have consistently shown that incomers are less successful than residents (Schoen *et al.*, 1986; Antonovics *et al.*, 1987; Kadmon and Shmida, 1990; Platenkamp, 1990; Platenkamp and Foin, 1990). Other evidence of specific local adaptation includes the findings that the direction in which selection acts on given phenotypic characters can vary among sites separated by only a few meters (Stewart and Schoen, 1987; Argyres and Schmitt, 1991), and that crosses within natural populations tend to be more productive between parents that are growing close together than between parents that are separated by a few tens of meters (Schmitt and Gamble, 1990; McCall *et al.*, 1991).

Though these results show that there is often some degree of microenvironmental specialization or local adaptation within plant populations, it can scarcely be concluded that individual plants have the potential to grow successfully only in a very narrowly circumscribed range of conditions. Many of the papers in this volume demonstrate the contrary view, that individual plants can maintain high levels of performance over a wide range of conditions. Given the relative immobility of individual plants and the ubiquity of environmental heterogeneity, it is reasonable to infer that plants will have evolved sufficient plasticity to grow successfully over a range of conditions. The available data, especially at small spatial scales (representing the distances traveled by pollen and seeds) and at small genetic scales (within populations of a single species), do not suggest that either specialization or plasticity predominates. Some balance of specialization and plasticity among individuals in plant populations, perhaps shifting in response to local environmental heterogeneity, seems more likely.

This is not an issue that we can resolve by studying the responses of groups of genetically uniform plants to carefully controlled environmental variation under laboratory conditions. The variation expressed in controlled-environment studies will depend on the range and nature of the test environments selected. Similarly, our assessment of the degree of variation among genotypes depends on the breadth of our sampling: Larger samples of genotypes, or tests across a wider range of environments, may always reveal unusual individual responses that substantially increase our overall estimates of variation in a trait. These problems may not be of much concern to breeders concerned with plant performance in managed environments (Mayo, 1987), but they can be seriously misleading in studies of plants in natural environments (Lawrence, 1982). The appropriate approach in studies of wild plants is to describe the patterns of environmental heterogeneity that exist and the responses that plants have evolved while growing under these conditions, that is, to study the nature and selection of plasticity under natural conditions. In this chapter we shall also suggest methods for carrying out such studies and point out some promising directions for research on the evolution of plasticity in natural populations.

II. Interrelated Concepts of Plasticity

Concepts of plasticity differ from discipline to discipline, leading to confusion when reading the diverse literature on plant responses to environmental heterogeneity. The usage of the word "plasticity" is not consistently well defined across disciplines, and there are often subtle differences in the operational definition of plasticity even within disciplines. Plasticity is gauged in one way or another by the degree of variation found in traits describing plant form or function. At the simplest level, the variance around the mean of replicate measures of a trait provides a measure of plasticity. Depending on the purpose and scale of the investigation, that variance may be estimated from measurements on single individuals repeated over time, from clonal replicates of individuals grown in different environments, or from individuals randomly sampled in one or more local populations and grown in one or more experimental environments. In all of these cases, a trait with greater variance would be considered, generally speaking, more plastic. This usage is rooted in the colloquial meaning of plastic, the idea that some basic stock is malleable and can readily take various forms. We emphasize, however, that plants that are able to make broad physiological adjustments to different environments, and that therefore express a large amount of plasticity with respect to physiological characters, will be able to survive and reproduce successfully across this range of environments, and will therefore express a small amount of plasticity with respect to characters related to fitness. Plasticity at the level of physiology is likely to be accompanied by stability at the level of fitness. This point is taken up again in Section IX.

Confusion in the literature of plasticity arises from three sources. First, the variation studied may range from individual plants varying over time to interspecific variation among geographically separated populations. Second, studies do not consistently distinguish, and sometimes do not even recognize, the genetic and environmental components of phenotypic variation. Third, depending on the purpose of the study, there can be substantial differences in the temporal scale at which traits are defined and variation is considered. Generally speaking, evolutionary ecologists tend to measure plasticity as the variation among plants in particular genetically defined stocks growing in different experimental test environments. They view plasticity in terms of the mean environmental variance estimated in experiments that allow the variation among individuals to be partitioned into genetic and environmental components. Physiological ecologists, on the other hand, tend to view plasticity more in terms of the adjustments that plants make in response to changing environmental conditions. They measure plasticity by changing form or function without particular regard to its genetic basis. It is useful to elaborate the nature of these contrasting points of view.

Botanists in general, and physiological ecologists in particular, usually use the term "plasticity" to suggest adjustments in the form or function of an individual plant subjected to changing environmental conditions (Jennings and Trewavas, 1986; Kuiper and Kuiper, 1988). Their concern is with phenotypic variability, not with the genetic processes that underlie that variability. More often than not, plasticity is studied and described as an attribute of the population or species rather than of the individual, despite the fact that the measurements are made on individual plants. In this sense, plasticity is illustrated by the heterophylly of submerged versus emergent leaves on an aquatic *Ranunculus,* or by the acclimation of the photosynthetic light response in *Trillium* leaves as the tree canopy closes in at the end of spring. Many of the ecophysiological processes discussed in this book emphasize such plastic changes, with the plasticity being variously referred to as physiological, developmental, or ontogenetic. In this botanical view, plasticity is characterized by the amount of change, the pattern of that change, and the temporal scale over which the change is expressed in individual plants (Kuiper and Kuiper, 1988).

Evolutionary ecologists are equally concerned with the amount and pattern of plasticity, but less attention has been given to the response time of plastic changes (Schlichting, 1986; Thompson, 1991). Evolutionary ecologists study the responses of individuals, but usually do not take into account plasticity in individual plants over time. Their focus is on traits that can be readily measured on individuals that characterize the longer-term functional adaptation of a plant to its environment. They favor traits closely linked to plant survival and reproduction, that is, to components of fitness (Lechowicz and Blais, 1988). For example, they might estimate seed production in preference to maximal stomatal conductance, or total vegetative biomass in preference to the partitioning coefficient for investment in leaf biomass. They use these data on individuals, together with techniques adopted from quantitative genetics, to analyze the genetic and environmental basis of the phenotypic variability among individuals of known genetic relationship sampled from a range of environments (see Lawrence, 1982, Via, 1987). Their focus is on the phenotypic and genetic variability *among* individuals in plant populations rather than on the plastic variation of individual plants (Schlichting, 1986, 1989; Sultan, 1987; Stearns, 1989; Thompson, 1991). There is more concern with spatial heterogeneity as a source of plastic response, or with temporal variation between generations, than with the effects of temporal variation over the life of individual plants.

Differences in the way that studies of plasticity approach temporal variation can be an especially perplexing source of confusion. In deciding how a plant response will be characterized when studying its plasticity, a clear conceptual line has to be drawn between dynamic regulatory processes such as stomatal opening and closing and characteristic variables that define the functional set points in a regulatory system. Most would consider the maximal stomatal conductance as a trait characterizing stomatal regulatory strategy and subject to plastic variation; conversely, most would consider the moment-to-moment changes in stomatal conductance as regulated dynamics, and not as plasticity. This can become a source of confusion because a trait that is chosen for a study of plasticity may vary on several different time scales. For example, an evolutionary ecologist may choose to investigate variation in photosynthetic capacity among plants in a population by assaying maximum photosynthetic rates under favorable test conditions at a single time period during the season. A physiological ecologist would be concerned that this estimate of photosynthetic capacity will vary seasonally, as well as with conditions immediately before and during the assay. In other words, the variability among individuals can be confounded with variability within individual plants if traits are not appropriately chosen and measured. Evolutionary ecologists strive to choose traits that do not vary much from time to time when measured on the same individual, so that they can better estimate the variability among individuals—variability within an individual plant is unwanted noise. Physiological ecologists, on the contrary, plan explicitly to study the variability of an individual as its environment changes on time scales from seconds to weeks or months—variability among individuals is unwelcome noise that makes the definition of representative response patterns more difficult.

III. Evolutionary Ecology of Plasticity

Natural populations may comprise not many highly specialized genotypes, but one or a few generalists able to cope with the modest amounts of environmental variance found at small spatial scales. Both extreme points of view, narrow genetic specialization and virtually unlimited phenotypic plasticity, are contradicted by the most mundane observations. A more reasonable approach is to recognize both genetic variation in performance and the environmental constraints within which this variation is expressed. We can then recognize that phenotypic plasticity is not merely an arbitrary attribute of plants, but instead a property that evolves through the natural selection of genotypes that determine a more or less plastic response to environments that vary in space and in time. The crucial concept that we wish to advance here is that the amount of environmental variance expressed by plants grown under different conditions is itself a character under genetic control, that is, plasticity is heritable. This fact has been recognized for a long time (Bradshaw, 1965; Schlichting, 1986), but without receiving the attention it deserves (Stearns, 1989; Thompson, 1991).

The concepts of plasticity and its heritability can be given precise statistical meaning in experiments where several genotypes or families are each grown and scored in several environments (Lawrence, 1982; Falconer, 1986). The variance of average scores among the genotypes is a genetic variance (s_G^2); The variance of average scores among the environments is an environmental variance (s_E^2). These two components sum to the total variance, provided that each genotype responds in the same way to differences among environments. However, if genotypes respond differently when the environment changes, a third source of variance will be contributed by genotype–environment interaction ($s_{G \times E}^2$). We consider the environmental variance, s_E^2, as a good measure of plasticity for studies in evolutionary ecology—the greater this environmental variance, the more plastic is the character. It should be noted that there are alternative opinions among evolutionary ecologists on exactly which variance components should be considered the measure of plasticity. Scheiner and Goodnight (1984) perfer the sum of s_E^2 and $s_{G \times E}^2$ as a definition of plasticity, whereas Thompson (1991) considers plasticity to be defined by the interaction effect $(s_{G \times E}^2)$ alone. We prefer to keep the concept of the mean phenotypic plasticity (s_E^2) separate from that of the genetic variance of phenotypic plasticity $(s_{G \times E}^2)$. This $s_{G \times E}^2$ can be thought of as the heritable portion of the total environmental variance. It has been discussed theoretically by Via (1987), Via and Lande (1985), and Lyman (1989, 1991), in an experimental context by Bell (1990a,b), and in the field situation by Via (1991).

To the extent that the quantity of environmental variance can be inherited, it can evolve under natural selection, if a greater or lesser sensitivity to environmental variation is advantageous. The environmental variance ties the phenotypic variation we observe to its environmental basis, and is therefore a good measure of plasticity. But if the phenotypic variation is due to s_E^2 alone, then there is no potential for differential selection favoring the degree of plastic response exhibited by the fittest individuals in the population. In that case, the plants do respond to environmental variation, but there is no opportunity for improving the magnitude or direction of this response through selection of greater or lesser plasticity. Without some genetic component to the plastic response, selection of that response is impossible. The environmental heterogeneity that a plant "perceives" and responds to from an ecological point of view will only be relevant from an evolutionary point of view if there is heritable variation in plasticity. This argument applies to any plant trait or response, but we shall be interested primarily in reproductive success, or in characters that are highly correlated with reproductive success. The concern with the evolution of plasticity, rather than merely its description, necessarily shifts the focus of investigations of plasticity from the individual plant to the population.

IV. Implications for Physiological Ecology

The preceding chapters have been concerned mainly with the mechanisms by which individual plants exploit gaps, patches, and resource-rich microsites. This emphasis is consistent with the long tradition in physiological ecology of studying plant responses to environmental gradients. The shift from early experimental definition of, for example, equilibrial photosynthetic rates along a gradient from 0 to 2000 μmoles photons/m²/s to the transient response to sunflecks is a natural one. It is a shift from an ecophysiological perspective focused on stable acclimation to experimental gradients to one that takes more account of the spatial and temporal heterogeneity of resources in natural environments, and leads to a view of plasticity gauged by changes in traits during the life of a single individual (Kuiper and Kuiper, 1988). This shift engenders studies of plant responses to experimentally manipulated resource patches, rather than along gradients. Our mechanistic understanding of plant responses has been much enriched by these investigations.

This new ecophysiological approach, with which this book has been largely concerned, is nonetheless incomplete. In most mechanistic investigations, individual variation is considered only as a nuisance that increases the number of replicates necessary to get a good estimate of the response. In fact, this variation among individuals in a population is a reflection of the effects of environmental heterogeneity at the population level and is of interest in its own right. Such variation among individuals in a plant population has important implications both for ecological patterns (Bazzaz and Sultan, 1987; Tilman, 1990; Turkington and Mehrhoff, 1990; Bazzaz,1991), and for evolutionary processes (Bradshaw, 1965; Antonovics *et al.*, 1987; Stearns, 1989; Bell, 1990a, 1991a; Cohen and Levin, 1991). It is also of interest to agronomists and plant breeders, who are often concerned with developing strains of crop plants that perform well over a wide range of environments, and may be willing to sacrifice exceptionally high levels of performance in particular environments in order to do so (Mayo, 1987). In this context, we can ask how variability in plant traits maps onto variability in the environment over the relevant spatial and temporal scales. Do plants respond to the same levels of environmental heterogeneity that we might measure instrumentally, or do they buffer or filter the variation in some way? Are the levels and patterns of plastic responses to environmental heterogeneity similar or different among traits? What is the balance of genetic and environmental effects that account for the range and pattern of phenotypic variation across a range of heterogeneous environments? The notion that phenotypic plasticity is a heritable trait exposed to selection gives us a powerful general principle for approaching these and other questions.

V. A Fine-Scale Survey of Maple Seedlings

In the body of this chapter we shall put forward a number of generalizations concerning environmental variation and how plants respond to it. We shall be particularly concerned with spatial pattern in native populations of plants growing in undisturbed natural environments at very small spatial scales, of the order of dispersal over a single generation. This is the scale that is most relevant to the evolution of plasticity, although it has been largely ignored by both ecologists and plant physiologists. Many of the suggestions that we shall make are speculative and are intended as hypotheses to be tested rather than as assertions of ascertained fact. The empirical basis for these speculations has been provided by work at the McGill Field Station at Mont St. Hilaire in southern Quebec, the published part of which will be cited in the appropriate following sections.

To provide a concrete example of phenotypic variation in a natural population, we shall analyze an unpublished survey of seedling sugar maples (*Acer saccharum* Marsh) growing in the understory of old-growth forest at Mont St. Hilaire. Most of these seedlings became established after damage from a glaze ice storm in December 1983 had increased insolation on the forest floor (Melancon and Lechowicz, 1987). Even in this single locality, which we refer to as the Lake Hill site, the seedlings are likely to differ from one another genetically (Perry and Knowles, 1991), and we can describe only the phenotypic variation among these individuals without partitioning its genetic and environmental basis. This is sufficient for our present purposes, but limits our discussion to a strictly ecological rather than an evolutionary perspective. We are essentially concerned with comparing the range and nature of phenotypic variation in different traits of the seedling maples growing at this forest site.

To quantify the spatial pattern in this population, we randomly harvested five seedlings in each of 198 meter-square plots randomly located within a 50 × 50-m grid. We aged each seedling by counting terminal bud scars and calculated its mean annual production of woody tissue (NAP: g/year); both aboveground and belowground woody tissues were included. The NAP is a good index of seedling fitness in this canopy tree that survives long periods of suppression in the understory (Canham, 1988). We also measured the area and mass of leaves on each seedling to calculate its specific leaf mass (SLM: mg/cm^2). The SLM is central to the functional organization of leaves (Gutschick, 1987), including those of sugar maple (Lei and Lechowicz, 1990; Ellsworth and Riech, 1992), and is known to respond strongly to light regime (Givnish, 1988). Figure 1 shows the spatial distribution of these two seedling traits within the sampling grid.

VI. Physical Environment in the Forest Understory

We are concerned to relate phenotypic variation among these maple seedlings to fine-scale environmental heterogeneity in this habitat. The simplest approach to measuring environmental variance is to record the readings of instruments that are sensitive to environmental factors, such as insolation or the concentrations of inorganic ions, at different points

Mean Annual Wood Production, g/yr

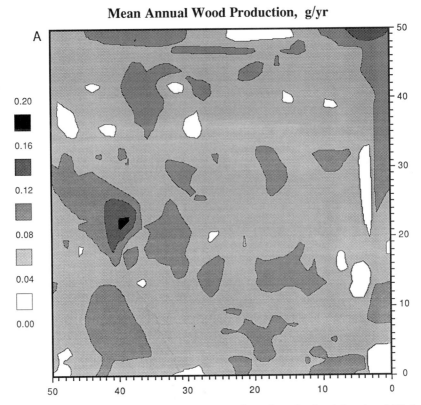

Figure 1 Contour plots of (A) the mean annual wood production (g/year) and (B) the specific leaf mass (mg/cm²) of sugar maple seedlings sampled in 198 random meter-square plots in the understory of a mature forest with *Fagus grandifolia* and *Acer saccharum* codominant in the canopy. The canopy at this Lake Hill site was heavily damaged by an ice storm in the winter of 1983–1984 but had largely closed in when these samples were taken in 1991 (Melancon and Lechowicz, 1987).

within the sampling area. By replicating measurements at each sampling point the error variance (confounding true measurement error with unanalyzable spatial variation at the sampling point) can be estimated; the remaining variance component is an estimate of the variance among sampling points. We do not yet have data of this sort for the Lake Hill grid, but we have found substantial variation among sites within a similar 50 × 50-m grid nearby on the mountain for soil pH, nitrogen, and potassium (Lechowicz and Bell, 1991). Even at very small spatial scales, the physical environment is not uniform, and we can expect selection for plasticity within this local population of sugar maples.

When the sampling area is extended, a greater range of recognizably

Specific Leaf Mass, mg/cm²

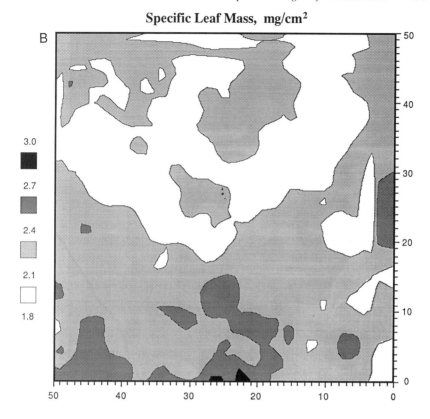

distinct habitats will be included, and it is reasonable to expect that the estimate of environmental variance will tend to increase. A plot of the environmental variance among a set of sites against the average distance separating the sites expresses spatial pattern (for technical discussion of the statistical procedures involved, see Burrough, 1983; Webster, 1985; Trangmar *et al.*, 1985; Robertson, 1987). If the regression has zero slope the sites are uncorrelated, and as much variance can be found within a smaller as within a larger area. If the regression has positive slope then nearby sites are positively correlated, and conditions of growth will become steadily less predictable as one moves away from any given point. The variance of physical features of the environment increases with distance even within 50 × 50-m grids at Mont St. Hilaire (Lechowicz and Bell, 1991), and similar results have been reported in less undisturbed environments (Palmer, 1990; Robertson, 1987; Robertson *et al.*, 1988; Moloney, 1988). Surveys at larger spatial scales have led us to propose the general rule that variance increases continuously with distance at all scales (G. Bell *et al.*, unpublished). This suggests two possible generalizations about the evolution of plasticity. First, species of plants that disperse their propagules more widely will display less phenotypic plasticity (or greater phenotypic stability) with respect to fitness when challenged with a range of deliberately manipulated environments. This is because their offspring, as a group, will encounter a greater quantity of environmental variance, and selection will therefore have favored physiologically plastic genotypes, which because they are able to survive and reproduce over a wide range of environmental conditions will express relatively little variation in fitness. Second, any given species will display less plasticity (greater stability) for fitness toward physical variables whose variance increases less steeply with distance, since the levels of these variables encountered by offspring will be less predictable. To the best of our knowledge, neither of these hypotheses has been tested.

VII. Plant Response: The Best Measure of Environmental Variance

The principal objection to using meter readings as measures of the environment is that their relevance to plant performance is often difficult to define adequately. In our survey of maple seedlings, we would like to describe the variance of the environment in which they are growing. We might begin by deciding that the most relevant environmental factor for NAP and SLM is insolation, but that would arbitrarily ignore covariant changes in soil nutrients, water, and similar factors (Bazzaz and Wayne, Chapter 13, this volume). Even if we accepted the simplification that plastic responses in NAP and SLM were dominated by variation in insolation, how could we quantify variation in insolation? Given the immense temporal and spatial heterogeneity of insolation in the understory of deciduous forests (Baldocchi and Collineau, Chapter 2, this volume), do we place a recording instrument at each of our seedlings throughout the growing season? If we had these numerous records second by second throughout the growing season, how would we summarize the variance in insolation? Given the effects of plant acclimation to previous insolation (Pearcy and Sims, Chapter 5, this volume), neither the simple variance nor even a power analysis of the time-series would provide a complete and biologically meaningful summary of the variance in insolation across all the seedling microsites. We are forced to conclude that the routine measurement of spatial and temporal heterogeneity in this single environmental factor, let alone all the covarying factors that can influence plant response, is at best impracticable, and may be impossible even in principle. We must turn to some alternative measure of environmental heterogeneity against which to gauge plant responses.

One alternative is the "phytometer" method, first introduced by Clements early in this century (Clements and Goldsmith, 1924) and recently revived by Antonovics and his associates (see Antonovics *et al.*, 1987). Clonally propagated plantlets transplanted into the natural environment serve as a bioassay of conditions throughout a study site. Bell and Lechowicz (1991) have referred to this design as an "implant" to distinguish it from "explant" trials, which have a similar purpose. In an explant trial, the bioassay is conducted using samples (soil cores, for instance) removed to a common test environment, in which they can be arranged randomly in order to destroy any purely spatial covariance. Implant trials estimate the full range of biotic and physical sources of environmental heterogeneity, but their interpretation can be complicated, or restricted to certain characters or test species, by high levels of transplant mortality. Explant trials circumvent this difficulty, but at the expense of assaying only some of the sources of environmental heterogeneity. Both approaches have the advantage that the environmental variation is measurable in units directly relevant to plant fitness: size or seed production, for example. These methods have been used to demonstrate the existence of spatial heterogeneity that influences plant fitness at scales from 10^{-1} to 10^2 m both in pristine environments at Mont St. Hilaire (Bell and Lechowicz, 1991; Bell *et al.*, 1991) and in more disturbed habitats (Antonovics *et al.*, 1987).

VIII. Separation as an Estimate of Environmental Similarity or Difference

Another method of studying environmental variance through plant response is to use *in situ* measurements of unmanipulated native plants. This approach uses the distance separating sampled plants as a basis for inferring the degree of difference in their microenvironments. We commonly suppose that two microenvironments close together are more similar than two that are farther apart, and this has been shown to be the case in forest understory (Palmer, 1990; Bell and Lechowicz, 1991; Lechowicz and Bell, 1991; Bell *et al.*, 1991) as well as more generally in other habitats (Robertson, 1987; Robertson *et al.*, 1988; Moloney, 1988). We can confidently predict that any two of our sampled seedlings separated by a meter should on average be more similar than two separated by 10 m. In other words, the variance of NAP (or SLM) among all pairs of seedlings separated by 1 m should be less than that among all pairs separated by 10 m. A graph of the variance of any factor on the distance apart of sampling sites is a description of the structure, or pattern, of the environment. This approach can readily be extended to partition phenotypic variance into its genetic and environmental components by analyzing the spatially patterned responses of replicate plants that are genetically characterized and planted out over the spatial grid in this natural environment. Such an analysis would provide measures of plasticity and its heritability unambiguously relevant to natural selection in this maple population.

There are some problems, however, with using variance as a measure of plant variability in comparisons of this sort. First, any measure of variance is dependent on the sampling interval on which it is based. If photosynthesis fluctuates from near zero to near maximum in response to sunflecks in a matter of seconds, but our infrared gas analysis system responds to changing rates only on the order of minutes, then we have artificially damped the real variance in photosynthesis. Only good technology and thoughtful sampling design can account for such effects of sampling interval on estimates of variance in a character. Second, because variance is calculated as squared deviations from the mean, any changes in units or measuring scales will change the variance, just as it will change the mean. If we measure nocturnal tissue pH of a CAM plant on a scale of pH units (logarithms of hydrogen ion concentration), the variance (and the mean) will be far smaller than if we measured hydrogen ion concentration itself. The same is true if we log-transform a measured variable to meet the requirements of a statistical analysis. To compare the relation of variance to distance among different characters, it is often useful to use the z-standardization (normalization) that adjusts the sampled values so that they have a mean of zero and a standard deviation of one (Zar, 1974). Unless the initial distribution of the data is very far from normal, this standardization removes the unwanted effects of scale and units of measurement without altering the pattern of plant variability over distance. We use the normalized phenotypic scores of the maple seedlings to describe the spatial pattern of NAP and SLM throughout this small area of forest floor (Figure 2).

The sampling design and our lack of genetic information about the sampled individuals place some limits on our discussion. Our estimate of the error variance is the variance among the five sampled seedlings within each square meter, so that we cannot analyze spatial variance at scales of 1 m or less. The remaining variance is our estimate of the overall environmental variance at scales exceeding 1 m within the 50×50-m sampling area. This estimate is substantial and highly significant for both characters: for SLM, environmental variance $s^2_E = 2.28 \times 10^{-2}$ (error variance $s^2_e = 2.59 \times 10^{-2}$, $P = 0.0001$ for H_0: $s^2_E = 0$); for NAP, $s^2_E = 2.80 \times 10^{-4}$ ($s^2_e = 6.91 \times 10^{-4}$, $P = 0.0001$). We emphasize again that these values, based on individuals of unknown genetic relationship, are not pure estimates of the environmental variance; they include

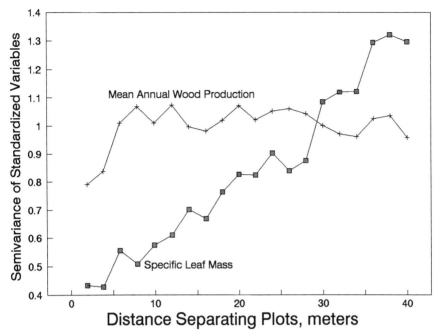

Figure 2 Semivariograms for standardized values of specific leaf mass and mean annual wood production for the seedling sugar maples sampled at the Lake Hill site (cf. Figure 1).

portions of the genetic and genotype–environment interaction variances, as well as genotype–environment covariance. However, experiments at Mont St. Hilaire using the understory herb *Impatiens pallida* from similar old-growth forests have shown that the environmental component of variance is much larger than the genotypic components (Bell *et al.*, 1991; Schoen *et al.*, in press), and we are confident that for our present purpose these estimates can be interpreted as a measure of plasticity.

The variance of SLM, which increases steadily with distance, illustrates the use of *in situ* plants separated by known distances to gauge patterns of environmental heterogeneity (Figure 2). The spatial response of this particular trait undoubtedly arises from the correlation between nearby points with respect to the physical factors to which variation in SLM represents a response—for example, nearby points will tend to have similar light regimes. However, it is the plant response itself that provides both the simplest and the most appropriate integration of these physical factors. Explant studies using genetically uniform material of *Arabidopsis* and *Hordeum* have likewise shown substantial small-scale environmental variance and pattern for similar areas at Mont St. Hilaire, in trials where genotypic sources of variance can be excluded (Bell and Lechowicz, 1991). However, these sources of variance were not absent: Different genotypes of the test species showed different patterns. A plant's "perception" of local variation in the environment depends on its genotype as well as on the trait being monitored. In either explant or implant trials, it is preferable to use a range of defined genotypes; the precision of estimates is not thereby compromised, and their generality is greatly extended. The existence of some heritable differences in plant response to this forest understory environment only emphasizes the potential for selection to act on plasticity within the range of conditions found in this small area.

IX. Nature of Plastic Responses to Environmental Heterogeneity

A plastic type varies physiologically in response to variation in the external environment. If plasticity has evolved under selection, plastic variation is likely to be functionally appropriate, in the sense of tending to increase expected survival or reproductive success, rather than being a mere passive fluctuation. Physiological processes may then vary as widely, or even more widely, as the environmental factors to which they are a response. However, we expect that the eventual output of these underlying physiological processes, lifetime reproductive success, will vary much less. This is because any lineage within a population that inhabits a heterogeneous environment will encounter different conditions in different generations. The environment perceived by the line descending from any given family will change through time, as members of the lineage are dispersed in successive generations to a succession of different sites. The long-term fitness of a lineage in these circumstances will depend on its geometric mean rate of increase and will therefore be greater when its environmental variance of fitness is less (Gillespie, 1977). Selection in a heterogeneous environment for *plasticity* of physiological response will therefore result in a *stabilization* of reproductive output.

This is little more than a simple principle of general homeostasis, but it is not quite as straightforward to test as might appear. It is not immediately clear how the data should be transformed to make different environmental and trait scores comparable. In a few instances a direct approach to the comparison of biotic and physical variation may be possible when the same variable can be meaningfully scored for both plant and environment. For example, we might predict that the rate of increase of variance with distance will be less for plant tissue potassium concentration than for soil potassium concentration. Whether any such comparisons are valid is questionable: Even if potassium is expressed in the same apparent units (g/g) in both plant and soil, there is no assurance that the mass of potassium relative to soil is truly comparable with the mass of potassium relative to plant tissue. The difficulty is compounded because the magnitude of the variance for untransformed scores will depend on the units of measurement, as explained earlier. The z-transformation, which we have used in Figure 2, forces the overall variances of all scores to be equal, which will not always be appropriate; there is no compelling theoretical rationale either for any particular scalar transformation or for scaling on the mean. However, environmental *pattern*, as expressed by the rate of increase in variance with distance, does not encounter these difficulties. We therefore suggest that in general the rate of increase of variance with distance will be less for functionally important plant responses than for the physical environment. The rate of increase of variance can be estimated as the slope of the regression of log variance on log distance.

Given that the rate of increase in variance over distance allows us to compare plasticity in different traits, we can ask whether different traits show different patterns of plasticity in the same range of environments. To show that this may happen, we have compared the amount of variation in NAP and SLM as a function of distance in the seedling maple population at Mont St. Hilaire (Figure 2). We might expect a priori that the amount of variation in both traits would increase with distance at nearly the same rate as more different microenvironments are encountered, but in fact this was not the case. The patterns of phenotypic variability in NAP and SLM of the sugar maple seedlings in this population are not alike. The variance of SLM increases steeply and continuously with distance, but NAP shows a quite different pattern: Variance increases from 0 to about 5 m separation, but thereafter remains at nearly the same level, so that the correlation between sites is nearly zero for distances in excess of 5 m. Put another way, both SLM and NAP are plastic traits, but the plasticity of SLM is directly coupled to increases in environmental heterogeneity and that of NAP is not. NAP is a less plastic trait, or a more stable trait, than SLM in the face of increasing environmental heterogeneity. This is to some extent an artefact of the arithmetic scale that we have used; if we were to sample plots several kilometers distant, there is little doubt that a greater variance in NAP would be found. Nevertheless, the two variables are clearly patterned in different ways, which we might express by saying that SLM measures a coarse-grained environment whereas NAP measures a fine-grained environment, or alternatively by saying that NAP is a more stable character than SLM. This difference in the environmental grain of the two plant responses is shown graphically by the contour plots of these data (Figure 1).

Since different characters will give different estimates of environmental variance and pattern, can we predict which characters will be more plastic and which less? It follows from the argument at the head of this section that characters that are more highly correlated with fitness will be less plastic, or more stable. The slope of log variance on log distance, which provides a convenient measure of plasticity, will be smaller for characters that are more highly correlated with fitness. The prediction is valid either for *in situ* surveys or for implant experiments using native plants. Insofar as selection favors plasticity, rather than leading to genetic specialization, it will favor genotypes with mediocre reproductive success sustained over a wide range of environments with different physical characteristics. However, this may require quite different physiological mechanisms in different circumstances. To give a hypothetical example, an umbellifer growing in full sun might be adequately defended against invertebrate herbivores by modest investments in photoactivated toxins but require high water-use efficiency to deal with its drought-prone water regime; conversely, at a nearby shaded site, water-use efficiency may be less important while investments in antiherbivore defenses may be much greater. The proximate physiological mechanisms involved in dealing with the physical environment would thereby retain much of the spatial correlation exhibited by physical variables, while selection would erode this correlation at the level of their eventual outcome, reproductive success. This is an evolutionary argument and applies only to native plants sampled in the environment they have evolved in; it would not necessarily apply, for example, to explant studies using exotic species as bioassays. The maple seedlings may provide an example of this principle, since the likelihood of eventual growth to reproductive maturity will be much more highly correlated with NAP than with SLM.

X. Artificial Selection for Phenotypic Plasticity

The most valuable technique for investigating these aspects of phenotypic plasticity in the laboratory is the selection experiment. Despite the recent increase of interest in plasticity, we know of very few attempts to manipulate the plastic response of plants by selection over many generations, using reproductive success itself as the character under selection; a partial exception, using plant height, is Jinks *et al.* (1977). A brief review of related work on morphological characters in *Drosophila* is given by Scheiner and Lyman (1991). The most straightforward experiment would be to select directly for plasticity by testing a range of genotypes in each generation and selecting those with the least environmental variance in fitness. This scheme might not work, because it would be likely to cause selection for genotypes that had low variance because they had very low fitness in any environment, as the result of deleterious mutations. It would be preferable instead to base selection on the geometric mean of fitness over environments. This is equal to $\bar{x} - \frac{1}{2} \sigma^2_E$, where $\bar{x}$ is the arithmetic mean and σ^2_E the variance over environments, for any given genotype. This design would address two fundamental questions.

First, what is the cost of plasticity? Since plasticity is limited, and reaches different limits in different species, it is presumed that acquiring a greater degree of plasticity entails some cost, but the nature and magnitude of this cost remain unknown. In the selection experiment, a cost would reveal itself as a correlated response of mean fitness: as the environmental variance of fitness fell, arithmetic mean fitness would also fall, if genotypes can express high levels of plasticity only at the cost of a reduction in mean fitness. Second, is plasticity dissociable? We have argued earlier that plants may display different levels of plasticity with respect to different physical variables; this assumes that selection can act independently on the responses to different variables, rather than simply reducing overall sensitivity to the physical environment. There is evidence for such differential plastic responses across key functional traits in herbaceous plants (Lechowicz and Blais, 1988; Macdonald *et al.*, 1988; Schlichting, 1989). In the selection experiment, this means that selection lines in which plasticity with respect to (say) temperature had been increased would not necessarily exhibit greater plasticity with respect to other variables, such as nutrient concentration. It is even possible that different plastic responses are antagonistic, so that increased plasticity with respect to temperature tends to be accompanied by greater sensitivity to variation in nutrient concentration; this would act as a further constraint on the evolution of plasticity with respect to any given physical variable.

A different but equally valuable approach is to select indirectly for greater plasticity. Since it is supposed that plasticity is an adaptation to variation in the physical environment, populations that are maintained over many generations in a diverse environment should evolve greater plasticity (with respect to the particular variables contributing to that diversity) than comparable populations maintained in a uniform environment.

XI. Conclusions

We have attempted to describe how plants respond to environmental heterogeneity not only as individuals but also more extensively on the scale of the local population. Both natural and managed environments often have great spatial and temporal heterogeneity with respect to factors that influence plant growth and reproduction at all scales, from less than that of the individual plant to greater than that of the habitat as a whole. The progeny produced by a plant, being dispersed away from the parental site, will encounter a range of environments, and any individual plant must therefore be able to grow successfully over a range of conditions. Selection will direct the evolution of an appropriate degree of plasticity. In studying the mechanisms by which individual plants adjust to temporal changes in their immediate microenvironment, we should not lose sight of the importance of larger-scale environmental heterogeneity in the evolution of plasticity.

Plant physiologists usually study the acclimation of individual plants to conditions of growth that vary in time. The reasons for this are largely methodological. In this chapter we have described a quite different perspective, which comes from studying populations of plants whose conditions of growth vary in space. Our arguments are statistical rather than deterministic, and evolutionary rather than physiological. Although these two points of view are quite different, and are normally pursued by different groups of scientists using different techniques, we see them as complementary rather than antagonistic; eventually, each will be necessary for the success of the other, as the validation of the evolutionary arguments requires physiologically informed experiments, while the generalization of the physiological work requires an evolutionary interpretation. The principal suggestions that we have made, all requiring further investigation, are as follows.

1. Phenotypic plasticity is a heritable trait that evolves under natural selection and that can be manipulated by artificial selection.

2. There is substantial physical environmental variance at the spatial scale of plant dispersal; selection will therefore act on plasticity even within local populations of plants.

3. Environmental variance increases with distance; the rate of increase of environmental variance with distance supplies a measure of environmental structure. Species that disperse their propagules farther will be more plastic. Any given species will be less plastic more stable with respect to physical variables whose variance increases less steeply with distance.

4. The most appropriate measure of environmental variance is plant response, as measured by implant, explant, or *in situ* studies. Plants vary less than their physical environment, at least for functionally critical traits.

5. Characters that are more highly correlated with fitness (reproductive success) display less environmental variance, as the outcome of the evolution of plasticity.

6. Selection experiments offer the best approach to the genetic architecture of phenotypic plasticity. The main objects of selection experiments should be to demonstrate and quantify the cost of plasticity, to investigate the independence of the plastic response to different environmental variables, and to test the hypothesis that plasticity evolves in heterogeneous environments.

The genetics of phenotypic plasticity have been neglected in the past, because geneticists prefer to work on highly stable characters, whereas physiologists and ecologists tend to ignore individual variation. This neglect accounts for the fact that many of the propositions that we have advanced are highly tentative, or even highly contentious. However, a Darwinian approach to the plastic response is long overdue, and our main object in writing this chapter has been the hope that evolutionary arguments will contribute more prominently to the next generation of research.

Acknowledgments

Work toward this chapter was supported by grants from the Natural Sciences and Engineering Research Council of Canada and from the Quebec Fonds pour la Formation de Chercheurs et l'Aide a la Recherche. We thank the Gault Board for permission to work at the Mont St. Hilaire Reserve and Jon Stewart-Smith and Sarada Sangameswaran for gathering the data on the maple population. The comments of Maureen Stanton and an anonymous referee improved the chapter substantially.

References

Addicott, J. F., Aho, J. M., Antolin, M. F., Padilla, D. K., Richardson, J. S., and Soluk, D. A. (1987). Ecological neighborhoods: Scaling environmental patterns. *Oikos* **49**, 340–346.

Antonovics, J., Clay, K., and Schmitt, J. (1987). The measurement of small-scale environmental heterogeneity using clonal transplants of *Anthoxanthium odoratum* and *Danthonia spicata*. *Oecologia* **71**, 601–607.

Argyres, A. Z., and Schmitt, J. (1991). Microgeographic genetic structure of morphological and life history traits in a natural population of *Impatiens capensis*. *Evolution (Lawrance, Kans.)* **45**, 178–189.

Bazzaz, F. A. (1991). Habitat selection in plants. *Am. Nat.* **137**, S116–S130.

Bazzaz, F. A., and Sultan, S. E. (1987). Ecological variation and the maintenance of plant diversity. *In* "Differentiation Patterns in Higher Plants" (K. M. Urbanska, ed.), pp. 69–93. Academic Press, San Diego.

Bell, G. (1990a). The ecology and genetics of fitness in *Chlamydomonas*. I. Genotype by environment interaction among pure strains. *Proc. R. Soc. London, Ser. B* **240**, 295–321.

Bell, G. (1990b). The ecology and genetics of fitness in *Chlamydomonas*. II. The properties of mixtures of strains. *Proc. R. Soc. London, Ser. B,* **240**, 323–350.

Bell, G. (1991a). The ecology and genetics of fitness in *Chlamydomonas.* III. Genotype-by-environment interaction within strains. *Evolution (Lawrance, Kans.)* **45,** 668–679.

Bell, G. (1991b). The ecology and genetics of fitness in *Chlamydomonas.* IV. The properties of mixtures of genotypes of the same species. *Evolution (Lawrence, Kans.)* **45,** 1036–1046.

Bell, G., and Lechowicz, M. J. (1991). The ecology and genetics of fitness in forest plants. I. Environmental heterogeneity measured by explant trials. *J. Ecol.* **79,** 663–686.

Bell, G., Lechowicz, M. J., and Schoen, D. J. (1991). The ecology and genetics of fitness in forest plants. III. Environmental variance in natural populations of *Impatiens pallida.* *J. Ecol.* **79,** 697–714.

Bradshaw, A. D. (1965). Evolutionary significance of phenotypic plasticity in plants. *Adv. Genet.* **13,** 115–155.

Burrough, P. A. (1983). Multiscale sources of spatial variation in soil. I. The application of fractal concepts to nested levels of soil variation. *J. Soil Sci.* **34,** 577–597.

Canham, C. D. (1988). Growth and canopy architecture of shade-tolerant trees: Response to canopy gaps. *Ecology* **69,** 786–795.

Clements, F. E., and Goldsmith, G. W. (1924). "The Phytometer Method in Ecology," Publ. No. 356. Carnegie Institute, Washington, DC.

Cohen, D., and Levin, S. A. (1991). Dispersal in patchy environments: The effects of temporal and spatial structure. *Theor. Popul. Biol.* **39,** 63–99.

Ellsworth, D. S., and Reich, P. B. (1992). Leaf mass per area, nitrogen content and photosynthetic carbon gain in *Acer saccharum* seedlings in contrasting forest light environments. *Funct. Ecol.* **6,** 423–435.

Falconer, D. S. (1986). "Introduction to Quantitative Genetics," 3rd ed. Oliver & Boyd, Edinburgh.

Gillespie, M. (1977). Natural selection for variance in offspring number: A new evolutionary principle. *Am. Nat.* **111,** 1010–1014.

Givnish, T. J. (1988). Adaptation to sun and shade: A whole-plant perspective. *Aust. J. Plant Physiol.* **15,** 63–92.

Gutschick, V. P. (1987). "A Functional Biology of Crop Plants." Timber Press, Portland, OR.

Jennings, D. H., and Trewavas, A. J., eds. (1986). "Plasticity in Plants." Cambridge Univ. Press, Cambridge, UK.

Jinks, L. L., Jayasekara, N. E. M., and Boughey, H. (1977). Joint selection for both extremes of mean performance and of sensitivity to a macroenvironmental variable. II. Single-seed descent. *Heredity* **39,** 345–355.

Kadmon, R., and Shmida, A. (1990). Patterns and causes of spatial variation in the reproductive success of a desert annual. *Oecologia* **83,** 139–144.

Kolasa, J., and Rollo, C. D. (1991). Introduction: The heterogeneity of heterogeneity: A glossary. *In* "Ecological Heterogeneity" (J. Kolasa and S. T. A. Pickett, eds.), pp. 1–23. Springer-Verlag, New York.

Kuiper, D., and Kuiper, P. J. C. (1988). Phenotypic plasticity in a physiological perspective. *Acta Oecol. Plant.* **9,** 43–59.

Lawrence, M. J. (1982). The genetical analysis of ecological traits. *In* "Evolutionary Ecology" (B. Shorrocks, ed.), pp. 27–63. Blackwell Scientific Publications, Oxford.

Lechowicz, M. J., and Bell, G. (1991). The ecology and genetics of fitness in forest plants. II. Microspatial heterogeneity of the edaphic environment. *J. Ecol.* **79,** 687–696.

Lechowicz, M. J., and Blais, P. A. (1988). Assessing the contributions of multiple interacting traits to plant reproductive success: Environmental dependence. *J. Evol. Biol.* **1,** 255–273.

Lei, T. T., and Lechowicz, M. J. (1990). Shade adaptation and shade tolerance in saplings of three *Acer* species from eastern North America. *Oecologia* **84,** 224–228.

Macdonald, S. E., Chinnappa, C. C., and Reid, D. M. (1988). Evolution of phenotypic plasticity in the *Stellaria longipes* complex: Comparisons among cytotypes and habitats. *Evolution (Lawrence, Kans.)* **42**, 1036–1046.

Mayo, O. (1987). "The Theory of Plant Breeding." Oxford Univ. Press, Clarendon, London.

McCall, C., Mitchell-Olds, T., and Waller, D. M. (1991). Distance between mates affects seedling characters in a population of *Impatiens capensis* (Balsaminaceae). *Am. J. Bot.* **78**, 964–970.

Melancon, S., and Lechowicz, M. J. (1987). Differences in the damage caused by glaze ice on codominant *Acer saccharum* and *Fagus grandifolia*. *Can. J. Bot.* **65**, 1157–1159.

Moloney, K. A. (1988). Fine-scale spatial and temporal variation in the demography of a perennial bunchgrass. *Ecology* **69**, 1588–1598.

Palmer, M. W. (1990). Spatial scale and patterns of species–environment relationships in hardwood forest of the North Carolina Piedmont. *Coenoses* **5**, 79–87.

Perry, D. J., and Knowles, P. (1991). Spatial genetic structure within three sugar maple (*Acer saccharum* Marsh) stands. *Heredity* **66**, 137–142.

Platenkamp, G. A. J. (1990). Phenotypic plasticity and genetic differentiation in the demography of the grass *Anthoxanthum odoratum*. *J Ecol.* **78**, 772–788.

Platenkamp, G. A. J., and Foin, T. C. (1990). Ecological and evolutionary importance of neighbors in the grass *Anthoxanthum odoratum*. *Oecologia* **83**, 201–208.

Robertson, G. P. (1987). Geostatistics in ecology: Interpolating with known variance. *Ecology* **68**, 744–748.

Robertson, G. P., Huston, M. A., Evans, F. C., and Tiedje, J. M. (1988). Spatial variability in a successional plant community: Patterns of nitrogen availability. *Ecology* **69**, 1517–1524.

Scheiner, S. M., and Goodnight, C. J. (1984). The comparison of phenotypic plasticity and genetic variation in populations of the grass *Danthonia spicata*. *Evolution (Lawrence, Kans.)* **38**, 845–855.

Scheiner, S. M., and Lyman, R. F. (1989). The genetics of phenotypic plasticity. I. Heritability. *J. Evol. Biol.* **2**, 95–107.

Scheiner, S. M., and Lyman, R. F. (1991). The genetics of phenotypic plasticity. II. Response to selection. *J. Evol. Biol.* **4**, 23–50.

Schlichting, C. D. (1986). The evolution of phenotypic plasticity in plants. *Annu. Rev. Ecol. Syst.* **17**, 667–93.

Schlichting, C. D. (1989). Phenotypic integration and environmental change. *BioScience* **39**, 460–464.

Schmitt, J., and Gamble, S. E. (1990). The effect of distance from the parental site on offspring performance and inbreeding depression in *Impatiens capensis*—A test of the local adaptation hypothesis. *Evolution (Lawrence, Kans.)* **44**, 2022–2030.

Schoen, D. J., Stewart, S. C., Lechowicz, M. J., and Bell, G. (1986). Partitioning the transplant effect in reciprocal transplant experiments with *Impatiens capensis* and *Impatiens pallida*. *Oecologia* **70**, 149–154.

Schoen, D. J., Bell, G., and Lechowicz, M. J. (1993). The ecology and genetics of fitness in forest plants. IV. Quantitative genetics of fitness components in *Impatiens pallida*. *Amer. J. Botany* (In press).

Stearns, S. C. (1989). The evolutionary significance of phenotypic plasticity—Phenotypic sources of variation among organisms can be described by developmental switches and reaction norms. *BioScience* **39**, 436–445.

Stewart, S. C., and Schoen, D. J. (1987). Pattern of phenotypic viability and fecundity selection in a natural population of *Impatiens pallida*. *Evolution (Lawrence, Kans.)* **41**, 1290–1301.

Sultan, S. E. (1987). Evolutionary implications of phenotypic plasticity in plants. *Evol. Biol.* **21**, 127–178.

Thompson, J. D. (1991). Phenotypic plasticity as a component of evolutionary change. *Trends Ecol. Evol.* **6,** 246–249.

Tilman, D. (1990). Mechanisms of plant competition for nutrients: The elements of a predictive theory of competition. *In* "Perspectives on Plant Competition" (J. B. Grace and D. Tilman, eds.), pp. 117–141. Academic Press, San Diego.

Trangmar, B. B., Yost, R. S., and Uehara, G. (1985). Application of geostatistics to spatial studies of soil properties. *Adv. Agron.* **38,** 45–94.

Turkington, R., and Mehrhoff, L. A. (1990). The role of competition in structuring pasture communities. *In* "Perspectives on Plant Competition" (J. B. Grace and D. Tilman, eds.), pp. 307–340. Academic Press, San Diego.

Via, S. (1987). Genetic constraints on the evolution of phenotypic plasticity. *In* "Genetic Constraints on Adaptive Evolution" (V. Loeschke, ed.), pp. 49–71. Springer-Verlag, Berlin.

Via, S. (1991). The genetic structure of host plant adaptation in a spatial patchwork: Demographic variability among reciprocally transplanted pea aphid clones. *Evolution (Lawrence, Kans.)* **45,** 827–852.

Via, S., and Lande, R. (1985). Genotype–environment interaction and the evolution of phenotypic plasticity. *Evolution (Lawrence, Kans.)* **39,** 505–522.

Webster, R. (1985). Quantitative spatial analysis of soil in the field. *Adv. Soil Sci.* **3,** 1–70.

Wiens, J. A. (1989). Spatial scaling in ecology. *Funct. Ecol.* **3,** 385–397.

Zar, J. H. (1974). "Biostatistical Analysis." Prentice-Hall, Englewood Cliffs, NJ.

Index

Physiological Ecology
A Series of Monographs, Text, and Treatises

Continued from page ii